T0075569

CONTENTS

PART THREE: THE RECIPES 327

GILLED MUSHROOMS
WITH DARK SPORES

HORSE & MEADOW MUSHROOMS & PRINCE

Pink (whitish) free gills turning chocolate brown from spores, with ring (pp. 298–305)

RINGED WEBCAP

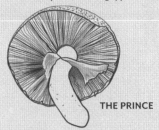

WEBCAPS, SHEATHED WOODTUFT & OTHERS

Rusty brown spores, attached gills, with ring or web (pp. 289–298)

THE PRINCE

CONIFER TUFT, FIELDCAP, WINE & WAVY CAPS & ROUNDHEAD

Cold dark brown, purple-brown, or purple-black spores; ringed; many on wood (pp. 310–320)

SWEETBREAD & DEER MUSHROOMS & ROSEGILLS

Pink-spored, no ring, on ground or on wood (pp. 283–289)

DEER MUSHROOM

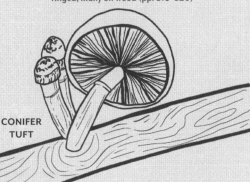

CONIFER TUFT

SHAGGY MANE & INKCAPS

Black-spored, no ring, elongated, often liquefying caps, on ground or wood (pp. 305–310)

SLIMESPIKE

SLIMESPIKES, PINESPIKES & GILLED BOLETES

Smoky gray to black spores (or yellow-brown for Gilled Bolete), decurrent gills, on ground (pp. 320–325)

SHAGGY MANES

FRUITS

OF THE

FOREST

A FIELD GUIDE TO PACIFIC NORTHWEST EDIBLE MUSHROOMS

DANIEL WINKLER

SKIPSTONE

Copyright © 2023 by Daniel Winkler

All rights reserved. No part of this book may be reproduced or utilized in any form, or by any electronic, mechanical, or other means, without the prior written permission of the publisher.

Published by Skipstone, an imprint of Mountaineers Books—an independent, nonprofit publisher
Skipstone and its colophon are registered trademarks of The Mountaineers organization.
Printed in South Korea
26 25 24 23   2 3 4 5 6

Copyeditor: Erin Moore
Design: Jen Grable
Cover photographs, top to bottom: American Blond Morel (*Morchella americana*), Pacific or Golden Chanterelle (*Cantharellus formosus*), Pale Oyster (*Pleurotus pulmonarius*), and King Bolete (*Boletus edulis*). All cover photographs by Daniel Winkler.
Frontispiece: Young Pig's Ear (*Gomphus clavatus*)
Illustrations on p. 27 and inside covers by Anna-Lisa Notter, p. 28 by Jen Grable, and p. 57 by Roo Vandegrift. All photographs by the author, except for the following: pp. 68 (*M. tridentina*), 173 (*Fistulina* tubes), and 177, and 186 by Alan Rockefeller; p. 72 by Joanne Schwartz; p. 78 (*Disciotis venosa*) by Wolfgang Bachmeier; p. 84 by James "Animal" Nowak; pp. 105 and 233 (top) by Rich Tehan; pp. 142 and 151 (*Leccinum manzanitae*) by Jonathan Frank; p. 173 (main image) by Bruce Newhouse; p. 230 by Tim Sage; p. 252 by Warren Cardimona; p. 253 by Libby Wu; p. 259 by Mike Potts; p. 290 by Yi-Min Wang; p. 302 by Thea Chesney; p. 352 (Bolete Quiche) by Daniel Kamberelis; p. 398 by Heidi Schor.

Disclaimer: Eating wild mushrooms is inherently risky: they can be tricky to identify, and people vary in their physiological reactions to mushrooms that they consume. The author has made every effort to ensure the accuracy of the information in this book; however, mycological research and knowledge are always evolving, and these recommendations are made without guarantee on the part of the author or publisher. It is incumbent upon the reader to assess their own skills and usage related to this guide and to mushroom foraging. Neither the author nor the publisher accepts responsibility for any mistakes in identification, personal reactions to consuming mushrooms, or other adverse consequences resulting directly or indirectly from information contained in this book. The author and publisher disclaim any and all liability.

Library of Congress Cataloging-in-Publication Data is on file at https://lccn.loc.gov/2022014430. The ebook record is available at https://lccn.loc.gov/2022014431.

Printed on FSC®-certified materials

ISBN (paperback): 978-1-68051-530-5
ISBN (ebook): 978-1-68051-531-2

Skipstone books may be purchased for corporate, educational, or other promotional sales, and our authors are available for a wide range of events. For information on special discounts or booking an author, contact our customer service at 800.553.4453 or mbooks@mountaineersbooks.org.

Skipstone
1001 SW Klickitat Way
Suite 201
Seattle, Washington 98134
206.223.6303
www.skipstonebooks.org
www.mountaineersbooks.org

LIVE LIFE. MAKE RIPPLES.

INTRODUCTION

When summer fades and the rains return in the Pacific Northwest, many outdoor enthusiasts become depressed, but mushroom lovers know that the rain sprinkles the forests with colorful, tasty offerings. While mushroom hunters elsewhere in North America are suffering in heat and humidity and worrying about poisonous snakes, nasty chiggers, and dangerous ticks, the Pacific Northwest is a mushroom hunter's paradise. No other region in the world has a comparable biomass of succulent chanterelles reliably popping up each year and such an abundance of other choice edible mushrooms. In addition, the Pacific Northwest is blessed with vast and open public lands. Between the Pacific coast and the Cascades, edible mushrooms can grow year-round! Familiarizing yourself with a handful of mushrooms completes the treasured trinity of flora, fauna, and funga, enriching your experiences in nature and making you a more aware member of your ecosystem.

The main focus of this book is to help you find, identify, and enjoy some two hundred edible mushrooms. The most dangerous toxic mushrooms (luckily, this region has few) are presented as well to avoid confusion with the edibles; it is better to know your villains so they cannot sneak up on you. The import-ant medicinal properties of select edible mushrooms are also highlighted. I have tried to make mushroom identification more accessible by avoiding complex myco-lingo; however, certain unique terminology for fungal characteristics is unavoidable, all of which is defined in the Glossary. What you will *not* find in this guide are descriptions of microscopic features or reactions to chemicals that mycologists use to identify fungi. Several great books include these tools and more to help you identify hundreds of regional fungi, and my favorite field guides and online sources are detailed in the References and Resources at the back of this guidebook.

This book focuses on edible Pacific Northwest mushrooms, covering the coastal areas of Alaska down to Northern California, as well as the mountainous slopes out West, including the inland temperate rainforest and drier eastern slopes of the Cascades and the Northern Rockies. Not directly addressed are edible mushrooms specific to the lowlands of California, the southwestern Rockies, and desert areas (yes, there are episodic desert mushroom fruitings!). The look-alike information is focused on the Pacific Northwest and mountainous Northern California, but many of the highlighted mushrooms in *Fruits of the Forest* radiate beyond these borders,

Shaggy Manes, Coprinus comatus, line up along a sidewalk.

and the knowledge shared in this book benefits foragers far beyond the Pacific Northwest.

Whether they're an ingredient in a peak culinary experience you might wax about years later, a wild-crafted source of protein, or possibly more of a fringe experience, the mushrooms presented here are memorable—and tasty.

WHAT ARE MUSHROOMS?

Mushrooms are mysterious creatures. Much of their life cycle unfolds invisibly to us. The reproductive organ of a fungus, they arise from a network of tubular, one-cell-wide strands of hyphae (roughly one-eighth the diameter of a human hair) known as a *mycelium* that grows wherever the organism finds nutrients and humidity—for example, in soil, in wood, or in rotting biomass such as dead leaves. Mycologists distinguish between mushroom and fungus. Fungus entails the whole organism, be it one cell as in some yeasts or a complex organism such as the "humongous fungus"—a Dark Honey Mushroom in Oregon's Blue Mountains with a mycelium that covers nearly 2,400 acres. A mushroom is the fruiting body of certain fungi, essentially a bundle of specialized hyphae of the organism ("fruitings" has been simplified to "fruits" in this guide's title).

A mushroom arising from the fungal mycelium is analogous to an apple growing on an apple tree. And similarly, more than thirty years of Swiss research showed that picking mushrooms is just as sustainable as picking apples, as long as the mycelium is not abused.[1] However, fungi do not have the capacity to turn sunlight into sugars via photosynthesis as plants and algae do; they have to feed on (or be fed by) other organisms. In this and other ways, fungi are more closely related to animals than they are to plants. For example, much of a fungus is made out of chitin, a tough polysaccharide and the same substance that makes up the exoskeleton of insects and the shells of crustaceans.

Often fungi are referred to as the decomposers or recyclers of the ecosystem, something that could be said of animals as well, whom we rather loftily label as consumers, while calling plants producers. But besides feeding on plant matter as decomposers, fungi have developed two symbiotic approaches to tap into plants' photosynthetic powers. The first is by providing a habitat for algae or cyanobacteria and living together to form lichens. The second is by living as mycorrhizae (*myco* "fungus" and *rhizo* "root" in Greek), in a symbiosis or mutually beneficial relationship with the feeder roots of a tree or other plant.

In mycorrhizal symbiosis, trees, shrubs, and nonwoody plants exchange their photosynthesized sugars for water and minerals absorbed by the mycelia of their associated fungi. Indeed, fungal partners connected to a tree's root system provide up to 80% of the nitrogen and nearly all of the phosphate the tree takes up, as well as a lot of water and other nutrients. In exchange, the tree provides 15% to 30% of its total photosynthates to the fungal symbionts.[2] Interestingly, most ectomycorrhizal fungal fruiting happens in summer or fall, after trees have grown new leaves and added new wood and now can spare sugars.

There are three main types of mycorrhizal associations—ectomycorrhizal, ericoid, and arbuscular (a type of endomycorrhizal). Many choice edible mushrooms are ectomycorrhizal, where the fungal mycelium connects to special roots on the outside (*ecto*), with the fungal hyphae covering these stubby root ends like a sewing thimble. Here the tree and fungus exchange photosynthetic sugars for soil minerals and nutrients. Chanterelles, boletes, corals, amanitas, brittlegills, milkcaps, and truffles are all ectomycorrhizal fungi. Being able to identify trees like pine, hemlock, fir, spruce, Douglas-fir, larch, birch, cottonwood, aspen, oak, tanoak, and chinquapin helps greatly in finding ectomycorrhizal fungi and their fruiting bodies. Ericoid mycorrhizal associations have developed between huckleberries (*Vaccinium* spp.) and other shrubby members of the Ericaceae, the rhododendron family. However, madrone trees, also part of the Ericaceae, are ecto-mycorrhizally associated with good edibles like hedgehogs, Western Matsutake, Black Trumpet, Madrone Scaberstalk, and Queen Bolete. The arbuscular association, where the mycelium branches into the root cells (*arbus*, Latin for "treelike,"; *endo*, Greek for "into"), is widespread, and this association exhibits the greatest diversity of hosts, although the fungi involved do not produce mushroom-like fruiting bodies.

Western red cedars, junipers, redwoods, maples, and fruit trees are

Freshly uncovered, hungry mycelia: they prefer dark, moist conditions, and grow in dead biomass.

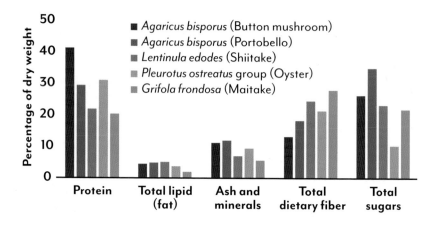

Figure 1. Nutritional components of several types of mushrooms. (Source: Friedman, "Mushroom Polysaccharides," Food, 2016.)

associated with arbuscular fungi; and while they do not produce mushrooms, tasty oyster mushrooms might grow on their dead wood, and shaggy parasols, blewits, and *Agaricus* in their leaf litter. These mushrooms are all decomposers, also called saprobes. The fresh mushrooms offered year-round in stores, such as the button mushroom, portobello, and shiitake, are saprobic and are all cultivated on wood, compost, and other biomass.

Another important non-mycorrhizal fungal lifestyle is parasitic. Parasitic fungi feed off living organisms; the Armillarias or honey mushrooms, for example, can kill living trees. The Lobster Mushroom (*Hypomyces lactifluorum*) digests and transforms its brittlegill (*Russula*) host, and *Cordyceps* like choice edible *Cordyceps militaris* turn insects into zombies.

Nutritional and Health Benefits

Eating mushrooms is not just a culinary delight. They yield incredible health and nutritional benefits that have not received enough attention. In addition, mushrooms are a great alternative to meat, with a much tinier carbon footprint. Morels are great as the pièce de résistance on top of a juicy steak, but mushrooms offer so much more than culinary refinement. They are rich in important minerals such as calcium, phosphorus, iron, copper, zinc, magnesium, potassium, and selenium[3] and offer a wide range of important vitamins like A, C, and all of the B complex,[4] though some B vitamins, like B_{12}, only in trace amounts.

Also, all mushrooms contain vitamin D_2, which, as in humans, is produced by exposure to the sun's UV light (see Collecting, Cleaning, and Preserving Mushrooms) when ergosterol, which is common in fungal cell walls, turns into vitamin D_2. Some commercial mushroom growers are exposing their button mushrooms to nanoseconds of very high doses of UV light to increase their vitamin D content. Marketers hope to have people as easily associate mushrooms with vitamin D as people now associate oranges with vitamin C.

Edible mushrooms like oysters or chanterelles have a water content of about 90% (woody conks can have a content as low as 70%), and the remaining 10% of weight is very high in protein, dietary fiber, and sugars, as well as minerals and some fats. By dry weight, gilled mushrooms like oysters, button mushrooms, and portobellos are up to 30% to 40% protein,[5] while firm-fleshed mushrooms like Maitake and Shiitake are around 20% protein (see Figure 1). Fungal protein contains all nine essential amino acids! Mushrooms are an excellent source of protein, as any vegetarian or vegan can attest.

Dried mushrooms have the highest dietary fiber content of all our main food sources.[6] Soluble and insoluble dietary fibers are very important for our digestive systems. If we have too little—a constant threat in our modern diets rich in highly processed, low-fiber

MUSHROOMS AS MEDICINE

How come fungi offer so many interesting compounds of known or suspected medicinal value? While mushrooms are usually protected from weather and predation by a skin-like tissue, the mycelium underground consists of unskinned tubular hyphae, or mycelial strands, one cell wide. The mycelium is constantly seeking out and growing into new food sources, supplying molecular "building blocks" to its growing tips while absorbing nutrients from the accessed environment. Visualize fungi digestion as a stomach turned inside out, releasing enzymes to break down biomass that the fungi then absorb as nutrients. In stark contrast to humans and other animals, which have protective skins and closely monitor what they ingest, fungi are open to the environment, though the hyphae are in control and will absorb only desired metabolites.

Excepting the aboveground fruiting bodies, the bulk of most fungi grow underground amid millions of other hungry microorganisms, many of them ready to devour the protein-rich fungal tissue. Fungi need a strong immune system to defend and hold their ground against hostile microorganisms. To do so they have developed antibiotic, antiviral, and antibacterial powers. These powers have long been harnessed for human health. Modern medicine focuses on *Penicillium* (a mold, also a type of fungi) and related allies, losing sight of more complex fungi. These often-overseen mushrooms offer a much richer cocktail of powerful compounds that aid the human immune system and are nutritious. This field guide shines some light on known medicinal benefits of species, although many of these insights are so far based only on in vitro lab work and lack human trials. Most of the information incorporated in this book is derived from publications of Chris Hobbs, such as *Medicinal Mushrooms*, and Robert Rogers, such as his extensive *Fungal Pharmacy* and the essential *Field Guide to Medicinal Mushrooms of North America*, which I had the honor to coauthor.

Reveling in the displays at the Puget Sound Mycological Society's annual Wild Mushroom Show in Seattle

ingredients—our intestines have a hard time moving material through the system. Insoluble fibers including fungal chitins offer a lot of surface to which toxins and other unwanted substances bind to be evacuated. Soluble fibers like beta-glucans are well researched due to their multiple functional and bioactive properties. Beta-glucans appear to lower the risk of heart disease and prevent the body from absorbing bad cholesterol from food. Glucans show bioactive properties such as immune-modulating, antitumor, antiviral, and liver-protective effects.[7]

Instead of drifting off into too many studies on the medicinal benefits of mushrooms, I would like to share two studies that show the benefit of regularly eating mushrooms. A recent study in Singapore by Lei Feng and others showed that seniors who consumed more than two standard portions of mushrooms weekly (300 g, or 10 ½ oz.) had 50% reduced odds of having mild cognitive impairment,[8] almost always the first step toward more serious cog-

nitive decline. A study by Min Zhang and others of breast cancer in more than one thousand Chinese women showed that daily intake of mushrooms (at least 10 g [⅓ oz.] fresh or 4 g [⅐ oz.] dried) reduced the participants' risk of recurrence by 40% (as did drinking green tea).[9]

Reducing our consumption of red meat contributes substantially to sustainability, since livestock production requires very high resource input, and vast areas of primary forest are destroyed each year to produce animal fodder and grazing areas. For our family, eating more mushrooms and less meat was never perceived as a sacrifice, but the opposite!

THE JOYS (AND POTENTIAL PERILS) OF FORAGING

Humans have foraged for wild foods including mushrooms since time immemorial. Edible fungi found in calcified dental plaque of Neanderthals in El Sidrón Cave, Spain, have been carbon-dated to nearly fifty thousand years ago. Growing your own food is, evolutionarily speaking, a very recent phenomenon, as is foraging in corner stores or supermarkets, where one could argue we are more likely than in nature to pick up questionable food, including the sometimes toxic cocktail of many highly processed, sugary, and fatty foods laced with chemical additives and residual traces of pesticides. Evolution has made sure that we are equipped with the right senses to succeed in foraging, but we need knowledge to do so safely. Although many of us do not often practice some of these senses, we all have the capacity to distinguish among different organisms. We might not have the terminology to describe the differences, particularly for organisms new to us, but we have an innate sense to see like any other animal and compare characteristics. These same skills are invaluable for identifying wild mushrooms.

While certain mushrooms are indeed deadly, the fear of dying from eating the wrong mushroom is blown wildly out of proportion. Years go by in the Pacific Northwest with no deaths from ingesting mushrooms, while deaths from *E. coli* food poisoning and automobile accidents are common. Please be extremely careful when driving to forage. Admittedly, I am playing with the irrational fear of mushrooms—fungophobia. However, even in a safe environment some dangers lurk. Not all edible mushrooms are easy to identify, and some species have toxic look-alikes. Most mushrooms need to be detoxified, accomplished by cooking them, and a

An enthusiastic forager picking Rainbow Chanterelles, Cantherellus roseocanus

APPLYING YOUR MUSHROOM KNOWLEDGE TO OTHER REGIONS

The foraging expertise you gain from the Pacific Northwest can be applied to most regions of the Northern Hemisphere where the habitat is similar. For example, edible mushrooms growing in mountainous spruce–fir forests share close similarities whether that forest habitat is in Colorado, the European Alps, or the Himalaya. The caveat is that you must also know what dangerous look-alikes might occur in these new areas; it is critical to be ever vigilant of genera known to contain toxic mushrooms. Be cautious when transferring knowledge directly for genera like *Amanita*, *Clitocybe*, and *Lepiota*, to name a few. Interestingly people who have learned to identy edible species in temperate Europe can apply their knowledge to the Pacific Northwest with great success!

There is nothing to worry about in the Pacific Northwest when foraging for common, safe choice edibles like porcini and chanterelles. Having said that, foragers need to be very careful trying to apply local Northwest mushroom knowledge to the tropics and vice versa!

tiny fraction of people have unexpected, mostly inexplicable, negative reactions, even to choice edible species. Luckily, a dozen or so of our region's choice edible mushrooms are fairly easy to identify (see Fourteen Fantastic Fungi in Part One), and a wise strategy is to start with them in your edible mushroom quest. Picking some of these mushrooms, like chanterelles and hedgehogs, does not require much more skill than identifying and picking edible wild berries. Many people pick berries based on what they learned from parents or friends. Mushroom identification is often learned in much the same way.

For a long time, a deep-seated irrational fear of mushrooms has protected the rich fungal resources in the Pacific Northwest. However, more and more people are overcoming their fungophobia to discover the joy of foraging. Some people argue that magic little blue-staining mushrooms may have contributed

to a shift in culture. But I think edible mushrooms have won people over with their rich flavor, high nutritional value, and the great fun and ease of foraging in Pacific Northwest forests.

Encountering the words "safe" and "wild mushrooms" in the same sentence may cause fungophobes to have painful heart palpitations. Such a reaction stems from ignorance. Yes, there are a dozen or so deadly mushrooms in the Pacific Northwest (see Truly Toxic Toadstools in Part One) out of possibly five thousand species estimated to occur here, but the key is to stick with the "safe edibles" like chanterelles, hedgehogs, and king boletes until you know your mushrooms much better. The world is full of people hunting wild mushrooms who don't know any scientific names or mycological terminology—they limit their foraging to a few easy but tasty mushrooms and live long, fungally enriched lives!

The traditional kickout at the annual Breitenbush Hot Spring gathering in Detroit, Oregon

IDENTIFYING MUSHROOMS

To the novice forager, mushrooms can look much the same. When relying on a field guide to identify an edible mushroom, it is crucial to make sure *all* characteristics match fully, including color, shape, size, spore color, odor, taste, and ecology. *Ignoring features can have fatal consequences.* Sometimes we are so keen to find a new choice edible that we can project its identity onto the wrong mushroom, forcing a false match. Such projections can get you into real trouble.

For example, if a description states that a mushroom grows on wood and you happen to find it growing in soil, that's a critical clue and should not be discounted; you should double-check your identification and—more important—*not* eat this mushroom. Yes, there may have been a piece of wood underground, but if you did not verify that, you do not have a match. Similarly, if you did not extract the whole stem base, or if your specimen is too old or young to display important characteristics, you do not have a match. Be careful. Harvesting very young mushrooms that do not yet clearly show their characteristics can be very dangerous; telling mushroom babies apart can be as hard as telling human babies one from another! Once you know a mushroom well, you can approach its identification in a more relaxed manner—but it takes some experience to get there. Remember this adage: *When in doubt, kick it out!*

FUNGAL TAXONOMY: WHAT'S IN A NAME?

Learning about mushrooms means attaching names to organisms you recognize. Many communities or families have their own common names for certain species, sometimes handed down over generations. There is nothing wrong with this approach: the challenge is communicating beyond your clan and confirming your mushroom finds using other sources, because common names vary. That is why universally accepted scientific names, Latin binomials like *Agaricus augustus*, are so helpful—they offer clarity. However, for real clarity, the binominal also needs an author, to connect it to the person who described the mushroom first, e.g., *Agaricus augustus*

TAXONOMY OF A KING BOLETE VS. A WOLF

DOMAIN: EUKARYA
All organisms whose cells have a nucleus and organelles

KINGDOM: FUNGI
All fungi, including yeast, molds, and mushrooms

PHYLUM: BASIDIOMYCOTA
All mushrooms with basidia, a club-shaped, spore-bearing structure, excluding mushrooms with asci

CLASS: AGARICOMYCETES
Most mushrooms, excluding Basidiomycetes such as smuts, rusts, and two orders of jelly fungi

ORDER: BOLETALES
All boletes, plus some families of gilled mushrooms, false truffles, and earth balls

FAMILY: BOLETACEAE
Boletes, mushrooms with spongelike hymenium, excluding *Suillus* (Jacks) and their relatives

GENUS: *BOLETUS*
Big mushrooms with soft tube layer, fat stems, non-bluing flesh; all edible

SPECIES:
BOLETUS EDULIS
King Bolete

DOMAIN: EUKARYA
All organisms whose cells have a nucleus and organelles

KINGDOM: ANIMALIA
All animals

PHYLUM: CORDATA
All animals that have a backbone

CLASS: MAMMALIA
All animals that produce milk to feed their young

ORDER: CARNIVORA
All animals that feed on flesh easily

ORDER: CANIDAE
All dogs and doglike animals

GENUS: *CANIS*
All domesticated and wild dogs

SPECIES:
CANIS LUPUS
Wolf

Figure 2. Taxonomic ranks: Example of scientific classification of a King Bolete

Fr., for Elias Magnus Fries (1794–1878), the great Swedish mycologist.

In this book each described mushroom has a name and taxonomy section, which explains the meaning of the scientific name and includes the author with the year of publication of the first description, e.g., Fries 1838.

Scientific names that we know need replacement are offered, for example, as *Agaricus* aff. *campestris*, *Agaricus* "*arvensis*," or, when there are several species involved, *Agaricus* "*arvensis*" group. This situation most frequently applies to European species names applied to American mushrooms that

FUNGAL BAR CODES AND THE DNA REVOLUTION

The successful application of molecular research starting in the 1990s allowed analysis of fungal DNA using tiny segments such as the internal transcribed spacer (ITS). Made of a sequence of around 700 genetic base pairs shared by all fungi, the ITS forms a sort of "primary fungal bar code." Within this bar code, changes caused by mutations that have occurred across millions of years of evolution allow researchers to draw conclusions regarding the degree of closeness or relatedness in the tree of life. However, different DNA regions are used for different purposes, and ITS is excellent for differentiating closely related species. Still, ITS is more informative for some species of fungi than others, and for some groups of fungi, like morels and boletes, researchers must evaluate three or four different genetic regions to draw conclusions about evolutionary, or phylogenetic, development.

Taxonomists apply complex statistical analysis to calculate the most likely evolutionary paths, which are depicted in a branching treelike scheme, grouping organisms together based on DNA similarities. On the phylogenetic tree, each branch (clade) contains only taxa (taxonomic units like order, family, genus, species) that are related.

we now know are distinctive species. They should have their own names, but their names have not been published yet. Also, several scientific names apply to the same species, which often include a European and an American name. In general, the first properly published name is valid; names published later are synonymous (expressed by = in the taxonomy notes). With a scientific name also comes a place in the fungal tree of life, which often allows us to see if we are dealing with a harmless family or if any toxic relatives are waiting in the bushes.

In the past, mycologists decided which traits best represent a group of fungi. Today, results of DNA analysis indicate closeness of relation and allow us to deduce relevant traits. Some morphological criteria established in the past have weathered the DNA revolution well, like the use of spore color as a main criterion for separating families of gilled mushrooms. Others, like the shape of spore-producing tissue (for example, grouping different "toothed fungi"), have turned out to be misleading, since the toothed hymenium evolved several times in what we now know are not closely related groups. Researchers using DNA sequencing and other modern methods are reclassifying species all the time; in this field guide, I've done my best to include the latest names and groups. That said, mycologists will likely continue to make new discoveries about classification. For most mushrooms featured in this guide, the species account mentions family and order to give an idea of degrees of relation to other mushrooms.

LEFT: *White Elfin Saddle,* Helvella crispa, *a favorite forest hang-out spot for elves;* RIGHT: *A secret subur-ban fairy portal made visible by* Marasmius oreades, *Fairy Ring Mushroom*

When several species in a genus are presented, only the first one includes this taxonomic information.

Taxonomy is used to bring order to the confusing and mesmerizing diversity of life. The very first taxonomists divided organisms into essential categories like edible, poisonous, dangerous, useful, and useless. Thus, basic taxonomy comes naturally to all organisms in some form. However, as a science, taxonomy became naming and classifying organisms based on their degree of relatedness (their shared ancestry), an important tool in the life sciences.

Fungal taxonomy has gone through many changes over the centuries, each stage refined with the arrival of new techniques. Taxonomists first used the morphology of fruiting bodies to name mushrooms and to group like with like. Microscopy was a revolutionary technology that brought to light similarities or differences in cellular structure and especially spore shape. Chemical anal-ysis and cultivation of mycelia were additional new tools to help show if two fungi were the same species or different. All of these methodologies left a lot of room for an individual taxonomist to decide which criteria to prioritize and which to ignore when grouping mushrooms. Then came the DNA revolution.

For fungi, the evolutionary tree coming to light is quite different from the evolutionary tree postulated before the DNA revolution. DNA analysis shows that many traditional mushroom taxa need serious adjustment. Whole orders and families have been dissolved because their members are less closely related than their looks would indicate. One of the first revolutionary rearrangements of the fungal tree of life was to split *Coprinus* (a single genus that had included Inkcap, *C. atramentarius,* Mica Cap, *C. micaceus,* and Shaggy Mane, *C. comatus*) into three different genera (*Coprinopsis, Coprinellus,* and *Coprinus,* respectively) in two different families. These conical-to-cylindrical-capped gilled mushrooms are all black spored and deliquescent. The new alignment allows us to reevaluate species' physical traits in light of the new genera.

This time around, it was not the taxonomists' picking the defining traits but the mushrooms' own DNA driving the analysis and inspiring the decisive characteristics—quite a revolution. Some of these DNA-inspired taxonomic changes are surprising, others long suspected. But before we had the genetic evidence, who was to say definitively what was a convergent trait versus what was homologous? For example, Shaggy Manes and Inkcaps used to be classified in the same genus, *Coprinus*, based on the assumption that their shared trait of a conical to cylindrical liquefying cap was inherited from one common ancestor, thus being homologous. However, it turns out that liquefying caps, a successful strategy for spreading spores, have evolved independently several times. Based on DNA evidence, *Coprinus* was divided up into several genera, some even nested in different families, meaning that their liquefying caps are an analogous trait of convergent evolution, like wings in insects and birds, which do not share a common winged ancestor.

Common names often are based on similarities to animals (think Deer, Oyster, and Coral mushrooms; Bear's Head), the shape of things (Shaggy Parasols, Black Trumpet), locations and season of appearance (Meadow Mushroom, Woodlovers, Spring King, Fall Skullcap), descriptive traits (Brittlegills, Sawtooth, Milkcaps), mythology (Witches' Butters, Elfin Saddles, Fairy Ring Mushroom)—the list goes on.

Often, we have to reach back in time or research other cultures to discover the inspiration for names such as chanterelle, morel, matsutake, and bolete.

In the eighteenth century, Carl Linnaeus introduced the binomial (two names) system of latinized genus and species names to positively identify a specific organism. Many of his animal and plant names were spot on and have held up well through the centuries. But Linnaeus's mushroom naming was less helpful. He created only ten genera to cover the entire fungal kingdom—one of which, *Agaricus*, contained every gilled mushroom then known. Another of his groups, *Boletus*, he applied to every mushroom with pores, be it woody conks or porcini. Though it was pointed out to Linnaeus that in Roman times *Boletus* referred to the Caesar Mushroom, *Amanita caesarea*, he did not bother to correct his error. Today, we recognize close to a thousand genera of macrofungi in the Pacific Northwest alone! Whenever a mushroom is given a new scientific name, it must be properly described and documented in a valid publication, and specimens, known as type specimens, must be deposited in an herbarium.

Common names, from a scientific perspective, are a headache, since they are not standardized. But when it comes to sharing and spreading scientific knowledge to the wider population, common names are extremely helpful. Most people do not have any association when hearing *Inocybe*, while the common name—"fiberhead"—points out an important trait for identification: the presence of fibers on the caps of these toxic LBMs (little brown mushrooms).

And guess what, the scientific name has the same meaning, from Greek *ino* "fiber" and *cybe* "head."

The absence of standardized common names for our mushrooms plagues the United States and Canada, as does the lack in these countries of a federal fungal "red list." Lists of endemic threatened and endangered fungi are generated by most other industrialized countries. The shortcomings are an expression of our society's fungophobia. Nearly all mushrooms have common names in European countries, including the UK, as well as in China, Korea, and Japan. In these countries, too, professional and lay mycologists work hard to make fungal knowledge more accessible to their fellow citizens. In the US and Canada, attempts of that nature have so far failed, be it a moat around the fungal ivory tower or simply lack of effort. We still have far to go to convey the diversity of North American mushrooms to the general public.

To this end, this field guide presents common names for Pacific Northwest mushrooms. Common names create common ground, even if there are regional differences; they make mushrooms more accessible. This guide also provides the Latin and Greek roots of species' scientific names. A good name we can connect to, be it scientific or common, and having the choice to learn either (and hopefully both) for the same organism, increases learning and recollection. In the past it was often argued that there are too many common names for the same mush-

room. The DNA-fueled taxonomic revolution has turned the table. Now, some common names are more stable than the scientific ones! With improved taxonomy, some common names should perhaps be updated to reflect new relationships.

LEARNING YOUR MUSHROOMS

Some people are lucky enough to grow up foraging for edible mushrooms with their families. The next best thing is having knowledgeable friends. A great way to meet such friends is by joining organizations focused on sustainable food and permaculture or by getting involved with one of the more than a dozen mushroom clubs or mycological societies that meet regularly all over the Pacific Northwest (look for a roundup in Resources at the back of the book).

When I moved to Seattle in 1996, I joined the Puget Sound Mycological Society after attending the annual fall mushroom show. For more than a quarter century, I have benefited immensely from participating in and volunteering to coordinate countless activities. Other members and the speakers invited to the monthly meetings have opened my mind in so many ways to appreciate mushrooms, often way beyond their culinary value. The best way to learn about mushrooms is to join in these all-volunteer activities and share your love for mushrooms with like-minded folks. However, don't expect strangers to share their favorite hunting spots— you'll need to find your own.

HOW TO USE THIS GUIDE

This field guide to edible mushrooms of the Pacific Northwest is organized into three main parts. Part One provides all the practical information you need to know to hunt and collect mushrooms, including details on habitat and seasonality; necessary gear and safety matters; where to collect (and where *not* to collect); instructions on preserving mushrooms; and more.

Part Two describes more than 170 species of commonly found edible mushrooms in the region, with notes on identification, look-alikes (some nonedible or even toxic), habitat and seasonality, any known medicinal uses, and name and taxonomy, as well as a photograph or two of each featured species, plus key look-alike species, to help you safely identify them in nature. In the species descriptions, symbols are used sparingly. For instance, ∅ stands for diameter in fruiting body measurements.

For some species, icons (defined below) that appear next to their common names indicate important information about known edibility, possible look-alikes, and toxicity. Consult the listing itself and other resources as you wish for details. Don't consume any mushrooms whose identity you are not confident about.

 Requires special preparation and/or caution; may cause digestive upset

 Difficult to identify; may have dangerous look-alikes

 Not well established as edible

 Toxic; may cause severe discomfort

 Deadly; high probability of organ failure

Select toxic or deadly mushrooms are also described because learning these is essential to staying safe when branching out in your culinary quest.

Part Three showcases 39 recipes and dishes for all manner of edible mushrooms to delight just about every palate. From soups, sauces, and salads to hummus, leathers, butters, and more, these savory recipes will encourage you to forage year-round in the Pacific Northwest.

This book offers what you need to know to identify many mushrooms with confidence. Space considerations limit how many photographs the guide can feature, so once you have identified

King Bolete, Boletus edulis *var.* grandedulis, *collected in early July in the Cascades*

a mushroom with a field guide, double-check your ID by searching for images online by the scientific name. You'll find hundreds of pictures for common edible species, enabling you to easily find a visual match if your identification is on target. Using social media and online forums can also be very helpful with identification or confirmation you have applied the right name. A caveat: never trust a stranger to confirm your ID. Mushroom enthusiasts can overestimate their knowledge. Always wait for several knowledgeable people to confirm your ID. Before you entrust anyone with your health, check their mushrooming background!

In the past, mycologists received their hard data regarding distribution and fruiting dates from fieldwork, catalogs of specimens deposited in herbaria, and information from trusted colleagues. Today, a lot of this info can be gleaned from extensive online mushroom databases—iNaturalist and Mushroom Observer are invaluable.

Simply upload photos of your finds, which helps the mushrooming community learn more about distribution and seasonality of mushroom fruiting. You also might receive precious help from other mushroom enthusiasts or mycologists. Citizen science findings shared by mushroom lovers is a key contribution to substantially improving our knowledge base. The databases iNaturalist and Mushroom Observer also allow for inquiries of what mushrooms are documented for a specific location or period. For example, you could search for all the mushrooms uploaded over a two-week period from Mount Rainier National Park. Equipped with such a compilation, you might have much more success identifying what you encounter during a foray.

For the Pacific Northwest, two free mushroom identification programs are extremely helpful. MycoMatch (formerly MatchMaker) built by Ian Gibson in Victoria, British Columbia, is an awesome ID tool containing more than

4,200 described species, 2,300 of them illustrated with 6,000 images. Using its matching tool, you can search, for example, for all gilled mushrooms that smell of garlic, or all smooth-capped, white-spored fungi growing in clusters on conifers—really useful ways of comparing!

Danny Miller's Pacific Northwest Mushroom Pictorial Key built on MatchMaker data is invaluable as well. The pictorial key is based on images with brief descriptions, and taxonomically, it is always up to date. The pictorial guide can be installed on a smartphone, as can MycoMatch. Michael Wood's MykoWeb (with a focus on California fungi) and Michael Kuo's MushroomExpert (with a focus on Midwestern species, albeit no information on edibility) are both helpful too. You'll also find detailed mushroom portraits of the most common Pacific Northwest edibles on my Mushroaming website.

Mushrooms don't grow without precipitation. Instead of driving hours hoping to find mushrooms, check weather patterns in the weeks before, or consult websites that show the precipitation history (or snowmelt patterns in spring). Some weather websites show precipitation history for specific locations. NOAA offers a great, though clunky, website (water .weather.gov/precip/) where you can see precipitation amounts overlaid on a map for recent weeks. Another website (iweathernet.com/total-rainfall-map -24-hours-to-72-hours) supplies data for the past three days. The comprehensive Resources in the back of this field guide contains a wealth of information.

PART ONE

THE HUNT

This part of the field guide provides advice on getting to know mushrooms, what makes a mushroom edible (or dangerous), safety issues for foragers, and information on Pacific Northwest habitat, seasonality, and fruiting. Also included are key details on how to collect, clean, and preserve and store mushrooms. Methods include drying, powdering, freezing, and pickling. Happy hunting!

OPPOSITE: *A basket of morels*

GETTING TO KNOW MUSHROOMS

When hunting mushrooms, you need to apply all your senses. Considering some key questions will help you in your quest: Where does it grow, and what are its look-alikes? What makes a mushroom edible? Successful identification comes through the practice of looking closely and comparatively at all kinds of mushrooms to become familiar with the deep diversity of shapes, features, and habitats. There is no substitute for time spent in the field, "getting your eyes and hands on" and applying all your senses.

APPLY ALL YOUR SENSES

Mushrooms demand all your senses. First come sight and feel. When learning to identify mushrooms, note the differences in spore-bearing tissues like pores, spines, tubes, and gills (see Figure 3). The gilled mushrooms are called agarics and feature many details you will learn as you go along. Check out how the gills connect to the stem. When you find a group of the same mushrooms, observe how they change color, shape, and surface texture from youth to old age. When you pick a familiar mushroom, take home a few unknown mushrooms (stored separately from your edibles)

and try to identify them. Even if you fail to pin down the species, don't despair! Just studying a mushroom, even without an identification, enriches your knowledge base, enhances your fungal terminology, and helps you to discern important details new to you.

Handling mushrooms is so important. A fungophobe may shout, "Don't ever touch a mushroom unless you know it is harmless—most are toxic!" Calmly inform the alarmists that touching a mushroom, even a deadly mushroom, will not kill anyone, just like holding hands will not get you pregnant! Feel the mushrooms' consistency. Do they seem light or heavy? Are their stems or caps sticky or dry? Smooth, warty, scaly, or hairy? Is the flesh fragile, brittle, firm, or tough?

If it is a gilled mushroom, check out the gills. How are they spaced? Are the blades broad (deep) or narrow (shallow)? Many and closely packed together, like leaves in a book, fewer and distant from one another, or somewhere in between? And how do the gills relate to the stem? Do they reach the stem or are they free of it? If attached, in what way: abruptly or with a notch or sinuous area? Do they run down the stem—are the gills decurrent? To understand this specific terminology, refer to Figure 4 showing the patterns of gill attachment.

Does the color change when the mushroom is handled or the flesh is exposed? Pay special attention to the stem base. Does it have extra tissues shaped like a cup, known as a volva? Often when a mushroom is scratched or sliced, there might be color changes in the flesh of the stem base not occurring elsewhere in the mushroom. Is there mycelium attached and, if so, what color? There might be a unique odor beyond fungal and earthy, like almondy, fishy, or of bleach.

When applying all your senses (including smell), don't forget taste! Tasting a mushroom is helpful—that is, if you dare. To be clear: tasting does *not* include swallowing—and I strongly discourage tasting little brown and deadly mushrooms and Ascos! Often just a tiny piece will divulge a spicy, mild, or bitter flavor, or the usual: fungal. Remember to spit the tiny sample out. Some people like to rinse their mouth, especially if they are unsure if the mushroom belongs to a harm-

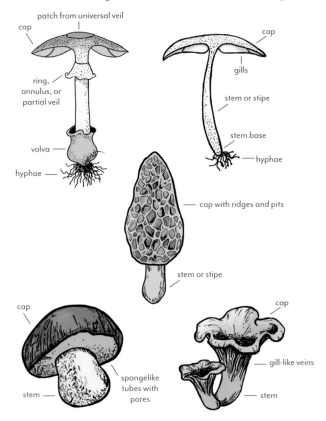

Figure 3. Parts of a mushroom: **TOP ROW:** *cross-sections of a cap and gills and entire mushroom to illustrate structure.* **MIDDLE ROW:** *Morel.* **BOTTOM ROW:** *Bolete, Chanterelles.*

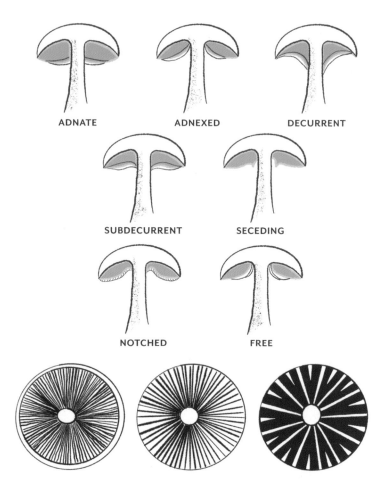

ADNATE   ADNEXED   DECURRENT

SUBDECURRENT   SECEDING

NOTCHED   FREE

Figure 4. Gills can be free or attached to the stem in several different ways. Figure 5. Gills are spaced evenly and generally fall into one of three categories: crowded, close, or distant.

less genera (harmless genera for tasting include brittlegills, *Russula*; field mushrooms, *Agaricus*; slippery jacks, *Suillus*; and corals, *Ramaria*) or one known to contain some really sketchy members (*Amanita*!). The amount of a poisonous mushroom that you have to ingest to cause ill effects depends on the species and its toxins. For the daredevils out there: many mushrooms are edible when cooked but poisonous when ingested raw. The specific presentations in Part Three: The Recipes include relevant details. Detoxification is also covered in Cooking with Mushrooms in the next section, Collecting, Cleaning, and Preserving Mushrooms.

As you expand your knowledge of edible mushrooms, you may want to collect your target edible species and carefully identify it at home. In general, I prefer to eat a mushroom new to me in the com-

FRAGRANCE AND MUSHROOM ID

When learning about mushroom odors, it is crucial to smell the mushrooms themselves to make the connection. We invoke a range of descriptive terms, but the actual fungal odor has its own characteristics. Mushroom odors can be alluring or repellent, and people's perceptions of these smells vary. Here are some helpful fungal odors, with PNW mushrooms from this guidebook as examples.

Almond, anise: The Prince, Almond Woodwax
Fruity: Wood Blewit, Golden Chanterelle
Spicy like Red-Hots cinnamon candy: Western Matsutake
Spicy veering off to skunky: Cloudy Funnel
Farinaceous (cucumber/watermelon): Sweetbread Mushroom, Cottonwood Mushroom, Leopard Knight (poisonous)
Chlorine: Smith's Amanita (poisonous), Bleach Cup
Fishy: Shrimp Brittlegill, Coccora
Phenolic (creosote-like, tarry): Buck's Agaricus (nonedible)
Radishy: Poison pies
Spermatic: Fiberheads
Musky: Oregon White Truffle

pany of someone who has eaten it many times before without negative results. Second best is when a person familiar with the new mushroom confirms my identification first. **Never eat an unidentified mushroom.** In my twenties I got talked into cooking up a superficially identified gilled white mushroom. I survived, but my digestive system quickly emptied its contents. To this day, I am not exactly sure what I ate, but I did not repeat this foolishness. Do not get involved with whitish gilled mushrooms unless you are absolutely certain of their identity. There are too many kinds, and a few are toxic and even deadly.

WHAT MAKES A MUSHROOM EDIBLE?

Sorry, there are no shortcuts. An old mushroomers' joke informs us that "Every mushroom is edible, at least once!" No overarching fungal characteristic will alert you to toxicity. You must learn to recognize each edible mushroom species based on its specific criteria. There are no helpful generalizations, and even a mushroom you have eaten many times before may turn out to be toxic, in the case of accumulated liver toxins or a newly developed allergic reaction. Though it happens extremely rarely, most edible mushrooms can cause adverse physical reactions in unlucky individuals. The good news is very few people are affected. Thus, eating only a few bites of any species new to you the first time you try it is prudent. In reliable field guides (such as this one) mushrooms reported to provoke personal reactions for some but not everyone are identified and marked as "Requires special preparation," indicated by an exclamation point.

In short, there is no widely agreed-upon approach to labeling a wild mushroom edible. There are so many factors, such as proper processing for detoxification, personal reactions including allergic reactions, freshness of the mushrooms, and also often a lack of precise, definitive taxonomy. To write this book, I consulted a wide range of sources, including commonly available North American field guides, a recent review of the edible mushroom species throughout the world published in *Comprehensive Reviews in Food Science and Food Safety*, and many European references. Many other cultures have a much longer history of appreciating wild mushrooms and much better funding for research on their edibility and toxicity.

Edible gilled mushrooms are generally much harder to identify than edible nongilled mushrooms (like the boletes or hedgehogs), and the consequences of misidentification are higher. Why? There are many more species of gilled mushrooms than nongilled, and species are often differentiated by minute details. Also, edible gilled mushrooms are more likely than edible nongilled mushrooms to have dangerous or poisonous look-alikes. It is much safer for beginner mushroom foragers to focus their study of edible mushrooms on the easy-to-identify, nongilled mushrooms. No wonder very few gilled mushrooms appear in the list of Fourteen Fantastic Fungi.

THE (UNRELIABLE) FREESTYLE APPROACH TO ID

During a hunt in an oak forest near Jakar, Bhutan, we were looking at a cluster of brown mushrooms seemingly sprouting from the ground. They looked familiar, but I could not place them right away. Jigme, a retired forest worker who had joined our foray, was unsure too, but he wanted to collect them for a meal, stating, "If I don't know a mushroom, I parboil them to render them edible." I was shocked that a person would reach his advanced age consuming unknown mushrooms and trusting in heat detoxification, a process that will not eliminate deadly amatoxins. I questioned him further and he shared a rule to his freestyle approach: never eat any gilled mushroom that has a ring on the stem!

Very interesting, I thought, since several deadly mushrooms have a ring. Still, this is not a reliable approach. I repeat: please do *not* start cooking up all ringless mushrooms! For example, the very toxic *Amanita smithiana*, the most dangerous matsutake doppelganger, has when fresh only flaky remnants of a ring that quickly shrivel up to nothing. The mushroom that evoked my exchange with Jigme turned out to be a honey mushroom growing from wood underground and sporting a fibrous partial veil on the stem. Cooking renders honey mushrooms edible for most people.

Lush coastal rain forest in Alaska

WHERE AND WHEN TO COLLECT

Mushrooms are members of ecosystems, and each has its preferred habitat. Just like you would never find me on a golf course, you will never find a chanterelle in a pure grove of western red cedar. The Golden Chanterelle is mycorrhizal with Douglas-fir and western hemlock, and that is where you will find them. The Pacific Northwest is not endowed with a great diversity of trees, but most of them are associated with edible mushrooms. Knowing your trees is instrumental to discovering them.

The habitat for all mushrooms presented in this field guide is described, since it is crucial information for finding them. However, not every Douglas-fir is surrounded by chanterelles in October and not every dead alder sprouts oysters in spring. It takes patience and time to discover collection spots. Don't expect others to share their beloved spots they might have acquired over years of trial and error. Still, many mushroom hunters will share insights into where you might get lucky, like general area, forest type, altitude, exposure, fruiting time, and other tips. These helpful tips—often generously shared in mycological society emails or social media posts—will help beginners a lot. But you could check the right habitat eleven months of the year and still not discover any mushrooms popping up. Timing is everything!

To fruit, mushrooms need sufficient humidity and a conducive temperature above freezing. The western, lowland Pacific Northwest is unique in having its prime mushroom season centered around October. Most fungi in the temperate Northern Hemisphere "fruit"—

FOURTEEN FANTASTIC FUNGI

Most of the safe easiest-to-ID mushrooms on this list are nongilled. Straightforward characteristics and few troubling look-alikes make them easier to identify with confidence. The only exceptions are the gilled Saffron Milkcap and Red-bleeding Milkcap, due to their unique, color-changing milk, the distinctive Shaggy Mane, and oysters growing in clusters on wood.

Bear's Head, Lion's Mane, Coral Tooth—*Hericium abietis, H. erinaceus, H. coralloides*

Black Trumpet—*Craterellus calicornucopioides*

Cauliflower Mushroom—*Sparassis radicata*

Chicken of the Woods—*Laetiporus conifericola* (edible with caution)

Golden, Rainbow, and Cascade Chanterelle—*Cantharellus formosus, C. roseocanus, C. cascadensis*

Hedgehogs—*Hydnum* spp.

King Bolete—*Boletus edulis* and other *Boletus* spp.

Lobster Mushroom—*Hypomyces lactifluorum* (edible with caution)

Morels—*Morchella* spp. (edible with caution, toxic when raw)

Oyster Mushrooms—*Pleurotus pulmonarius, P. populinus, P. ostreatus*

Pig's Ear—*Gomphus clavatus*

Puffballs—*Lycoperdon perlatum* and *Calvatia booniana*

Saffron Milkcap and Red-bleeding Milkcap—*Lactarius deliciosus* group, *L. rubrilacteus*

Shaggy Mane—*Coprinus comatus*

produce their mushrooms—in midsummer or late summer, while in the summer drought climates, the fruiting occurs in fall and winter, or as it seems with climate change in California, whenever enough rain falls. Summers in the Pacific Northwest can be bone-dry, which is a total turnoff for fungi to grow their reproductive organs. They want rains plus conducive temperatures.

Mushroom hunters pay close attention to precipitation: no humidity, no mushrooms. Mushrooms growing on wood often react first to precipitation, since dead wood retains water well. Mycorrhizal fungi often take two or three weeks to grow their mushrooms after the first soaking rains arrive, ending summer drought. If the soil is already humid, mushrooms will react much faster to showers. While saprobe fruiting can be correlated to substrate availability, temperature, and humidity, mycorrhizal fruiting patterns are much more complex, since they are completely dependent on the amount of sugars their symbiotic trees can provide. After a crippling summer drought, generous fall rains may not trigger a great fruiting flush since stressed trees simply have few resources to share. Even the previous year's weather extremes can impact fruiting.

Fruiting depends greatly on sites. Weather, altitude, and slope exposure influence soil or substrate temperatures as well as water availability. Spring mushrooms often react to soil temperature and will start fruiting earlier in more southern latitudes, in milder climates, at lower elevations, and on sun-exposed sites. Northern latitudes, continental climates, higher locations, and shady exposure delay fruiting. In dry years, shady slopes or densely forested habitat provide more favorable microclimates. Unusual summer rains might kick off the main mushroom season much earlier. Eastern Pacific Northwest areas that have summer rains and thunderstorms or high-altitude areas where frost comes early often experience summer fruiting.

Climate change is mixing up the fruiting patterns known to previous generations of mushroom hunters. Many spring mushrooms fruit several weeks earlier now, and the fall season is extended as well. Long-term predictability has become much more uncertain, a real challenge for people organizing mushroom events: the emergence of our fungal friends can be assumed but never guaranteed.

WHERE *NOT* TO COLLECT

Fungi are specialists in taking up nutrients needed for growth and rare compounds and minerals. Like a specialized chemical factory, each species breaks down complex and sometimes toxic compounds into harmless molecules, and at time produces modified toxins as well as other nourishing, healthy compounds. This makes fungi top contenders for

Wild collected Fire Morels and Spring Kings next to cultivated mushrooms at Pike Place Market in Seattle

effective soil remediation and toxic site restoration. It also makes them concentrators of substances that bioaccumulate in their tissues. And though many toxic compounds can be rendered harmless by cooking—either breaking down or evaporating these compounds—others cannot: these mushrooms are labeled as poisonous. Even good edible mushrooms can contain and concentrate toxic compounds and trace elements, including heavy metals, radioactive isotopes, and manufactured toxins like pesticides.

Starting in the 1970s, researchers began quantifying heavy metal accumulation in vegetables and grains. Mushrooms next piqued the interest of food scientists. A warning regarding heavy metals in edible mushrooms was issued in 1978 by concerned German food safety officials; Germans consume a lot of wild mushrooms. Currently Germans are advised to limit their intake to less than half a pound per week. Yet even this warning closes with a chipper note: "However, there are no concerns about the occasional consumption of larger quantities."[10] Radioisotopes like cesium-134 and cesium-137 were released from

TRULY TOXIC TOADSTOOLS

In any discussion of edible mushrooms, fungophobes will jump right away to deadly mushrooms. Sadly, the following mushrooms have killed people in the Pacific Northwest. (This is not a comprehensive list of deadly mushrooms featured in this field guide.)

Death Cap and Western Destroying Angel—*Amanita phalloides, A. ocreata*
Deadly Skullcap—*Galerina marginata*
Fatal Dapperling—*Lepiota subincarnata*
Red-pored Beauty Bolete—*Rubroboletus pulcherrimus*
Chicken of the Woods—*Laetiporus conifericola*
Wavy Cap—*Psilocybe cyanescens*

Amanita phalloides is perhaps the deadliest of these. It is an introduced and invasive mycorrhizal species spreading in our area. A man from Spokane died after ingestion of Red-pored Beauty Bolete by infarction of the midgut. *Psilocybe cyanescens* killed a young child; magic mushrooms can cause convulsions and high fevers in young children. Strangely, the choice culinary mushroom Chicken of the Woods, eaten safely by most when cooked well, is linked to a case of a deadly anaphylactic shock. In addition, there have been many close calls from *Amanita smithiana* and *Gyromitra esculenta*. Ringed Conecap (*Pholiotina rugosa*) and Deadly Webcap (*Cortinarius rubellus*) are potentially deadly, but no deaths have occurred in the Pacific Northwest.

Never eat a mushroom you cannot identify with absolute certainty. As a beginner, stick to the known, safe edibles described in this book and, before venturing out, get to know the villains!

HEAVY METAL CONCENTRATORS

Many beloved edible mushrooms have been indicated as heavy metal concentrators, accumulating high amounts of these elements in their tissues. These include the Prince, Meadow Mushroom, Horse Mushroom, and other species of the genus *Agaricus* as well as some members of the family Agaricaceae, such as Shaggy Mane (*Coprinus comatus*), shaggy parasols (*Chlorophyllum* spp.), and several puffballs (including *Calvatia* spp. and *Lycoperdon* spp.). However, species of other families of gilled mushrooms—such as the Wood Blewit (*Lepista nuda*, Tricholomataceae) and Fairy Ring Mushroom (*Marasmius oreades*, Omphalotaceae)—have also been implicated.

Furthermore, members of other orders—such as boletes (e.g., King and Bay Boletes, *Boletus edulis* and *Imleria badius*, Boletales) and brittlegills (*Russula* spp., Russulales)—have been shown to accumulate serious amounts of toxic metals in polluted sites. For example, near a smelter in southern Poland, peak lead pollution for the Bay Bolete reached thirty times the admissible European Union value, and for cadmium it was ten times higher.[11] Interestingly, the highest cadmium accumulations are present in spore-producing tissues like gills or pores, sometimes over ten times higher than in the stem, which usually contains the lowest heavy metal concentration in fruiting bodies, with the cap levels ranging in between.[12]

nuclear bomb tests in the 1950s and in even greater amounts from nuclear disasters like Chernobyl in April 1986, which spawned dozens of studies published in Europe on wild edible mushrooms. After Chernobyl, my family, along with countless others in Europe, quit collecting mushrooms for many years. We are lucky that, according to research by mycologist Matt Trappe, the Fukushima nuclear disaster in March 2011 did not dump a serious toxic load on the edible mushrooms in the Pacific Northwest.[13]

Other concerns include a wide range of industrial pollutants and toxins such as pesticides, herbicides, and fungicides, all suspected of harming the health of humans and other organisms. A Polish study looked at the residue amounts of persistent chlorinated hydrocarbons in Chanterelles and King and Bay Boletes and revealed that these organochlorine pesticides were present at rates several times higher than in local vegetables. Another Polish study showed DDT accumulation in fungi matches accumulated quantities found in mammals and fish similarly exposed to DDT.[14] Fortunately, cooking mushrooms can substantially reduce the amounts of pesticides and insecticides; the latter are also often used on nonorganic cultivated mushrooms. The longer you cook mushrooms, the more they will detoxify as the heat degrades and evaporates some of the toxins, as demonstrated for button mushrooms.[15] Areas receiving chemical treatments, such as golf courses, conventional farmed agricultural areas, and toxically managed backyards laced in herbicides and pesticides are best shunned by pot hunters.

Most relevant for foragers are warnings of heavy metal concentrations in wild edible fungi. Some species are better than others at taking up and accumulating heavy metals like cadmium, zinc, mercury, arsenic, and copper to a concentration in their fruiting bodies several times higher than the substrate they are growing in. High heavy metal concentrations can be found naturally in soils as a result of the local geology, especially in serpentine soils, but more dangerously are caused by pollution from sources like mining and resulting mine tailings, downwind from still-operating or former smelters, along highways, and in forest areas that are "fertilized" with biosolids—an industry euphemism for sludge from wastewater treatment facilities. Spreading this stinky sludge is unfortunately a common practice in some areas ("dissolution is the solution"). Better to heed the warning signs and keep a wide berth! A study from Eatonville, Washington,[16] reported that some of the mushrooms collected in conifer forests treated with sludge had extremely high concentrations of arsenic and lead: Western Amethyst Deceiver (*Laccaria amethysteo-occidentalis*), Rosy Slime Spike (*Gomphidius subroseus*), Orange Milkcap (*Lactarius luculentus*), and the White Coral Fungi (*Clavulina coralloides* and *C. cinerea*).

It is common sense to avoid picking wild edible mushrooms from old mining or smelter sites or from superfund sites in your area. Do your homework to map them out. It's also wise to avoid known concentrator species (see Heavy Metal Concentrators) unless you are collecting in a pristine environment. Urban and suburban foragers should be selective and accept that certain edible mushrooms are a bad choice from a health perspective. However, what is missing are studies that look into the health impacts of heavy metal bioaccumulation on consumers of wild edible mushrooms.

FORAGING GEAR LIST

Here are must-have items for any successful mushroom foraging trip:

- Mushroom basket, overflow bag, and mushroom knife
- Sturdy shoes
- Warm clothing (depending on the season)
- Rain protection (year-round, just in case)
- Water (more than you might think, just in case)
- Emergency whistle and mirror
- Emergency snacks like nuts, dried fruit, or protein bars
- Matches and/or a lighter
- Map, compass, and/or a GPS device
- Spare batteries
- Charged cell phone and portable charger
- Orange hat or bright clothing during hunting season

ADVICE FOR STAYING SAFE WHEN EATING MUSHROOMS

Here are a few suggestions that will help you to enjoy wild edible mushrooms and avoid unpleasant results.

- Only eat mushrooms known to be edible and that you can identify with absolute certainty.
- Always cook your mushrooms; overcooking is better than undercooking!
- Don't eat any mushroom raw, unless you have verified that they are safe eaten raw.
- After confidently identifying specimens, eat just a few bites at first of a species that is new to you to see if it is agreeable.
- Eat any culinary mushrooms new to you early in the day.

- Go easy: do not overindulge in any mushroom, especially the eat-with-caution species.
- Don't pressure other people to eat mushrooms.
- Don't mix several species that are new to your system into a single meal.
- Be acutely aware of habitat, such as polluted soils, mine tailings, pesticides, etc.
- Avoid rotten food—brown or gnarly flesh; old, formerly edible mushrooms can make you sick.

One study from Elinoar Shavit and Efrat Shavit in 2010 analyzed lead and arsenic levels in American Blond Morels (*Morchella americana*, formerly *M. esculenta*) collected in old apple orchards in New Jersey and New York.[17] The study discusses longtime mushroom hunter Mr. Bob Peabody of New Jersey, who died despite receiving chelation therapy once it was determined that he had heavy metal poisoning. Following investigations ruling out factors in his living and work environments, Peabody's physicians concluded that the lead and arsenic entered his system through regular consumption of apple orchard morels, which Bob had picked and eaten for many decades. Neither Peabody nor the mushroom hunting community were aware that for decades (until the practice was finally

outlawed in the late 1960s) a high percentage of these orchards had been treated with lead arsenate pesticides. Shavits' study confirmed these toxins linger in the soil and, alas, in the morels as well.

The lack of reliable studies on what could be devastating health impact of heavy metals intake from edible mushrooms indicates to a certain degree that there seems to be no major health risk in enjoying wild edible mushrooms. A German study tested human fecal matter for cadmium and copper after a three-day *Agaricus* diet.[18] It concluded that the hard-to-digest fungal chitin with its cadmium load "mostly passes through the intestinal tract unscathed without resorption." Hence, *Agaricus* fungi may not cause cadmium intoxication in humans.

In juxtaposition, an animal study showed that fungal cadmium was well absorbed in the liver, the kidneys, and the body overall by lab rats.[19] Still, the researchers closed their study with a statement about human impacts:

While it is concluded that the bio-availability of cadmium contained in mushrooms is not less than that of cadmium contained in other feeds and foods, most mushroom species do not contain more cadmium than other vegetables. In humans the average cadmium intake from the total diet is considerably below the tolerable level. Generalized warnings against consumption of mushrooms are therefore not considered necessary.

Despite this conclusion, the clear evidence of heavy metal accumulation in general in many choice edibles screams for more studies on the possible health effects. For this reason, I decided to have my heavy metal levels evaluated, especially since I have been enjoying substantial amounts of mushrooms collected in suburbia (shaggy parasols; king, birch and suede boletes; blewits; oysters, etc.) for several decades. Luckily, my test results did not show high levels of the heavy metals that mushrooms actively accumulate like arsenic, zinc, mercury, and cadmium. However, my lead values were way too high, but since mushrooms do not accumulate lead but rather reflect the concentration of lead in the environment. I attribute my lead levels to lead plumbing removal work I did. Chelation therapy is removing the lead out of my system, but sadly "my" lead might end up being sprayed into our woods as biosolid "fertilization."

STAYING SAFE WHILE FORAGING

One of the biggest challenges when mushroom hunting in the Pacific Northwest is not getting lost. It's easy to get

Entrance area of the Oregon Mycological Society Mushroom Show in Portland

disoriented in the woods in pursuit of our precious friends. It takes a certain awareness to simultaneously track your zigzagging through the woods while scanning the area for the next mushroom. Some people have an innate gift at keeping track of their orientation, but many do not. Know your limits—a night in the woods is no fun, and not every mushroom hunter has lived to tell the tale! It is easier to keep your bearings when foraging on a slope—no one confuses up and down—but it gets a bit challenging when crossing several ridges. The most challenging environment is a flat area or one with many little hills.

It is absolutely advisable to have a compass and/or a GPS device with you! However, both are useless if you do not know how to use them. If you are using these functions via cell phone, be aware that many forest areas have no reception (or only along major highways), although GPS satellite signals work without reception. Check your battery level before venturing out; a dead cell phone or GPS unit will not be of any help directing you back to your car.

The Pacific Northwest is blessed with having few dangerous and annoying organisms taking the fun out of mushroom hunting. Yellow jackets and ground-nesting hornets are two of those few, but they are usually not active in peak fall season. Most dangerous can be encounters with wild predators like mountain lions and black bears (and sometimes aggressive, mostly male hominids), but luckily most encounters are harmless. Still, take precautions, especially when foraging alone; it is always much safer to venture out with a buddy. Inform yourself about the possible dangers of your mushroom hunting area. West of the Cascades, there are no venomous snakes, and east of the Cascades, rattlesnakes prefer sunny, exposed places, not moist mushroom habitat. Learn to identify poison oak and poison ivy, and where they grow in the region.

Thankfully, there are not any chiggers in the Pacific Northwest, but ticks can transmit Lyme disease, Rocky Mountain spotted fever, and a few other tick-borne diseases in our region, though instances of such diseases are generally rare in the region. Lyme disease can cause decades of serious health issues. If you're bitten, it is essential to treat the infection right away. Usually an infection will cause a red swollen ring around the bite (or maybe not). Mainstream medicine will prescribe extended doses of antibiotics; herbal medicine recommends taking teasel root extract (*Dipsacus* spp.) immediately after infection.

COLLECTION PERMITS

In general, most public lands in the Pacific Northwest allow mushroom hunting. After all, what would be the point of public land that could be logged but that the public could not use to enjoy mushroom hunting? Limited mushroom collection is also allowed in many state parks and national parks in the United States. But many areas in the Pacific Northwest require permits for mushroom collection, whether commercial or for your own use. It is helpful to know the ownership and managing agency so you can check the collecting rules before you go.

The rules do not just differ between Canada and the United States, but

LEGALLY TRADED WILD MUSHROOMS IN WASHINGTON

The Washington State Department of Health limits wild edible mushroom trade in stores, farmer markets, and restaurants to the following wild harvested mushrooms: Golden, White, Winter, and Blue Chanterelles, Black Trumpet, Hedgehogs, King Bolete, Lion's Mane and Bear's Head, Cauliflower Mushroom, Saffron Milkcap, Matsutake, Oyster, Lobster Mushroom, Oregon White and Black Truffles, as well as all *Morchella* species. Note that Shaggy Parasols, Pig's Ears, and puffballs are not included. Similar limiting regulations have not been issued in Oregon or British Columbia.

also between states and provinces and county and local forest reserves. For example, in British Columbia, mushroom hunting is permitted on many Crown lands without a permit, but it is illegal to pick mushrooms in a provincial or national park. In the United States, in the Pacific Northwest, many state forests have restrictions on how many mushrooms you can pick, sometimes with differing quantities for regular people and people who have purchased a commercial collection permit. Privately owned forests also have their own rules, and some offer individual permits or sell commercial collection permits.

Be warned: the permit system is a jungle. Figuring out the rules and requirements is challenging. Each jurisdiction has their own permits and regulations; there is no one-size-fits-all permit, and the regulations change regularly. While in recent years several jurisdictions have made recreational collection permits available online, others still require an office visit, many of them closed on weekends or possibly hours away from where you would like to collect. Some collection permits are free, while others cost anywhere between $20 and $150, the latter for commercial picking. The funds collected by the US Forest Service go into the special forest products program. They are being used, for example, for expenses in connection with fire morel collection events, to produce multilingual informational materials, clean up commercial collector campsites, support collection permit programs, and manage environmental impacts on fish and wildlife, and people. To stay current on the permit system for the region would be its own job. The best approach is to talk to a local ranger or visit the website of the specific state forest, state park, national park, national forest, or private forest owner when the season starts and check for mushroom or forest product permits.

COLLECTING, CLEANING, AND PRESERVING MUSHROOMS

To get your mushrooms from forest to table requires several steps: picking and transporting mushrooms, cleaning and preparing them, and finally cooking, drying, pickling, storing, or otherwise preserving foraged wild mushrooms for your enjoyment.

PICKING AND TRANSPORTING MUSHROOMS

A firm container such as a basket or a bucket (the latter troubling my old-world-conditioned soul) protects mushrooms from being crushed. Also, a basket facilitates spore dispersal post-picking. In a pinch, plastic bags work during an outing, but not for storage. Keeping your mushrooms in a *closed* plastic bag too long renders them inedible, turning them into a stinky mush overnight. Cloth tote bags are a much better alternative. (I usually have paper shopping bags in my backpack for overflow: their firm material offers better protection than a thin plastic bag.)

Mushrooms react differently to transport: while rubbery chanties or firm king boletes don't mind a bouncy ride in a basket, fragile edible Amanitas or Brittle Milkcaps are quickly rendered unattractive and arrive in a thousand pieces. Having separate containers or bags—especially when picking small and big or firm and fragile mushrooms at the same time—and keeping species separate guarantees you'll arrive home with a much nicer bounty.

When you collect unknown mushrooms or specimens for identification, you have to **pick them with the complete stem, including the stem base**. Often the stem base features some essential clues to a mushroom's identity. Once you know an edible mushroom well, there is no point in collecting the soil-covered stem base. Whether you pluck your mushroom first or cut the

CUT OR PULL?

Each mushroom season, discussions on social media heat up regarding the proper way to harvest mushrooms. Should you cut them, leaving the stem base in the ground, or should you pull or pluck your mushroom, often with a small twist out of the ground? I prefer plucking so I can minimize waste by carefully cleaning (and keeping) the stem base. Many people have strong opinions, and it is nice to see that at the root of the discussion is the long-term health of the mycelium from which the mushrooms grow.

Portland's Oregon Mycological Society conducted a study at the foot of Mount Hood from 1986 to 1997 to research chanterelle growth patterns and the impact of chanterelle harvest on mushroom productivity. Nearly 5,500 chanterelle fruiting bodies were recorded. Of special interest were harvest comparisons of cutting versus plucking and the impacts on overall production. Three main types of multiple plots were established and monitored: in one group of plots the chanterelles were left untouched, in another group they were plucked, and in a third group they were cut. Overall, there was no statistical difference in fruiting body biomass production in areas where mushrooms were cut versus plucked. Interestingly, over the study's twelve years' duration, areas that were harvested produced slightly more chanterelles than nonharvest areas.[20]

stem without uprooting your mushroom, the key is to collect the mushroom as cleanly as possible. Soiled stems will smear perfectly clean gills and caps, requiring tedious unnecessary cleaning. While picking mushrooms can be ecstatic, cleaning sullied and broken mushrooms for hours at home is rarely experienced as such. The cleaner the mushroom is when it lands in your basket, the easier it will be for you to process. I love using my mushroom knife; its integrated brush helps me clean on-site.

Mushrooms past their prime usually suffer badly in transport, and you will probably discard them at home. Best practice is to leave them standing in the woods—they might still have a few days or weeks to produce and spread their spores. When field dressing my mushrooms, I like to leave no trace so as to not give my spots away. Also, I do not appreciate encountering other people's fungal debris (although I love examining it to learn what has been growing there). Nor do I like finding other people's flagging marking "my" mushroom spots; I love removing such plastic pollution (and other junk) from the woods, but I *do* respect and never mess with research or forestry marking.

CLEANING AND PREPARING MUSHROOMS

Foragers form strong opinions about washing mushrooms, but as always in life, things are seldom black and white. Mushrooms contain about 90% water and grow when there is plenty of rain. They have evolved to live with it. However, the nature of a mushroom's more delicate, spore-producing private parts makes a big difference in how well a mushroom handles washing. Boletes, with their spongy tube layers and densely gilled mushrooms, can retain surprising amounts of water, affecting cooking and especially drying. Yet carefully rinsing the cap surface or stem does not affect most mushrooms. Most of the time, brushing or wiping with a wet cloth or paper towel is more efficient.

Mushrooms growing in sandy soils require more thorough washing or scraping. Rain-soaked, dirty chanterelles, sandy morels, debris-loaded Cauliflower Mushrooms, or soil-sullied corals handle washing or spraying very well, which speeds cleaning enormously. Some people like to add a bit of vinegar to the washing bowl to break the surface tension of the water and help dislodge dirt and debris. When cleaning really sticky mushrooms, like jacks or slimespikes, remember to peel the sticky layer or use a veggie brush while rinsing the cap. Once you've washed your mushrooms, allow them time to dry on a net or towel before storing them or further processing them. In fall, you can place them outside for several hours or overnight (beware of garden slugs); when it is still warm, it's best to store them in the fridge. And of course, you can fry the freshly washed mushrooms right away, but dry fry them first—that is, start the frying process without oil, adding oil only after they lose their moisture and stick to the pan.

After cleaning, most mushrooms are best readied for further processing by slicing. Most commonly, mushrooms are sliced to about $1/8$ to $1/4$ of an inch (3 to 6 mm) thick. When processing boletes for drying, a slicer device that cuts a bolete into ten even, wide slices in one push will revolutionize your efficiency! Mushrooms that are already bite-size,

Fire morels: Note the fused, hollow cap and stem, creating one big cavity, typical for Morchella.

like Winter Chanterelles or Fairy Ring Mushrooms, do not need to be sliced. Sometimes I pinch stems from the caps so they fry better. Instead of slicing, many fibrous mushrooms like oysters and chanterelles can be easily ripped or pulled apart into nice pieces, an enjoyable process. Some mushrooms are much better enjoyed with minimal (or without any) slicing, like when barbecuing porcini, matsutake, or the Prince. Washing the cap surface protects them from burning when barbecuing them gill-side up. Whole caps of some gilled mushrooms like shaggy parasols, grisettes, and young oysters turn very tasty when the caps are fried whole, gill-side down.

COOKING WITH MUSHROOMS

Many people rely on their knowledge and habits of cooking vegetables or meat when first cooking with mushrooms. But one quickly realizes what mycologists have long known: mushrooms are not part of the plant kingdom; they are their very own creatures. Unlike plants that are structurally made of indigestible cellulose, mushrooms consist of the polysaccharide chitin (fascinatingly, also common in insect exoskeletons, crustacean shells, and fish scales). This tough, fibrous, glucose-based polymer is also hard to digest. Fungal cell walls are full of beta-glucans, one of the intriguing medically active components in mushrooms and commonly used to combat heart disease and high cholesterol.

No worries: chitin, glucans, and other huge, complex sugars contained in mushrooms do not pose any problem for diabetics, as they are not proper sugars but rather dietary fibers. Interestingly, chitin's structure was first revealed by Albert Hoffman—the same Swiss chemist who stumbled upon LSD produced by the ergot fungi and also revealed psilocybin in magic mushrooms. Luckily for consumers, mushrooms did not figure out how to strengthen their cell walls with sclerotin or calcium carbonate as insects and crustaceans do, which creates their incredibly tough outer shells. However, the more chitin present in fungal tissue, the tougher it is.

ADVENTURES IN PORCINI PREP

During a Mushroaming adventure in Tibet, we brought the fresh porcini we had picked that day in the mountain spruce–fir forest to a Chinese restaurant. We asked the Sichuanese cooks to prepare them and suggested that they first be browned in some oil. However, the cooks were used to preparing porcini in a soup without browning them first. The dish we got was far from the expectations we had built up that day while foraging.

We talked to the cook again and asked to have the porcini fried first. This idea was so outlandish to the chef that it took another round of sending the mushrooms back until we finally received beautifully browned porcini slices. They were very surprised that porcini could be prepared this way and would develop such an intriguing aroma and nice consistency.

A vegetable slicer living up to its potential—porcini slicing!

The tough stems of Shiitake contain much more chitin than the softer caps. And oyster mushrooms can have some of the lowest chitin levels of the well-researched major edibles, which makes them easily digestible.

Cooking mushrooms helps render them digestible. Heat is a powerful agent to help break apart chitin, giving the digestive system access to all the nutrients locked up in fungal cells. Enzymes that can break down chitin, called chitinases, occur in plants like common beans and tomatoes, and in fruits like banana, kiwi, and avocado to fight back insects. Chitinase is also produced in omnivorous animals like chicken and pigs, and humans have some as well. Results from a study in Italy showed that 80% of the tested participants had fungal chitin-digesting chitinase in their gastric juices.[21] One could speculate that the number of people lacking chitinase would be higher in the fungophobic UK and lower in countries that feed not only on mushrooms but also on insect larvae.

Not only does cooking improve the nutritional value of mushrooms, it is essential for rendering them edible, part of detoxification for many species. In general, don't eat any mushroom raw, unless you are sure it is harmless without heat detoxification. Many people can enjoy raw, edible species of *Agaricus*, such as the Prince, Meadow Mushroom, King boletes (*Boletus* spp.), Matsutake, culinary truffles, Beefsteak Polypore, Coccora (*Amanita calyptroderma*), and apparently some jelly fungi and puffballs. However, the likelihood of an adverse reaction is much higher when indulging in raw versus cooked mushrooms. It is best to first eat small amounts of an edible mushroom that is new to you to see how your system digests it uncooked.

Getting back to cooking, it is always better to overcook than to undercook your mushrooms. *Porcini ain't no zucchini!* The heat softens the tough polysaccharides, which makes the nutrients contained in the cells more accessible. And while many foods benefit from precise attention to internal temperature and cooking time, mushrooms with their chitin-reinforced cell walls are remarkably forgiving. They are still very enjoyable whether you cook them for ten, twenty, or forty minutes.

Another factor when processing mushrooms is the size of pieces or slices. The more surface exposed (as in smaller or thinner pieces), the stronger the impact of processing (and detoxifying). When enjoying mushrooms that are easier to digest, like oysters, chanterelles, or boletes, this is a moot point. However, when working with tough mushrooms or their tougher stems, thinner slices make them more palatable.

Most people regard mushrooms as just another vegetable, which is not that surprising, since most people hunt their mushrooms in the produce section. But mushrooms deserve specific treatment. A crucial process in cooking mushrooms is reducing their water content, usually around 90%—the softer the mushroom, the higher the water content. Reducing the water content firms up mushrooms nicely. It can be done by simply exposing them to heat in an oven or a frying pan. Defying veggie-informed cooking, mushrooms become firmer when boiled in hot water or other liquids. The heat will make the fungal cells release some water. However, water content in mushrooms is not the enemy. Sure, soggy mushrooms need to be dried out quite a bit to render them enjoyable, but juicy morsels of King Boletes or Shaggy Manes, or the caps of Shaggy Parasols are extremely delicious.

Sautéing, grilling, or roasting your mushrooms enhances their flavor substantially. Two important chemical processes take place when cooking: Maillard and caramelization reactions. Both improve flavor, especially to bring out a food's umami nature. The Maillard reaction is a nonenzymatic process that happens to amino acids when heated to 280 to 330°F (140 to 165°C) and thus browns onions, potatoes, steaks—and also mushrooms. Caramelization is specific to sugars in a similar temperature range and adds another enticing flavor component. If you take it a bit further, pyrolysis kicks in, the final heat-induced breakdown of organic matter leading to burning—something to avoid!

To optimize these processes when frying or baking, it is crucial that your mushrooms are not drowning in liquid. Overfilling the (frying) pan is the usual reason. When mushrooms piled too high release all their water, these processes get drowned out. Many people regard mushrooms fried in a single layer and turned when browned as the best-tasting mushrooms. When sautéing other vegetables, it is best to cook them separately from the mushrooms. The end result of golden, meaty, browned-outside, juicy-inside mushrooms will speak for itself! Mushrooms are versatile and diverse, and they can also shine their culinary light quite well without the browning. There are endless recipes to bring out the best qualities of each species of mushroom. (See Part Three for my favorite mushroom recipes.)

PRESERVING MUSHROOMS

Wild mushrooms are highly seasonal, so preservation and storage have long been of concern and many preservation

Fire morels drying in the sun to increase their Vitamin D load

techniques have evolved. The heyday of brining and canning is over; modern technology in the form of dehydrators, refrigeration, and freezing has revolutionized preservation and made it much easier to enjoy wild mushrooms year-round. Techniques are often informed by personal choice, tradition, and experience, but also by the species itself. For example, while king boletes and morels lose very little attractiveness when being dried, chanterelles are poorly suited for drying, unless it is done as a first step before processing them into mushroom powder. Chanterelles' fresh, firm-rubbery consistency hardens when dried, and rehydrated chanterelles maintain a tough consistency even when soaked for several days. Cooking chanterelles before freezing them is the way to enjoy your harvest in the future! Species-by-species processing and cooking requirements are included in the mushroom presentations in Part Two.

Drying Mushrooms

The most ancient approach is drying your mushrooms in the sun, under a cliff or roof, or over the fire, but that has changed since we discovered electricity and created appliances like ovens and dehydrators. These new gadgets let us control the temperature much better. To dry mushrooms, the recommended temperature is up to 130°F (55°C); however, many people like to use lower temperatures like 110°F (44°C), or even 95°F (35°C). Fungophile Mark Todd, the "Cheese Dude," swears by using such a low temperature, since it will prevent the white flesh of King Boletes from browning and allow you to store them in a state closer to their uncooked nature. For slices of bigger King Boletes, it takes two drying phases: first around six hours until they are dry on the outside, then a rest of about eight hours, during which the moisture still retained in the core spreads out. The second drying (three to four hours) should render the mushroom slices crisp throughout.

Mushrooms are done drying when they crack with an audible snap. The more interior surface that is exposed, the quicker mushrooms dry out. Increase the surface area by slicing your mushrooms—the thinner the slices, the quicker they will dry. Be aware that the cap and stem surfaces will resist drying the longest, having evolved an outer layer to protect the fruiting body from drying out. Drying is optimized by slicing pieces to an even thickness and using an electric dehydrator. Foolishly, I dragged my feet for years reluctant to get another big gadget, but efficient drying of mushrooms and fruit has enriched our lives! Try these three methods:

A fan. I'm often traveling when mushrooming, and I have found that setting a simple fan next to your sliced mushrooms makes an enormous difference.

An oven. While not the most energy-efficient way to dry mushrooms, it works fine in a pinch. Set the oven at the lowest temperature, ideally 130°F (55°C), but for sure below 200°F (93°C), with the oven fan on if available. Keep the door open a crack by wedging in a chopstick or wooden spoon to allow moisture to escape. Spread mushrooms on trays and move them at least every hour so they dry evenly. It can take anywhere from two to six hours depending on moisture content and slice thickness (ideally around or below ¼ inch). Set a timer and monitor frequently—you do not want to burn the whole bunch!

A dehydrator. Much more convenient is a dehydrator, whether home built and powered by old-style incandescent lightbulbs or bought used

POWDERING MUSHROOMS

Powdering mushrooms goes way beyond spicing up your popcorn! It allows fungal aroma to really infuse soups and sauces, from subtle extra umami to bold fungal flavor. In addition, powders make excellent butters (see Bolete Butter in Part Three). Turning them into powder allows you to use mushrooms that may have good to great aroma but poor consistency or flavor when fresh: too soft or slimy, like some jacks (Suillus spp.); too tough, like big, old oyster mushroom caps or the stems of honey mushrooms, Shiitake, or shaggy parasols; are on the bitter side like Hawk's Wings (Sarcodon imbricatus); or have a special flavor like Candy Caps or Pepper Boletes. In my family, as kids, we tediously hand-cut Hawk's Wings into tiny pieces to produce a mushroom "powder" that we gave away as Christmas gifts. These days we have plenty of appliances to grind dried mushrooms in seconds.

Keep in mind that by powdering, you substantially enlarge the surface area of the ground mushroom, which can speed up loss of aroma and decay. Thus, it's best to grind up smaller amounts that will be used up quickly. However, we have had some mushroom powders sitting on our spice shelf for years and they are still useful. Boletes make a particularly delicious powder.

or new and ready to go. Dehydrators come in many shapes and sizes. Their important features are temperature controls and timers. Mushrooms dry in anywhere from four to twenty-four hours, depending on many factors such as quantity of mushrooms and trays, moisture content of the mushrooms, and slice thickness. Dehydrators deliver the most evenly dried mushrooms.

In the Pacific Northwest in summer, drying mushrooms outside is easy during hot sunny days, but in fall, when many mushrooms are harvested, the hours of intense sunlight dwindle and the air is cooler and more humid. Drying your mushrooms might take several days, which could invite mold. Also, you need to move the mushrooms indoors overnight or cover them so that they do not absorb condensation during the night. It's best to spread your mushrooms out on a clean surface, an old bedsheet (no fuzzy material, as that will stick to your mushrooms), or screens, such as window screens. Screens or sheets off the ground allow air circulation above and below, speeding the process. Nylon screens are better suited than metal, which might rust or leave rust prints on your mushrooms. Rolls of nylon screen are relatively cheap, and you could take a frameless piece with you when backpacking for several days should you discover some mushrooms you want to dry.

Minimalists can fall back on stringing up mushrooms (literally, with needle and thread) for drying. Especially well suited are whole or halved morels, but bigger, more voluminous mushrooms like king boletes need to be sliced—and you better make sure squirrels can't raid your operation. They actually dry mushrooms themselves too, by taking them up in trees to dry and later stashing their dried mushrooms in a tree cavity!

The mushrooms are fully dried when they are no longer pliable—they will snap when crunched. Put them in an airtight container as soon as possible. Ziplock bags work fine for extended freezer storage, or for a short time in a pantry, but many insects such as meal moths can easily bite through plastic bags. Glass jars are much more reliable for long-term airtight storage. Preheating your jars before filling them will generate a tight seal once the glass cools off. Then store your prizes in a dry, dark place; sunlight can degrade dried mushrooms. Some morel lovers insist that their morels, when properly stored, simply get better with the years like a fine wine. In our basement we still have some decade-old Telluride Red Kings that work great, freshly powdered, to flavor bolete butter.

When drying your wild mushrooms, you want to increase their vitamin D_2 content. Wild mushrooms naturally contain some vitamin D_2, but exposing them to more UV light, the easiest source being sunlight, increases that content. Irradiation with UV light turns the sterol ergosterol, which is present in all mushroom cell walls, into D_2 and will remain stable in dried mushrooms for many years, according to research by mycovisionary Paul Stamets. The more of the surface that is exposed to UV light, the more vitamin D will be produced. So slicing your fresh mushrooms thinly and turning them over while exposing them to sunlight will maximize their vitamin D content.[22]

ON MAGGOTS AND LARVAE

Who doesn't despise finding a lively litter of larvae or murder of maggots when slicing open a stately King Bolete? It feels so violating, so devastating! Having many times discovered an internally ravaged mushroom, I thought it was time to turn the tables. Knowing that larvae are eaten and highly esteemed in many cultures did not change my attitude. However, informing myself that these worms (larvae) are made up of 100% fungal compounds, I thought I owed it to the mushroom and my readers to take the leap. The mushroom larvae I sampled turned out to be quite tasty! My favorite were the ones that floated on top of the brine when I parboiled a Cauliflower Mushroom for pickling. When biting on the larvae, one is treated to a rubbery little crunch and a white paste with a sweet taste!

Larvae are rich in protein and fats. The larvae of fungal gnats, mushroom fly specialists, are no different. However, for most fungivores and especially boletivores, it is challenging to keep an open mind about this "lowly" life-form, since it is usually guilty of destroying one of our favorite foods and sometimes rendering it gut-wrenchingly disgusting. While I cannot see myself getting excited about picking larvae out of rotten mushrooms as a nutritious delicacy, as some of our ancestors might have done, I will not let a few fresh larvae in my favorite mushrooms spoil my meal. As long as there is no bacterial decay, a little extra protein never hurt anyone.

Maggots have rendered this King inedible!

Sliced White Kings, Boletus barrowsii, *ready to go into the dehydrator*

Be aware when drying your mushrooms inside, especially when using a device like a dryer with a ventilator, because zillions of spores might float in the air. Some people experience flu-like reactions from such spore loads, which can last for several days. It's better to move such operations outside or into the garage. Similarly, driving for hours with fresh sporulating mushrooms can bring on such allergic symptoms; it's best to cover your catch!

Reconstituting Dried Mushrooms

As many ways as there are to dry a mushroom, there are even more ways to reconstitute them. Different mushrooms reconstitute very differently. The small, tasty Fairy Ring Mushroom rehydrates in minutes; sometimes it's possible to just drop them into a sauce while cooking. King Boletes can be ready to use in five to ten minutes. Other mushrooms like to take their time, sometimes several hours. For many recipes, you can allow dried mushrooms to finish rehydrating while cooking. Using hot water or bringing dried mushrooms to a quick boil can considerably speed up reconstituion. Some people swear by rehydrating their mushrooms in wine, vermouth, milk, or even cream—all interesting options to play with! Reserve any liquid that is not reabsorbed. It is enriched with fungal fragrance and might be as important as the meat of the mushrooms for the cooking process.

When rehydrating, make sure that your mushrooms are clean enough that you feel fine using the liquid. Sometimes they may need a quick prerinse or bath, especially for sandy morels or dirty Black Trumpets. Discard this wash water before immersing your mushrooms in the liquid you are using to reconstitute them.

Dried clean mushrooms are easy to handle; dirty ones are trickier. A tall, clear canning or mason jar that can be closed and shaken allows heavy sediments to settle. Some people use a French press dedicated to mushroom rehydration: the plunger allows you to push the mushrooms down and use just as much liquid as necessary. When removing your mushrooms, be careful to leave the sediment at the bottom. When using the liquid for cooking, pour carefully to leave the gunk at the bottom of the jar. If you want to sauté reconstituted mushrooms, express some extra liquid by squeezing. When working with rehydrated mushrooms, it might be best to start out by dry frying your mushrooms to boil away some of the extra liquid before adding butter or oil.

Freezing Mushrooms

A few great mushrooms, like king boletes, morels, or matsutake, can be frozen uncooked. When freezing fresh mushrooms, it is best to freeze young or firm mushrooms after having cleaned them well. A compromised consistency will take a serious hit from freezing and thawing, and such older or softer mushrooms hold up better if they are dried or cooked before being frozen. Slices, with their large, exposed surfaces, are predestined for freezer burn, a deteriorating process that includes dehydration and oxidation caused by air exposure; the resulting discoloration and change in flavor does not render them inedible but might kill the joy! Use freshly frozen mushrooms rather quickly (don't forget them in the back of the freezer). Vacuum seal them for longer freezer storage times with much less quality loss. Matsutake freeze well immersed in water, an approach that retains their aroma quite well. Chanterelles do not react well to being frozen fresh; they often become bitter.

Chanties are predestined for cooking before freezing, one of the best ways to preserve most mushrooms. At minimum, fry or bake your mushrooms in the oven to the point when they start sweating, or, more dryly stated, until they release their water. At this point you can stop the cooking process and let them cool. The cooking has changed the chemistry of the mushroom, and all the enzymatic processes in the mushroom are stopped. Decay has been slowed considerably, and you can store them for several days in the fridge. However, they do not get better over time in the fridge. The sooner you freeze them, the better. Pack them in ready-to-go portions for freezing; the right portion size is helpful for easy future use. Store them in wide-mouthed containers and the icy chunk of mushrooms will slide out easily for meals later. In years of total bounty, I oven-sweat chanterelles on trays, then let them cool before filling small freezer bags half full and rolling them up.

One of my favorite techniques is freezing fully cooked mushroom duxelles (finely chopped and cooked mushrooms; see Part Three) or mushroom sauces for use later with pasta, rice, polenta, and dumplings or as fillings for puff pastry. Often after a mushroom hunt, we cook all the mushrooms, enjoy a good portion in a meal, and freeze the leftovers. Our freezer is filled with a lot of ready-to-go, well-portioned containers of a wide variety of mushroom dishes. I call them our TV dinners, although we haven't had a TV in decades.

When suspended in a sauce, soup, or any other liquid, or finely chopped as in a duxelles, mushrooms freeze much better than bigger pieces that allow air pockets, which facilitate unwanted freezer burn. I press a little bit of plastic wrap on top of the mushrooms before closing a container to reduce exposure and limit freezer burn. Be sure to label all your frozen mushrooms, especially the species, date, and location—it might bring back nice memories—and also some of the ingredients like dairy, grains, or meat, so you can tailor your servings to people's food preferences or limitations.

Yes, fresh chanterelles are divine, but mushrooms that have been frozen a year or two do not raise an eyebrow in our home. I have enjoyed frozen mushroom dishes that have sat in our freezer for way too many years. Their quality does diminish over time, however, and I would not serve some of the "ancient" freezer mushrooms to guests. I also spare my wife these fungal time capsules, as she has a much more sensitive digestive system than I do.

Canning, Brining, and Pickling

Before fridges and freezers, there was canning, brining, and pickling to preserve mushrooms in a hydrated state. While canning uses heat treatment of boiling water and a pressure cooker to sterilize mushrooms and jars, brining relies on the antimicrobial power of salt, and pickling with varying amounts of salt enlists the acidity of vinegars. These methods are often combined. All require that the fresh mushrooms are first quickly parboiled, then cooked or fried until they release moisture. Spices like thyme, bay laurel, rosemary, and onion and garlic can be integrated.

The culinary outcome of the processes is quite different.

Straight canning of mushrooms for preservation without adding spices is a dated technique, and famous mushroom chef Jack Czarnecki regards canned mushrooms as the least flavorful. Also, since mushrooms contain some protein, if canned mushrooms are stored for an extended period and not completely sterilized, deadly botulism can develop, although it is extremely rare. To preserve mushrooms for months or years after initially boiling or frying them, sterilize the filled canning jar in a pressure cooker or in a boiling water bath for another 30 to 45 minutes. Brining requires anywhere from 5% to 10% salt saturation in the brine, and the outcome is very salty mushrooms. The saltiness can be ameliorated to a degree by first washing the brined mushrooms and, if they are still too salty, placing them in fresh water to let the salt leach out. Firm mushrooms like milkcaps and brittlegills brine best, and it's best to use brined mushrooms in dishes that handle the extra salt input.

Brined and pickled mushrooms are very popular in Eastern Europe, and there is the well-loved tradition of eating them straight out of the jar and washing them down with vodka! Interestingly, traditional mushroom brining has not really caught on with the Pacific Northwest mushroom community. Most popular in the West is pickling mushrooms in vinegar and with spices (see recipe, Pickled Chanterelle Buttons).

PART TWO
THE
MUSHROOMS

The selection of mushroom species introduced in this section is based on the overarching theme of edibility. However, mushrooms cannot simply be divided into black and white, poisonous and edible. We have plenty of great, safe edibles in the Pacific Northwest, but there is also a culinary fungal twilight zone, a fascinating field that does not get enough exposure.

OPPOSITE: *A strong, young King Bolete, Boletus edulis*

LEFT: *Thick Cup;* RIGHT: *Lobster Mushroom;* OPPOSITE: *Basket of Fire Morels*

ASCOMYCETES

Most of the fungi in the Ascomycetes, the sac fungi, do not produce mushrooms we would consider eating, but a few do and they are an interesting and important group. Ascomycetes belong to the division of Ascomycota (the other being Basidiomycota) of the "higher fungi" (the subkingdom Dikarya) within the kingdom Fungi. They are often referred to as "Ascos."

Spring is the season when most edible mushrooms belonging to sac fungi fruit. Though most "Ascos" do not produce mushroom-like fruiting bodies, in spring these mushrooms fruit, including morels (*Morchella*), Spring Morels or Thimble Caps (*Verpa*), and Brain Mushrooms (*Gyromitra*). In summer and fall, we search for the illustrious Lobster Mushroom (*Hypomyces lactifluorum*), and in late fall, elfin saddles (*Helvella* spp.) and Oregon White and Black Truffles (*Tuber oregonense* and *Leucangium carthusianum*), which fruit through winter. The most commonly encountered Ascos in the woods are a great variety of cup fungi, growing in all sizes, many quite minute, some showing off delightful colors.

An important aspect of Ascos are yeasts used to make wine and brew beer as well as to bake, the most important being *Saccharomyces cerevisiae*. These yeasts cause alcoholic fermentation by feeding on starch and sugar and turning them into carbon dioxide and ethanol. The CO_2 carbonates beer and leavens bread. The alcohol is retained in beverages but evaporates while baking. Another famous Asco genus is *Penicillium*. Some species produce bacteria-killing antibiotics; others such as *P. camemberti* and *P. roqueforti* produce tasty soft cheeses.

MORELS, *MORCHELLA*

What is it about morels that stirs such passion? Perhaps it is that they symbolize the end of the dark winter—spring has sprung! Hunting for elusive natural morels camouflaging as pine cones and finding them nested among blooming Calypso orchids is one of the most exciting ways to immerse yourself in the powerful return of the life force. Similarly, finding flushes of fire or burn morels poking through burned soil in a devastated ashen landscape attests to the tenacity of life. Admittedly, picking landscape morels from a monotonous, beauty-bark bed in suburbia does not have the same glory, but a precious fungus emerging unexpectedly in a biodiversity desert makes our hearts jump.

Morels are mysterious creatures, and our understanding of their ecology is still spotty. While many of them grow in connection with specific tree species, we do not understand the nature of the association. The association does not match any of the known mycorrhizal symbioses. Furthermore, a few species like fire morels[23] and the landscape morels (*Morchella importuna* and *M. rufobrunnea*) grow without tree-root associations. Landscape morels are now being cultivated on a large scale, especially in China.

Another mystery is the toxicity of morels. Raw morels seem to cause serious poisoning in most people, but even well-cooked morels can cause minor to serious gastrointestinal distress with bloating, abdominal pain, vomiting, and diarrhea. The actual toxin or toxins are still unknown, but they seem limited to freshly cooked morels and appear not to be related to hydrazine (also known as a component of rocket fuel) as contained in the Gyromitras. Reported neurological problems include dizziness, disorientation, a staggering gait, fine tremors, and blurred vision; these symptoms can occur within a few hours, and are not limited to undercooked morels, but are usually gone by the next day.[24] However, a mushrooming friend of mine suffered through half-body paralysis. After falling in love with freshly collected morels, he ordered more online, and then he apparently overindulged, and ended up in intensive care for several days. Thankfully, he regained control over his nervous system after a week. Others who have eaten morels for years can develop a reaction. If you are a part of the unfortunate "morelly" challenged 10% and show signs of an allergic reaction, you are well advised not to ingest morels.

Morels are used in traditional Chinese medicine to tone the GI tract, reduce phlegm, and treat shortness of breath. Morels have demonstrated in vitro activity against *E. coli* bacteria, and their polysaccharides have shown anti-tumor and anti-inflammatory activity.

Plenty of reports suggest that mixing alcohol and morels could cause trouble, yet many people enjoy morels with red wine, and as mycotoxin expert Denis Benjamin reports in *Mushrooms: Poisons and Panaceas*, there is no scientific evidence for alcohol being an aggravating factor yet. The German mycological society refuses to include morels as a recommended edible, but acknowledges that their advice is mostly ignored, since people love morels too much and the tradition of morel enjoyment is deeply ingrained. The name "morel" and its scientific name, *Morchella*, are derived from its German name, *Morchel*, which itself is a reference to *Möhre*—the carrot or beet—due to similarity in shape. German immigrants brought morel mania to this continent, a joy still intensely celebrated each spring in the Midwest and spreading like wildfire everywhere else!

Germany, Switzerland, and Austria, followed by France, are the world's biggest importers of morels, which are sourced from Turkey, Pakistan, India, and eastern Tibet. A lot of fire morels are sourced from western North America.

For several years in the late 1990s, I observed how eastern Tibetan morel resources were developed. Traditionally, morels were mostly ignored as edibles in Tibetan culture. Chinese Muslim mushroom buyers visited Tibetan areas with the right conifer forest habitat, showed local Tibetans what they were interested in, and offered very high prices for morels. However, once people learned to collect

Delighted Fire Morel huntresses

A Tibetan morel hunter holding Khukhu Shamo

photos show up on social media, but to be ahead of the curve, you want to have a more ecosystem-based approach. In the mountains, morels start fruiting several weeks after the snow melts: the soil has to warm up first. The most straightforward scientific approach is probably checking for 53°F (12.7°C) soil temperature. One year I armed myself with a thermometer and was getting readings at different altitudes and different soil depths. If you check the soil, it's important to measure between 4 and 6 inches—an important detail I did not know back then. However, that day I found my naturals without the help of a thermometer and never bothered to bring it again. ("Naturals" are morels that emerge annually without needing a fire to trigger their fruiting.)

them and brought in good quantities, the buyers dropped the price dramatically. The Tibetans felt betrayed and some ugly confrontations ensued. Also, locals figured out how to insert stones into hollow morels to receive "fair prices." The first few years were a bit rough, but the market mushroomed quickly; ten years later, even European buyers were showing up in the Tibetan hinterland. Tibetans call morels "Khukhu Shamo" ("cuckoo mushroom"), since it appears at the same time in spring as the high-flying bird that craftily outsources child rearing by laying its eggs in other species' nests before migrating back to overwinter in mild India.

WHEN AND WHERE TO FIND THEM

Morel hunters have all kinds of indicators and theories for when and where they will pop in spring. Nowadays you can just pay attention to when morel

Some people swear by using plants as indicators, be it their height after emerging or their flowering times. You might even learn to recognize perfect habitat and timing by a specific plant. Showy sun-flowered balsamroot (*Balsamorhiza*) announces morel season east of the Cascades. Most iconic is the fairy slipper (*Calypso*) orchid with its exquisite purple-pink flower. Different regions and habitats will have different indicators, and correlating common, easily visible plants with morel fruiting is definitely a strategy worth pursuing, especially if you can correlate the appearance of your morels in the mountains with, let's say, the lilac flowering in your yard.

Good morel years will have plenty of soil moisture from spring rain or snowmelt and mild air temperatures that are gradually warming, not jumping from nightly lows in the 30s to daily highs in the 80s. Heat dries out the soil and the morels. Morels grow first on sunny south-facing slopes, which warm faster and earlier, and fruit a bit later, but often for much longer, on shady north-facing slopes.

Looking for morels, especially the naturals in unburned forests, is often tricky business; they are just so darn hard to spot. You will find yourself looking intensely at a forest floor not seeing a single morel. And then surprisingly you see your first morel—and suddenly a whole bunch! The wild thing is, these morels were there the whole time you looked intensely and saw zilch! Clearly, perception is generated in the mind, and you need to adjust your filters.

The more often you hunt morels, the easier it becomes to see them. Finding naturals is much harder than finding burn morels, since seeing their fruiting bodies in a regular landscape with all its distractions is much, much harder than seeing morels on burned, ashy ground covered in fallen needles after a wildfire. Many morel fungi respond to habitat disturbance or destruction by fruiting. Logging, thinning, or road construction will induce fruiting the next spring. Morels also often appear in spring in places where concrete was recently poured or sheet rock pieces were left behind. It seems like the extra influx of minerals common in lime and ash, such as calcium carbonate and others, enhances or even triggers fruiting.

A group of fire or burn morels, given sufficient soil moisture and precipitation, will fruit in impressive numbers the first spring or summer after a forest fire. The second year, a few morels will grow but in just a tiny amount compared to the first postfire flush. Morel hunting in burns is easier for several reasons: morels fruit in much higher numbers, they are much more visible, and the burned habitat is easy to recognize. However, if a fire burned too hot, it likely destroyed all sclerotia. Sclerotium is the overwintering fungal equivalent of a potato, the nutrient storage organ

Burned mountain conifer forest in the Cascades

Northern Morel, Morchella norvegiensis: *Note the cavity created by the fusion of stem and cap, seen here in profile,* typical for all Morchella *species.*

that—unless toasted in intensely burned areas—allows morels to grow rapidly in spring.

The main challenges facing fire morel pickers are overeager administrators and a lot of other biped morel predators. The easy harvest attracts mushroom hunters from near and afar, but usually there are plenty of morels for everyone. Easily accessible, online wildfire maps let every morel lover precisely locate burns. A real challenge can be when the Forest Service closes off burns and access roads, making it much harder to get into perfect habitat. However, a plus is that since fire morels fruit basically only once and in such abundance, fellow mushroom aficionados are generous when it comes to sharing burn information.

COMMONLY FORAGED PACIFIC NORTHWEST MORELS

The combination of a fairly thin, hollow fruiting body made out of chewy, firm flesh is unique to morels. This great structure and the fabulous rich fungal flavor of most species render them very versatile in the kitchen. When harvesting morels, cut the stem at ground level. Often there is a lot of good biomass in stem bases, so it pays off to cut close and carefully, but make sure there is no dirt attached. Sometimes morels can be quite sandy or splashed with dirt or ash after heavy rains, so they may need to be washed. Washing also encourages millipedes, ants, and other insects hiding inside the mushroom to abandon the sinking ship. However, these squatters are fairly rare, especially in burn morels. Usually I do not wash my morels, and if there is a little bit of sand after drying, it comes off easily during rehydration and settles at the bottom of the jar. Discard these dregs; do not use them to enhance your sauce like the rest of the precious soaking liquid.

There are some general features that all true morels share, though their lifestyles and habitats can differ in the Pacific Northwest. Identifying morels

PROCESSING AND COOKING MORELS

The hollow nature of morels allows for easy drying, including simply stringing them up. Using the sun, often plentiful during morel season, facilitates vitamin D production (see Drying Mushrooms in Part One). Morels can be dried whole, halved, or in strips depending on size and personal preferences. Drying morels will reduce the toxicity that is common to all fresh morels. However, cooking morels thoroughly is an important part of detoxifying them. Most troublesome experiences seem to derive not from dried morels but from undercooking fresh morels.[25] Dried morels are easily rehydrated within fifteen to thirty minutes, and using hot or boiling water will speed things up (see Collecting, Cleaning, and Preserving Mushrooms in Part One). Many morel enthusiasts insist that when properly stored—airtight, protected from light—dried morels improve with the years just like a fine wine.

As long as the moisture is not trapped inside their hollow bodies, fresh morels store well. They keep best in a paper bag in the fridge; like all mushrooms, they will rot in an airtight container or plastic bag. Stored properly, fresh, larvae-free morels will easily last for several days to a week or more. Still, it is best to check your treasure once in a while to remove mature morels gone bad.

Morels offer juicy, firm-textured morsels of fungal goodness, making them extremely versatile! When preparing for cooking, it is nice to leave small ones whole or slice them into beautiful halves. Midsize morels are easily sliced into strips or gorgeous wheels. The biggest morels are perfect for stuffing (see Stuffed Morels in Part Three). Morels are great in wine or cream sauces, as well as in duxelles, enjoyed with pasta or rice. They are excellent fried and paired with all kinds of greens like peas and asparagus, or used with eggs in omelets or quiches, or simply fried in butter. When powdering morels, detoxify them by cooking them first; frying them with onions or shallots can spice them up nicely.

Morel wheels

Comparison of main types of morels: Left to right are a Brown Morel, Morchella brunnea; *Spring Morel or Wrinkled Thimble Cap,* Verpa bohemica; *and a Brain Mushroom or False Morel,* Gyromitra esculenta

is much more straightforward than the health advisory around eating them mentioned earlier. Their fruiting bodies range from cone-shaped to round, reaching 5–20 cm in height, and their stem and cap are fused together. Fertile caps of black morels are honeycomb-like, with vertically arranged ridges, lighter-colored pits, and fairly pointed caps. Blond morel caps have more irregularly shaped ridges and pits and often rounder caps. Morels fruit mostly in spring, or early in summer at higher elevations. Their habitat varies by species and ranges from beauty-bark beds, orchards, riparian forests, mixed forests, and conifer forests to recently burned forests.

The Spring Morel, *Verpa bohemica,* and the False Morels, *Gyromitra esculenta* and *G. montana,* are somewhat similar, but slicing your morels in half will quickly reveal if you have a true morel or not. In true morels the stem and cap are fused, forming a single cavity; in *Verpa* and *Gyromitra* they are not.

All true morels are in the family Morchellaceae, order Pezizales. True morels are easy to identify as a genus, but identifying individual species is challenging. Species do not exhibit clear-cut features, and even specimens within a species can vary in shape, color, and size. It took advanced phylogenetic DNA studies to distinguish our Pacific Northwest species in recent years. Here is what we know: there are three different sections in *Morchella*: black morels (*M. elata* clade, "Distantes"), yellow or blond morels (*M. esculenta* clade, "Morchella"), and white morels (*M. rufobrunnea* clade, "Rufobrunnea").[26]

Currently there are more than a dozen species in the Pacific Northwest, most of them black morels. A bit of an oddity is the Half-Free Morel, *Morchella populiphila* (formerly *M. semilibera*), the common name referring to the lower half of the cap being free while the upper half is fused to the stem. The scientific name informs us that this morel grows with black cottonwoods. Our common blond morel is *M. americana*, but there is also a second blond morel in the Pacific Northwest, the Contorted Blond Morel (*M. prava*). The white morels encompass globally only two species, and here in the Northwest we find the Blushing Morel (*M. rufobrunnea*).

 **AMERICAN BLOND MOREL**

Morchella americana

The irregular shape of the ridges and pits of *Morchella americana* makes it easy to recognize. As in all true morels, the fertile cap tissue is fused to the stem and creates one continuous hollow space inside. The color of the ridges varies from white to pale yellow when young, remaining whitish or turning to brownish-yellow, sometimes with some reddish staining. Pits can be grayish-brown to nearly black when young, becoming pale brownish-yellow.

In the Pacific Northwest, the American Blond Morel typically grows around black cottonwoods along rivers. They are also found around ornamental ash trees and in apple orchards. The very similar but often smaller (height below 10 cm) and seemingly deformed Contorted Blond Morel (*M. prava*) can be found, according to retired Evergreen State College mycologist Michael Beug, in orchards and along rivers as well, but also in vineyards and conifer forests. Avoid morels growing in old apple orchards that were sprayed for decades with toxic lead arsenate pesticides.

NAME AND TAXONOMY: *Morchella americana* Clowez and C. Matherly 2012; *M. esculentoides* M. Kuo et al. 2012. Before often referred to by the name of its Eurasian cousin, *M. esculenta* Fries 1801.

Mature (left) and young (right) American Blond Morel, Morchella americana

Mountain Black Morel, Morchella snyderi

Mountain Blond Morel, Morchella tridentina

! MOUNTAIN BLACK MOREL
Morchella snyderi

Snyder's Morel is one of our "naturals," meaning that it fruits annually in the same location without needing a burn to trigger it. Its most striking feature is the pocketed or buttressed stem. The ridges are arranged vertically as is typical for black morels. When young, the cap is pale yellowish, but ridges mature to smoky-brown or black. Preferred habitat is mountain conifer forests including Douglas-fir, ponderosa pine, and white fir, but sometimes Snyder's will fruit in beauty bark and also in big clusters. Depending on altitude, naturals fruit between April and early June, usually in the same location year after year if the weather allows.

Very similar are two other natural species: the Northern Morel, *Morchella norvegiensis* (see photo, p. 64), a conifer-associated morel occurring from sea level to tree line in the Northern Hemisphere,[27] and the often smaller, natural Brown Morel (*M. brunnea*; see the earlier photo of the three types of morels),

which has a thinner, more fragile, and simpler nonconvoluted stem and which you will notice sports hairless pits when you examine it with your loupe. It has a preference for alder and aspen groves, but is also found with conifers. Alan Rockefeller has observed both Brown and Mountain Black Morels fruiting on burns several years after a fire has occurred.

NAME AND TAXONOMY: *Snyderi* honors Leon Carlton Snyder (1908–1987), who studied morels in Washington; *brunnea* is Latin for "brown;" and *norvegiensis*, latinized Norwegian. *Morchella snyderi* M. Kuo and Methven 2012; *M. brunnea* Kuo 2012. *M. norvegiensis* Jacquet. ex R. Kristiansen 1990 = *M. eohespera* Beug, Voitk, and O'Donnell 2016.

! MOUNTAIN BLOND MOREL
Morchella tridentina

Recognizable as a black morel by its vertically aligned whitish ridges that darken only when old, this is the palest of the black morels, though others can start out very pale when not exposed to light. Mountain Blonds usually fruit after Mountain Black Morels. Though not a fire morel, it is frequently observed on the edges of burns. Interestingly, our Mountain or Western Blond Morels are essentially genetically identical to mountain morels described in the Southern Alps in Trentino, Italy, hence *Morchella tridentina*.

NAME AND TAXONOMY: *Morchella tridentina* Bres. 1898 = *M. frustrata* M. Kuo et al. 2012.

! FIRE MORELS
Morchella eximia, M. sextelata, and *M. exuberans*

These three species of morels are easily recognizable as fire morels, since they fruit in burned-over areas and, like other black morels, are dark with longitudinally arranged ridges, often running continuously from the base to the top of the cap. While DNA analyses clearly differentiate between these three burn species, telling them apart is challenging. Some claim that the Luxuriant Morel (*Morchella exuberans*) can be differentiated by an internally folded and chambered stem, while the Exceptional Morel (*M. eximia*) and Sexy Morel (*M. sextelata*) both have simple stipes. Furthermore, *M. eximia* seems to like fruiting close to creeks (as well as elsewhere on burns), but otherwise so far *M. sextelata* and *M. eximia* cannot be distinguished by their appearance. *M. eximia* was originally described from a burned area in France more than a hundred years ago and has also been found in burned areas in Asia. There are claims that *M. exuberans* can be greenish ("pickles"), but pinkish and greenish tones do not seem to be specific in fire morels.

NAME AND TAXONOMY: *Eximia*, Latin "exceptional," *Morchella eximia* Boud. 1909; *M. sextelata* (Latin "sixth" type of *M. elata*

LEFT: *Young Luxuriant Morels,* M. exuberans; **RIGHT:** *Most likely an Exceptional Morel,* M. eximia

Fruitings from two types of fire morels

Chesnaux 2012, *M. septimelata* (Latin "seventh" type of *M. elata* group) M. Kuo 2012.

 ## GRAY FIRE MOREL
Morchella tomentosa

Of all our fire morels, this species is most easily recognized due to its black velvety stem when young (hence its nickname, fuzzy foot morel). With aging, and alas we know this phenomenon, thinning of hair turns this fuzzy foot into a gray morel. This citizen of burned mountain conifer forests is beloved for two reasons: it can be quite meaty (many specimens have a double-walled stem), and they flush on burns after most people are done collecting the early fruiting bunch of fire morel species. In high-altitude burns they can fruit into August with sufficient rains.

Young and mature Gray Fire Morel, Morchella tomentosa

group) M. Kuo 2012; *M.exuberans* (Latin "luxuriant") Clowez, Hugh Sm, and S. Sm. 2012. The following species are all synonymous with *M. eximia*: *M. anthracophila* (Latin "coal-loving") Clowez and Winkler 2012, *M. carbonaria* (Latin "charcoal") Clowez and

Laddered Landscape Morels, Morchella importuna

NAME AND TAXONOMY: *Tomentosa* Latin "densely matted." *Morchella tomentosa* M. Kuo 2008.

> ## (!) LADDERED LANDSCAPE MOREL
> ### *Morchella importuna*

The Laddered Landscape Morel is the most common black morel growing around human habitations and in disturbed areas in the Pacific Northwest. In the western lowlands in bark mulch landscape beds, it often fruits as soon as early March. It also grows around burn pits and in disturbed and logged areas, often when trilliums start to bloom, according to Michael Beug, who regards them as very good to eat. It is fairly easily recognized by its close, ladderlike arrangements of ridges and by its early fruiting. Individual mushrooms can reach 20 cm in height.

You can count on this motto in March: "I do not go anywhere to look for morels; but wherever I go, I look for morels!" The joy of finding it in unexpected locations easily outweighs its lack of aroma. Blandness is also an issue when buying these morels from big greenhouse cultivation operations now established all over Sichuan, China, and slowly catching on in the United States.

NAME AND TAXONOMY: *Importuna*, Latin for "troublesome," since some ignorant homeowners were distressed that this morel invades their gardens. This morel is genetically extremely close to the European *Morchella elata* (Fries 1822). However, the holotype (the original mushroom used for the first description) is lost. A substitute in the same location matching the original description needs to be selected to be used as a neotype for DNA analysis. Confusingly, Fries's short description matches several Swedish black morel species. *M. importuna* Kuo, O'Donnell, and T. J. Volk 2012.

BLUSHING MOREL
Morchella rufobrunnea

Much more common in California than in the Pacific Northwest, this fully saprobic morel usually grows in disturbed areas close to human settlements and often on wood chips, hence its other name, the Landscape Morel. It has very light-colored, vertical ridges with often darker pits standing in stark contrast. The flesh of these morels also stains red in age and with handling (untouched, these morels do not show this characteristic). This cultivatable morel is the blandest tasting of all.

NAME AND TAXONOMY: *Rufo-brunnea* Latin "reddish-brown," for the color change from bruising. *Morchella rufobrunnea* Guzmán and F. Tapia 1998.

Blushing Morel, Morchella rufobrunnea

FALSE MORELS AND CUPS

Under the common name "morel," two genera besides *Morchella* are invoked, *Verpa*, Spring Morels, and *Gyromitra*, False morels. Whereas *Morchella* and *Verpa* are in the same family (Morchellaceae) and thus similar in many regards, False morels or Brain mushrooms, *Gyromitra* are less closely related and in the family Discinaceae. This distinction is also reflected in the presence of different toxins in some *Gyromitra*. To emphasize the difference between *Morchella* and *Gyromitra*, we use the common name "Brain Mushrooms," which steers clear of the word *morel* for *Gyromitra*. Also included in this section are the Elfin Saddles, closest to morels in stature, some disc-shaped *Gyromitra* species reminiscent of cuplike fungi, and a few other types of true cup fungi.

 ## SPRING MOREL
Verpa bohemica

The first messenger of the return of fungal fruits is the Spring Morel (also known as the Wrinkled Thimble Cap). Some curmudgeons insist that being the first fungus is the only thing Verpas have going for them, but that does not do justice to these beacons of early fertility. Culinarily they compete with landscape morels (admittedly a low bar because of the bland flavor), but that should not take away from the joy of hunting fruitfully on a beautiful spring day and returning home with an enjoyable meal.

Luckily most people do not have to travel far to find Spring Morels. They are associated with omnipresent black cottonwoods that even release their sweet balsamic scent to remind us to look for their fungal friends. Cottonwoods like their feet wet and often grow in wetlands or along creeks and rivers. Don't bother looking where it is soggy wet, however: thimble caps do not like their mycelium flooded, and they often hide in the deepest blackberry brambles.

Some *Verpa bohemica* reach 20 cm; being taller than many true morels makes them easier to spot. The free, wrinkled cap (the cap is not fused with the stem, but sits connected atop the stem) has the firmest texture and offers the most interesting bites. The hollow stem is rather fragile and open; it can be stuffed with cooked mixtures and turned into a surprisingly satisfying dish. Just like true morels, Spring Morels have to be cooked very well to be rendered

LEFT: *Spring Morel, Verpa bohemica;* RIGHT: *Bell Morel, Verpa conica, with Calypso orchid*

harmless. In the past a lot of people advised mushroom pickers to parboil Verpas, but that is unnecessary. A good, thorough frying will detoxify them easily. They are in no way more troublesome than true morels.[28]

IDENTIFICATION

- Yellow-brown or tan cap (Ø 2–5 cm), darkening with age, deeply furrowed and wrinkled.
- Bluntly conic to bell-shaped, free cap only narrowly attached to stem top.
- Stem hollow but stuffed with whitish cottony pith, equal to club-shaped, 6–15 cm (up to 25 cm) tall.
- Flesh rather thin and fragile with indistinct odor.
- Fruits solitarily or in groups around cottonwoods in early spring.

LOOK-ALIKES: The Bell Morel or Smooth Thimble Cap (*Verpa conica*) is a "relaxed" Verpa—the cap is hardly wrinkled! It has a smooth or slightly wrinkled, ochre-brown to reddish-brown, bluntly conic to bell-shaped cap. It also fruits on the ground in spring and is edible after cooking. The stuffed white stem attaches only at the top (like *V. bohemica*) and may exhibit transverse belts or ribs from fine brown granules.

NAME AND TAXONOMY: *Verpa* Latin colloquial "male member or penis," and *bohemica*, since it is beloved by Bohemians, or rather was first described in Bohemia, Czech Republic. *Verpa bohemica* (Krombh.) J. Schroet. 1893. Morchellaceae, Pezizales.

 **BRAIN MUSHROOM**
Gyromitra esculenta

This distinctive mushroom is also called the False Morel, but if we insist on "morel" in the common name it should be the Toxic Morel. Though the scientific name *esculenta* translates from Latin as "edible," for some people, it was only edible once due to a fatal outcome! The symptoms of the tragic poisoning are quite similar to those caused by the notorious amatoxins present in Death Caps and allies. However, the Brain Mushroom's toxic agent, gyromitrin, is very volatile and can apparently be neutralized, a crucial difference to the heat-stable amatoxins.

Not surprisingly, differences in opinion on the edibility of Brain Mushrooms is the fungal equivalent to political polarization in the United States. While myco-conservatives fearfully focus on the lethal toxicity, myco-progressives point to the transformative power of proper processing, which is less challenging than correctly gutting fugu, the also relished, potentially deadly Japanese blowfish. While most mushroom hunters shun Brain Mushrooms for good reasons, since plenty of foragers have been seriously sickened, others praise its flavor and structure and claim it superior to regular morels. Some myco-progressives

invoke Finnish mushroomers who dryly state: "If False Morels were deadly, there would be no Finns!" However, many people don't like the idea of ingesting a mushroom that contains traces of the carcinogenic agent gyromitrin when raw, which the human stomach turns into monomethylhydrazine, or rocket fuel. Gyromitrin was recognized as the actual toxin based on symptoms encountered by a rocket scientist, and these mushrooms may contain traces of this substance even after they have been cooked.

Not surprisingly, this is a highly volatile chemical, but it dissipates easily when dried or cooked. When Brain Mushrooms were still en vogue in Europe and eaten by the ton, most were consumed after being dried. However, recurring fatalities have resulted in the outlawing of their trade in the EU. In Finland, *Gyromitra esculenta* are still legal, but they must include a note that instructs

Brain Mushroom, Gyromitra esculenta

LEFT: *Thick Cup,* Gyromitra ancilis; RIGHT: *Bleach Cup,* Disciotis venosa

popping in the frying pan. Finally, (3) several similar species can be told apart to a certain degree by color and size, but really, microscopy is needed to check spore details, since to the naked eye they can be dead ringers. Research published by Alden Dirks in 2023 surprisingly revealed that the Orange Thick Cup, *Gyromita leucoxantha* contains gyromitrin, while all other closely related Thick Cups, classified in the subgenus *Discina* of *Gyromitra,* which also contains *G. montana,* are lacking gyromitrin and hence are regarded as edible if well cooked. Subgenus *Discina* does not include *G. esculenta,* which contains the deadly toxin gyromitrin.

David Arora did not offer keys for these species, listed as *Discina* in his seminal book *Mushrooms Demystified,* with the promising common name of "pig's ear" (which is more commonly used for the unrelated *Gomphus clavatus*). This group of tough, wrinkled, disc- or cushion-shaped, round Ascos includes the Olympic Thick Cup, *Gyromitra olympiana,* which is usually a bit smaller with a less pronounced stem (more a point of attachment than a stem). It is brighter, with tones of yellow, than *G. ancilis.* The very rare and toxic Orange Thick Cup, *G. leucoxantha,* is the lightest in color of the bunch, with a yellow-brown to orange-brown cap that dries to bright red-brown. The Black and White Thick Cup, *G. melaleucoides,* with its gray-tan to dark brown upper fruiting body, stands out with its ivory to off-white underside that lacks prominent ribs and quickly turns a starker white as it dries.

Shape, size, and colors of mature specimens of these species can overlap. Their variability suggests that some daredevils may have already eaten some, and survived, and that they all might be safe, at least after they have been cooked. Still, I would stick with mushrooms that match the description of *G. ancilis!* The taxonomic challenges explain why very few people eat this tasty spring mushroom, although many who do eat it are probably not sure of the species.

IDENTIFICATION

- Irregular cup to disc shape (Ø 4–12 cm) with a depressed center, edges turned downward when young, flattening or turning upward with age.
- Wrinkled or convoluted upper surface usually dark cinnamon to red-brown, underside pale white, somewhat furrowed.

- Convoluted stem (Ø 1–2 cm) short or lacking, smooth or pitted.
- Tough but brittle flesh.
- Grows alone or in small groups (sometimes clustered) on dead conifer wood or in humus near rotten wood.
- Fruits in spring from southern British Columbia into California.

LOOK-ALIKES: The also edible Bleach Cup (*Disciotis venosa*) has a chlorine odor. The crust-like cup *Rhizina undulata* is marked by a lot of rootlike rhizomes attaching the fruiting body to the ground. Several, probably harmless after cooking, brown cups like *Peziza badia* have thinner, often fragile, multilayered flesh.

NAME AND TAXONOMY: *Ancilis* refers to a sacred shield that fell from heaven and became the protective palladium of ancient Rome. *Gyromitra ancilis* (Pers.) Kreisel 1984 = *Discina ancilis* (Pers.) Sacc. = *Discina perlata* (Fr.) Fr.

BLEACH CUP
Disciotis "venosa"

While the very similar Thick Cups (*Gyromitra ancilis*) and their allies can be unsettling due to their challenging ID and presence of toxic relatives, the Bleach Cup is easy to ID due to its clear stench of chlorine, which thankfully cooks away. In addition, being closely related to true morels (and hence also known as the Cup Morel), it is an excellent edible. Unfortunately, Bleach Cups seem to be rare in the Pacific Northwest; maybe too few people pay attention to them to report them. Molecular studies have revealed that North America has several species of *Disciotis* that need to receive their own scientific names.

- Cup- or bowl-shaped (Ø 5–20 cm) with upturned edge when young, later almost flat.
- Inside surface at first light brown to yellowish-brown, later dark brown; branched, radially arranged veins or folds, always lighter on the slightly rough outside.
- Stem, when present, is short (up to 1.5 cm), thick, ribbed, and usually buried, often only indicated with mycelium network.
- Flesh thick, firm, brittle, and double layered; outer layer brownish, inner whitish.
- Grows alone or in a small group in spring in damp soil or humus in conifer and hardwood forests, often along rivers, in non-fertilized sites.

LOOK-ALIKES: See the Thick Cup, *Gyromitra ancilis*, above.

NAME AND TAXONOMY: *Disci-otis* Greek for "disc-ear" (*otis*), *venosa* Latin for "veined." *Disciotis venosa* (Pers. ex Fr.) Boud. 1906. Morchellaceae, Pezizales.

BLACK ELFIN SADDLE
Helvella vespertina

Black Elfin Saddles are as gothic as a mushroom can be, and they fruit just in time for Halloween! Even the scientific name seems to suggest a "hell of a mushroom" but actually means "small pot vegetable" in Latin. While the size and cap shape are quite variable, the stems always have that unique "lacunose" look, brandishing holes and hollows, as if

Black Elfin Saddle, Helvella vespertina

Elfin saddle affected by Hypomyces cervinigenus

designed by Antoni Gaudí. Foolish people will scare themselves and mess up their digestive system by eating them raw or undercooked. Smart but wary people will use them as table decorations, since *Helvella* might contain traces of toxic gyromitrin (see Brain Mushroom, *Gyromitra esculenta*). And then there are the more daring mushroomers who enjoy elfin saddles well cooked for their crispy nature and morel-like flavor. Others like to dry and powder them. Watch out for elfin saddles affected by the whitish to pinkish mold of *Hypomyces cervinigenus*; nothing is known about the mold's edibility or toxicity. Elfin saddles fruit in fall and are often one of the last mushrooms to fruit before hard frosts end the mushroom season.

Recent DNA studies reveal that *Helvella lacunosa* is a misapplied name for our Black Elfin Saddle, since *H. lacunosa* proper grows on the East Coast of the US and in Europe. This study revealed several Western species but only described two species, *H. vespertina* and the very similar and also edible-with-caution *H. dryophila*, its name meaning "loving oak" in Greek. The Oak Elfin Saddle is best differentiated by its association with oak.

IDENTIFICATION

- Smooth to irregularly lobed cap with margins turned down and mostly fused to the stem.
- Cap light to dark gray, young underside bald and pale, darkening to pale gray.

- White to gray, sometimes black, stem is furrowed and contorted lengthwise and sometimes forming holes.
- Stem interior highly complex and chambered.
- Occurs in groups and clusters, primarily in fall and into winter.
- Grows with conifers on the ground or, rarely, on wood.

LOOK-ALIKES: Mushrooms coming somewhat close to these pieces of art are Brain Mushrooms (*Gyromitra* spp.), which are easy to tell apart by color, and less so true morels (*Morchella* spp.) and spring morels (*Verpa* spp.), which have more conventional stems and regular caps made of ribs and pits.

NAME AND TAXONOMY: *Vespertina*, Latin "evening," according to Vellinga—"western, where the sun sets." *Helvella vespertina* N. H. Nguyen and Vellinga 2013. Helvellaceae, Pezizales.

Orange Peel Fungus, Aleuria aurantia

ORANGE PEEL FUNGUS
Aleuria aurantia

This common Pacific Northwest cup fungus is easily recognized by its eye-catching bright orange color, often in stark contrast to the barren soil it grows from. Its flavor is not tongue-catching, but it offers harmless fungal protein and adds beautiful color to your food. It can even be enjoyed raw in salads. Some people praise it as delicious when fried in butter.

IDENTIFICATION

- Bright orange, cup- to saucer-shaped, wavy or irregularly contorted fruiting body (Ø 1–10 cm).
- More or less smooth and thin fleshed with downy, paler, or whitish underside.
- Grows on barren or disturbed soil.
- Fruits mostly in fall, but also in spring.

LOOK-ALIKES: The probably poisonous Snowbank Orange Peel (*Caloscypha fulgens*) is a spring-fruiting, spherical cup that usually displays traces of blue, at least on the cup's upper edge.

NAME AND TAXONOMY: *Aleuria* from Greek *aleurone* for "flour," for the mealy, downy underside; *aurantia* Latin "golden." *Aleuria aurantia* (Pers.) Fuckel 1870. Pyronemataceae, Pezizales.

SCARLET ELF CUP
Sarcoscypha coccinea

Red as red can be! This striking cup fungus brightens up dark winter days and invokes spring. It is often quite well hidden since it likes to grow from partially buried hardwood branches. Surely not a great culinary revelation, but it is so easy to identify, and it fruits during winter and early spring when low expectation secures great joy. Staying true

Scarlet Elf Cup, Sarcoscypha coccinea

to color, its flavor is reminiscent of red beets, but unfortunately the gorgeous color cooks away to gray. My Norwegian mushroom friend Pål Karlsen suggests boiling it briefly in cherry liqueur or some other tasty red liquid (using equal parts liqueur to mushroom) and letting it cool off in the liquid before removing it for drying. Then it is the perfect edible decoration. Recent DNA findings suggest that we will soon need at least one new name for our Western Scarlet Elf Cup.

IDENTIFICATION

- Bright red cup (Ø 2–6 cm, widening with age) with a whitish edge.
- Whitish, finely hairy exterior, usually with a white stem.
- White flesh, not really brittle, no particular flavor or odor.
- Grows alone or in small groups on dead or buried hardwood branches.
- Fruits mostly in winter and early spring.

LOOK-ALIKES: More orange than red, *Pseudaleuria quinaultiana* lacks the stem and cup shape and is more disc-shaped. It fruits in spring, usually in old-growth conifer forests, and its edibility is unkown.

NAME AND TAXONOMY: *Sarco-scypha* Greek "flesh cup," *coccinea* Latin "scarlet." *Sarcoscypha coccinea* (Jacq. Ex Gray) 1887. Sarcoscyphaceae, Pezizales.

TRUFFLES AND LOBSTERS

Besides the fact that they are both Ascomycetes, culinary Truffles and Lobster mushrooms are quite different from other mushrooms; they both happen to offer tasty, solid fruiting bodies. The Lobster Mushroom is the outcome of a hostile fungal takeover. *Hypomyces* parasitizes a Short-stemmed Brittlegill and turns an edible, middle-of-the-road white mushroom into a shocking orange-red fungal mass. The truffle lifestyle is much more ordinary and benign; it is straightforward mycorrhizal, with a fruiting body that grows underground.

When their spores are mature (and taste is most appealing), truffles release pungent odors to alert passersby, so the attracted creatures will excavate and devour the fruiting bodies, spreading the spores far and wide. Mycologists call fruiting bodies that form at least partially embedded in soil or humus "hypogeous" (underground) fungi. Culinary truffles are merely a tiny group of truffles; the majority of species are not eaten by humans and are known as "false truffles," a label rejected by other animals who eat them all with relish.

Bowl of Oregon Black and Oregon White Truffles

Author enjoying the aroma of truffles

Any classy glutton starts salivating when sensing truffles, and I am not just thinking of humans, but of boars, bears, squirrels, and spotted owls—all are bona fide truffle fans! As a colleague once stated, you say "truffle" and people reach for their wallets. Mysteries surround truffles, some of them preventing their collection. It starts out with trying to describe their flavor. People allude to earthy, musky tones with hints of garlic, mature cheese, nutmeg, chocolate, gasoline, and so on. Tasting one is the only way to make sense of that. But don't culinary truffles grow only in Italy and France? No, indeed. Some very tasty truffles are endemic to the Pacific Northwest, and in my opinion they are on par with the old-world delicacies.

WHERE TO FIND THEM: LIFE UNDERGROUND

Truffles fruit belowground, and finding them is a challenge. How do you find a mushroom that is basically invisible? My first few attempts using a rake felt invasive and destructive, and I lost

interest in truffles for a while. Going on truffle hunts with a trained truffle dog made all the difference, and such dogs are available these days for a fee. Collecting mature truffles that are invisible to the human eye may seem impossible, but to a dog with its discerning nose that can detect their fragrance, it is a whole different game! For commercial truffle outfits, collection should be restricted to assistance by trained dogs, since it is gentler and much more sustainable. However, truffle hunts do not have to be assisted by dogs.

If you happen to be in the right habitat at the right time—that is, on a young Douglas-fir plantation in winter—you can search for Oregon White Truffles by removing (then replacing) the duff layer, in small sections by hand, and looking for the light brown pearls that sit most often right at the interface of mineral soil and organic duff layer. If you get the habitat and season right, this crude approach will expose truffles in all stages, tiny to bigger, immature to mature.

PROCESSING AND COOKING TRUFFLES

Once they have tasted truffles, many people fall in love with their rich, earthy aroma, which is also historically claimed to be highly aphrodisiac. Scientifically inclined people point to the compound androstenol, a musky steroidal pheromone. Truffles are not enjoyed for their meat, like chanterelles, but as a spice. Thin shavings of carefully washed raw truffle are spread over prepared food just before or when serving. Luckily, since a pound of fresh prime Oregon Whites starts at $500, a little goes far, and an ounce will transform your cooking.

The best way to store your truffles is in the fridge, a temperature close to the forest floor, in a closed container lined with a moisture-absorbing agent, like dry rice grains or paper towels. When truffles sit in their own evaporated moisture, they quickly turn soft and rot! Sometimes the unspoiled parts of a truffle can be saved by cutting away the rotten parts. Be thorough. Once, I felt sick after enjoying *pasta al burro con tartufo nero*—I thought I had successfully removed all the rotten parts of my Oregon Black Truffle, but I had not—and ended up throwing up my linguini with butter and truffle, a very sad day.

There are always many more immature fruit bodies to each mature one, and their development needs to be protected. The Corvallis-based North American Truffling Society calls on you to replace needle duff you moved aside, not search too closely to tree trunks, and avoid collecting in steep, erosion-prone areas. Dog-assisted truffle hunters despise humans who hunt truffles without dogs, especially when using destructive rakes, which is illegal in most places. Access to truffle sites can be a challenge, and there is stiff competition and a lot of hostility from landowners whose truffles have been poached—thousands of dollars' worth of truffles can be stolen in a night. Thus, truffle hunting is a whole different ball game. So far, no truffles are known to be poisonous! However, "bluffballs" (a.k.a. *Amanita* mushroom eggs) can be seriously toxic.

Most groups of macrofungi have members that "went underground"; their fruiting bodies are basically folded up with the spore-producing tissue encapsulated inside. Culinary truffles like Oregon White Truffles are folded-up cup fungi; Oregon Black Truffles are origami morels. *Rhizopogon* truffles are boletes gone belowground. Their interior has the spongy nature of the tube layer, reminiscent of their aboveground family members. The very common pine associate, the Blushing False Truffle (*Rhizopogon rubescens*), is consumed in East Asia and known in Japan as *shoro*.

Truffles are most often plant-associated mycorrhizal fungi. Fruiting belowground protects the fruiting body from extreme temperatures, allowing them to fruit in winter in warm-temperate climates—including Mediterranean climates like the maritime Pacific

HOW TO MAKE TRUFFLE BUTTER AND OIL

Infusing butter with rich truffle aroma is easy. Basically, you simply have to be patient for a day or two. Place your truffles in an airtight jar big enough for the food item to be infused next to it. Use dry truffles (stored with rice or paper towels). Bars of butter can remain wrapped in paper, and hard-boiled eggs can stay in their shells, but remove plastic wrap from cheese. The more surface area exposed, the better the aroma is absorbed. When infusing olive oil, it is best to use wider jars, not narrow ones, and hang the truffles in a tea bag or a cheesecloth above.

Truffle butter and oil

Northwest. The soil and humus also protect them from drying out. Interestingly, the highest diversity of hypogeous fungi seems to be in Australia, where rains often followed by hot, drying winds might desiccate an aboveground mushroom before it sheds spores, a challenge truffles evade by fruiting underground. However, fruiting underground means that truffles depend on non-aerial agents for spore dispersal. To attract critters, most of them have developed intense scents effused when mature to signal to mycophagous animals, "Come and get me!"

OREGON WHITE TRUFFLES
Tuber gibbosum and ***T. oregonense***

White truffles of the Pacific Northwest are some of the most well-known truffles on Earth. Belonging to the genus *Tuber*, Oregon Whites are very closely related to all the most famous European truffles, be it the Black (Périgord), White (Alba), or Brown (Summer) Truffle (*Tuber melanosporum*, *T. magnatum*, and *T. aestivum*, respectively). Genetic sequencing has revealed that Oregon White Truffles include at least four species. Of those four, two are collected commercially and are the most commonly enjoyed: the early-fruiting paler *Tuber oregonense* and late-fruiting, thin-skinned *Tuber gibbosum*. The collective common name reflects the region where they are most abundant, specifically Western Oregon. However, they also grow west of the Cascades, from southern Vancouver Island south through Washington and Oregon and into northwestern California.

- Round when young, turning irregularly lobed and furrowed, peanut to large walnut size.
- Whitish to pale brown (after washing) with a tinge of olive to dull orange-brown, surface finely downy and often cracking.
- Solid whitish to brown interior with white marbled veins; cracks when squeezed.
- Odor at first mild, becoming strong and complex with notes of garlic, spices, cheese.
- Fruiting season for *T. gibbosum*: January–June, for *T. oregonense*: October–February.

- Grows with young (five-year-old) to early to mature Douglas-firs west of the Cascades, from sea level to 1,400 feet elevation.

NAME AND TAXONOMY: *Tuber* Latin for "knob," *gibbosum* Latin "humped, protuberant." *Tuber gibbosum* Harkn. 1899; *T. oregonense* Trappe, Bonito, and Rawlinson 2010. Tuberaceae, Pezizales.

OREGON BLACK TRUFFLE
Leucangium carthusianum

This mushroom generously emits a uniquely fruity but musky aroma. The common name Oregon Black Truffle seems to be a bit of a misnomer, since it was first described growing near a Carthusian monastery above Grenoble, France. However, in "tuberphile" Europe, this truffle is ignored, although its cool musky pineapple aroma with a chocolate finish is not found in any other culinary truffle. It can be used like any other truffle—served raw, thinly sliced on a dish, or crushed, as truffle hunter James "Animal" Nowak with Terra Fleurs taught me, and be

Oregon White Truffle, Tuber gibbosum

Oregon Black Truffles, Leucangium carthusianum

used as a nifty anchovy replacement in a Caesar salad. The fruitiness invites embellishing desserts.

Most truffles collected in the coastal Pacific Northwest north of Oregon are actually *Leucangium carthusianum.* Maybe it should be called the Washington Black Truffle; however, Oregon State University truffle research set the pace for discovering truffles in the Pacific Northwest and left its stamp on naming them. Since the Oregon Black Truffle fruits deeper in the soil, a dog-assisted hunt is really the way to collect these culinary gems. Their extended fruiting season will keep a truffle dog busy most of the year.

IDENTIFICATION

- Round to lumpy and irregular, blackish (purple hued), minutely warty but sometimes nearly smooth. Hazelnut to walnut size, sometimes bigger.
- Flesh solid, with greenish-gray to brown pockets of spore-bearing tissue separated by amorphous to marbled veins.

- Odor at first mild, becoming a fruity musk with notes of pineapple.
- Fruiting season September–April, sometimes into June.
- Growing with more mature Douglas-firs, west of Cascades, often 10–20 cm underground.

NAME AND TAXONOMY: *Leuc-angium* Greek for "white" (*leucos*) and "capsule" (*aggeion*), and *carthusianum* since it was first found close to a monastery of the Carthusian order. *Leucangium carthusianum* (Tul.) Paol. 1889. Morchellaceae, Pezizales. It is regarded as synonymous with *Picoa carthusiana* Tul. and C. Tul. 1862. However, DNA analyses place other species of *Picoa* in Pyronemataceae, so *L. carthusianum* would be a misfit.

 LOBSTER MUSHROOM
Hypomyces lactifluorum

The bright orange of the Lobster Mushroom seems to scream "come and get me and spread my spores far and wide"—

and most mushroom hunters gladly oblige! The Lobster is a unique mushroom, as idiot proof, or rather, as safe as the collection of wild mushrooms can be. Is it a mushroom or a fungus? Actually, it's both! The bright orange to red and smoothly pimpled surface of the fungus *Hypomyces lactifluorum* parasitizes the Short-stemmed Brittlegill (*Russula brevipes*). In addition, the Lobster often fruits much earlier than

Lobster Mushroom, Hypomyces lactifluorum: *Pimply surface showing the top of the perithecia, in which the fungus produces its white spores*

other fall mushrooms, handles drought very well, and resists decay for weeks.

The abused mushroom is blocked from producing spores by the hostile takeover. Its whole fruiting body, usually already infected as an innocent baby primordium, is pervaded by *Hypomyces*, which not only changes its appearance, but also improves its flavor without undermining the firm structure of the *Russula*. This structure— which works fine fried or cooked—the unique fishy flavor of the Lobster Mushroom, and its impressive color make it a choice edible, often used with seafood or as a seafood substitute.

However, be warned that some people's systems do not appreciate this fungal *frutti di mare*. Lobsters are frequently implicated in gastrointestinal distress, especially after undercooking or overindulging. Ingesting spoiled specimens is also troublesome; remove deteriorated areas before cooking well. Poor souls bothered by lobsters could perhaps donate them to mushroom dyers, who love its powerful dye, which stains orange, red, purple, and blue.

IDENTIFICATION

- Smooth surface when immature; white turning light to deep orange or red and with pimple-like bumps when mature; purple when old.
- Fruits on Short-stemmed Brittlegill (*Russula brevipes*), transforming the Basidiomycete's gills into smooth, blunt ridges.
- Begins fruiting in July into November, usually in small groups, often hardly reaching above the duff layer in conifer forests.

LOOK-ALIKES: Daring foraging harvesters have eaten the closely related Yellow-green Gillgobbler (*Hypomyces luteovirens*) and the first white to yellow to brick-red Ochre Gillgobbler (*Hypomyces lateritius*). The former parasitizes mostly brittlegills (*Russula* spp.), the latter prefers milkcaps, (*Lactarius* spp.), overgrowing their gills. However, there are no reports of anyone eating *Hypomyces chrysospermus*, the "Bolete Eater" (see photos on pp. 129 and 131). It

Two healthy Short-stemmed Brittlegills, Russula brevipes, *and a third one infected by* Hypomyces lactifluorum

first overgrows boletes as a white mold, then turns golden when mature. Sounds like the Lobster playbook, but the Bolete Eater renders porcini and many other boletes into an unattractive heap of soggy mush. And while *Hypomyces* species are found all over the Northern Hemisphere, our Lobster is endemic to North and Central America.

NAME AND TAXONOMY: *Hypo* for "below or under," *myces* for "fungus"—both Greek, for "getting under the skin of its host." *Lactifluorum* Latin "pertaining to milk flow"—a reference to milkcaps infested in other regions. *Hypomyces lactifluorum* (Schwein.: Fr.) Tul. 1860.

Delightful, colorful Basidiomycetes: Cauliflower Mushroom (white), Red Corals, Golden Chanterelle (both photographs), Woolly Vase (orange), and Pig's Ear (purple)

BASIDIOMYCETES

Most fungi that grow mushrooms are Basidiomycetes, belonging to the division of Basidiomycota, one of the two large divisions (the other being Ascomycota) of the "higher fungi" within the kingdom Fungi. Basidiomycetes include chanterelles and hedgehogs, boletes, gilled mushrooms, puffballs, stinkhorns, earth stars, corals, polypores, jelly fungi, smuts, and a few other groups. This group includes many good edibles, and many of its species fruit in the Pacific Northwest in fall.

JELLY FUNGI

Let's kick off the Basidiomycete mushrooms and their edibles with the jelly fungi (followed by the chanterelles) because, interestingly, these orders are the most ancient branches of the Basidios. All other orders of Basidiomycetes (including Boletales, Agaricales, and Russulales) are evolutionarily more recent.

Jelly fungi are an interesting bunch of mushrooms with a gelatinous texture, flabby consistency, and some cool colors. Jellies are grouped based on appearance, not taxonomy, and are found in four orders of the basidiomycete, three of them—Auriculariales, Dacrymycetales, and Tremellales—represented here. This category has a handful of good edibles. Their preferred habitat is decaying wood.

AMERICAN WOOD EAR
Auricularia americana

Wood ears are easily recognized by their flabby, rubbery brown fruiting bodies. The gelatinous but crunchy consistency offers a distinctive culinary experience, and wood ears are an important part of Chinese and other East Asian cuisines, where they are cultivated in immense quantities. Older field guides call our American Wood Ear *Auricularia auricula*, but DNA research suggests that this European species does not grow in North America. There are two Pacific Northwest wood ears: *A. americana*, found on dead conifers, and *A. angiospermarum*, on hardwoods. They are both edible, as are all wood ears around the world. *Angio-spermarum* is Greek for "contained seeds," since all hardwoods belong to the flowering plants known as angiosperms, their seeds being encapsuled. I have found most of my conifer-fruiting American Wood Ears in spring while looking for morels in the Cascades.

IDENTIFICATION

- Fruiting bodies ($\varnothing$ 2–10 cm) of varied shapes, but commonly ear-shaped or forming a shallow inverted cup, often irregularly ribbed or veined.
- Brownish colors from tan to dark brown, sometimes tinged yellow, red, or olive.
- The upper convex surface features a dense, silky covering of tiny hairs; lower concave fertile surface is smooth; margins are clear and smooth.
- Fruiting body laterally attached (with or without a short stem).
- Flesh thin-rubbery to flabby-gelatinous; mild flavor.
- Grows on dead conifers potentially year-round, in the Cascades and Rockies especially after snowmelt.

American Wood Ear, **Auricularia americana**

- Distributed widely in the Pacific Northwest and beyond, in both North and Central America.

MEDICINAL: In Chinese medicine, wood ears are considered nourishing for the lungs and helpful for improving blood circulation and fighting throat inflammation. In Europe, wood ears have a history of use for eye problems and against inflammation and angina. They are antioxidants and anticoagulants (blood thinners), so do not overindulge when bleeding or using aspirin or warfarin.

LOOK-ALIKES: Quite similar in color are the also edible Leaf Jellies (*Phaeotremella foliacea,* formerly *Tremella,* and *P. frondosa*) both growing on wood, sometimes in big clusters, and parasitizing False Turkey Tails (*Stereum* spp.). Leaf Jelly lacks the silky upper surface and usually grows more clustered and only rarely in a simple ear shape. Some brown

Leaf Jelly, **Phaeotremella foliacea**

cup fungi (especially *Peziza* spp.) can have similar colors but are brittle and not jelly-like and will grow on soil or rotten wood rather than from firm dead wood.

NAME AND TAXONOMY: *Auricularia* Latin "little ear." *Auricularia americana* Parmasto and I. Parmasto 2003. Auriculariaceae, Auriculariales.

CAT'S TONGUE
Pseudohydnum gelatinosum

Pseudohydnum gelatinosum is a beautiful soft, rubbery, translucent jelly fungus with spines on its underside. A multitude of common names refer to these unique features: Toothed Jelly Fungus, Jelly Tooth, or Cat's Tongue. Growing from dead wood, they are sometimes spoon- or fan-shaped, while other times they are more oyster-like with a short lateral stem or no stem. The cap is usually white, more rarely gray to gray-brown. Researchers are currently using only one scientific name for Cat's Tongue wherever it occurs, from Alaska down to the Amazon. Initial DNA work indicates that there may be more than one jelly tooth species, which might explain the differences in color and shape.

Cat's Tongue, Pseudohydnum gelatinosum

As a provocatively textured edible, it is more novelty than nourishment. The absence of a distinctive flavor can be seen as a disappointment or as the base for unlimited culinary possibilities! Cat's Tongue can be eaten raw or dried without cooking. Marinating it to turn it into candy, possibly after partially drying, can add

PROCESSING AND COOKING WOOD EARS

Wood ears (*Auricularia* spp.) are the most exciting of the jelly fungi. Even though they have been eaten in East Asia since ancient times, wood ears have been largely ignored as edibles in Western cuisine outside of hot and sour soups dished out in Chinese restaurants. Dedicated foragers have applied their creative minds to bring some jellies like wood ears into the fold of edibles.

Wood ears store very well, up to several weeks when refrigerated. They also are easy to dry—sometimes they're found already dried in nature—and rehydrate quickly in warm water. Organic mushroom farmer and mycologist Tradd Cotter suggests rehydrating them in broth or teriyaki-infused sauce (see Wood Ear Mushroom Jerky in Part Three), then dehydrating them again for a crunchy snack. Frying gives them a distinctive crunchy structure preferable to the flabby soup ear piece (which has never won my heart). Whatever method you use, keep in mind that wood ears benefit from added flavor. Frying them in goose fat and/or deglazing the fruiting bodies in soy sauce or tamari makes a delicious dish.

appealing flavor and aromas as a candy. First partially dehydrate it, then marinate it in fruit and/or lemon juice, roll it in sugar, and dehydrate to the desired consistency. It can also be used as a garnish or in salads.

- Fruiting bodies are fan-shaped to bracket or oyster-like (Ø 2–6 cm), whitish to sometimes light gray to gray-brown, with a rough surface and wavy margins.
- The underside, the hymenium, features soft white teeth.
- Stem attaches to the side of cap; sometimes fruiting body is stemless and attaches directly to substrate.
- Flesh is gelatinous, soft, translucent, usually white.
- Grows solitarily, in small groups, or densely clustered on decaying wood.
- Common in Douglas-fir forests from late summer into winter.

LOOK-ALIKES: Hedgehogs (*Hydnum* spp.) have teeth as well, but are yellowish and much firmer and do not have a jellylike consistency.

NAME AND TAXONOMY: *Pseudo-hydnum* references *Hydnum* (hedgehog mushrooms), and *gelatinosum* is from Latin *gelatus* "frozen, gelatinous." *Pseudohydnum gelatinosum* (Scop.: Fr.) P. Karst. 1868.

WITCHES' BUTTERS
Tremella mesenterica,
Naematelia aurantia, and
Dacrymyces chrysospermus

These yellow-to-orange jelly blobs are quite a bright sight—and they get really slippery and wobbly when saturated after rains! Many a mushroom hunter apparently under the spell of the buttery witch happily bags the blob without the notion that there are three comparable jelly fungi in the dark forest. They can be told apart in the field by the identity of the tree they are growing on, the presence of another fungus they are parasitizing, and the color of their attachment point.

Orange and Yellow Witches' Butters haunt hardwoods such as oaks and alder, but only if they are already infected by other fungi, since both are mycoparasitic. Yellow Witches' Butter (*Naematelia auran-tia*) parasitizes the orange-brown False Turkey Tail (*Stereum hirsutum*). The Orange Witches' Butter, *Tremella mesenterica*, lives off parchment crusts (*Peniophora cinerea*, *P. aurantiaca*, and *P. incarnata*), wood-decaying fungi that cover their victims in gray, orange-brown, or pinkish-orange crusts, respectively. Conifer Witches' Butter (also called the Orange Conifer Jelly), *Dacrymyces chrysospermus*, grows from dead conifers and only rarely from hardwoods. Moreover, *Dacrymyces* shows white tissue around the point of attachment, which is missing in the other witches' butters. So if you want to know what species you have collected, you have to pay attention to the substrate out in the woods. Once you are in your kitchen, it might be impossible to get a clear ID.

All three basically bland jellies are harmlessly digestible. They can even be enjoyed, be it as a colorful decoration to a salad, as a base for candy, just like Cat's Tongue, or floating in a soup. Some people advise blanching them quickly before using them for culinary purposes. Channel their gelatinous texture into sauces by blending them. They can also be turned into a tasty fruit leather (see recipe, p. 337).

TOP: *Conifer Witches' Butters,* Dacrymyces chrysospermus; *note the white attachment point on the inverted specimen.* **MIDDLE:** *Yellow Witches' Butters,* Naematelia aurantia, *with False Turkey Tail,* Stereum hirsutum. **BOTTOM:** *Orange Witches' Butter,* Tremella mesenterica, *with* Peniophora *spp.*

People in China have traditionally used Orange Witches' Butter as a demulcent to soothe inflammation of the throat and upper respiratory tract and to foster mild expectoration. Orange Witches' Butter has been scientifically recognized as an expectorant and anti-inflammatory for bronchial inflammation and asthma. One of its active ingredients, tremellastin, has been shown in animal studies to significantly lower triglycerides and blood sugar. *Tremella mesenterica* induces interferon production and prevents leukopenia—the reduction of white blood cells after radiation and chemotherapy. Other species of witches' butter have been researched less intensely.

IDENTIFICATION

These particular characteristics apply strictly to *Dacrymyces chrysospermus*, or Conifer Witches' Butter.

- Fruiting body is a bright orange, flabby or brain-like blob (Ø 2–8 cm) made of many longish irregular lobes originating from a single base.
- Loses color from the base up and often liquefies in age; upon drying, turns orange to red and bone hard.
- Has either a stout stem or is stemless; attached to wood by tough, white rooting base.
- Grows sometimes in a solitary blob, but more often in big groups, on dead conifer wood.
- Common in late fall and mild winters, but can occur year-round.

LOOK-ALIKES: Orange and Yellow Witches' Butters, *Tremella mesenterica* and *Naematelia aurantia*, consist of convoluted lobes, not longish parts. There are some other yellow and orange jelly fungi, but

Apricot Jelly Mushroom, Guepinia helvelloides

most are smaller (except *Tremella mesenterella*); none are known so far to be toxic. The also edible, but bland, gorgeous Apricot Jelly Mushroom, *Guepinia helvelloides* (formerly *Phlogiotis*) is peach to salmon-pink and grows in a fan, spatula, or funnel shape. It has no look-alikes and, if not too old or tough, can be prepared like Cat's Tongue and witches' butters.

NAME AND TAXONOMY: *Trem-ella* Latin "little shaking," as in tremor, and *mes-enterica* Latin "middle intestine," for its similarity to the human mesentery. *Naema-telia* Greek "gelatinous-wrap," and *aurantia* Latin "orange." *Dacry-myces* "tear-fungus," for its shape, and *chryso-spermus* "golden-seed," a reference to spore color—all Greek. *Tremella mesenterica* Retz.: Fr. 1769; *Naematelia aurantia* (Schwein.) Burt 1921 = *Tremella aurantia* Schwein.: Fr. 1822. Tremellaceae, Tremellales. *Dacrymyces chrysospermus* Berk. and M.A. Curtis 1873 (formerly also *D. palmatus* Bres. 1904). Dacrymycetaceae, Dacrymycetales.

JELLY ANTLER
Calocera viscosa

The unique, beautiful, yellow-to-orange Jelly Antler, *Calocera viscosa*, is a very small, tough-fleshed bifurcating coral look-alike growing on conifer wood. The Jelly Antler is harmless enough to be regarded by some people as edible, but most people do not consider it food—though it might make a neat edible decoration.

Jelly Antler, Calocera viscosa

CHANTERELLES, HEDGEHOGS, PIG'S EARS, AND MORE

Fruiting bodies with veined or toothed spore-producing surfaces developed multiple times in the fungal tree of life. However, for ease of presentation in this field guide, edible mushrooms with these features are combined into one group. Some of the best toothed fungi, hedgehogs (*Hydnum*) happen to be closely related to chanterelles; however, Bear's Head and Lion's Mane (*Hericium*) are not closely related to hedgehogs. Veined fungi have a spore-producing tissue that can look similar superficially to broad gills. Often their veins or folds fork, are interconnected, or form a wrinkly surface. Chanterelles (*Cantharellus* and *Craterellus*) are the most famous veined mushrooms, one of the most important groups of edibles on Earth. However, a few other mushrooms have similar veins, e.g., Pig's Ears (*Gomphus clavatus*) and Blue Chanterelles (*Polyozellus* spp.), although they are not closely related.

TRUE CHANTERELLES AND WOULD-BE CHANTIES

It is downright depressing to imagine a mushroom-blessed existence without chanterelles! They are the perfect wild edible mushroom for so many reasons: bright color, abundance, fruity aroma, and firmness. They are bug-free, they are easy to ID, they store and transport well—and the list goes on. And we are so lucky in the Pacific Northwest. There is no other region on Planet A blessed with such a reliable annual abundance of chanterelles. Yes, the Californian oak-associated *Cantharellus californicus*, known as Mud Puppy, might grow a bit bigger than our Pacific or Golden Chanterelle (*C. formosus*) and the Rainbow Chanterelle (*C. roseocanus*)—and Eurasia's *C. cibarius* might be more fragrant—but I prefer picking ten times the amount of PNW chanterelles, even if each ounce has only half the aroma of the more celebrated ones. I have picked chanterelles on four continents, and I never wished for another species in my meal!

Admittedly, having tiny, perfectly formed chanties on your plate is more attractive than cut-up pieces of a bigger fruiting body, but out in the woods, collecting these tiny mushrooms feels like fungal infanticide. I cover such babies up with moss or duff and return in a week to pick the matured mushrooms, whose biomass has multiplied! Sometimes, when they are the first of the season, I let myself harvest young chanterelles to pickle for antipasto or chutney—both make great presents. Their rich, fruity aroma paired with a firm structure makes chanties extremely versatile in the kitchen, and their seasonal abundance allows for endless experimentation.

Chanterelles make it so easy for us! Their joyous golden color is a snap to spot. The often chalice-like shape of the smoothly surfaced cap, with the strongly decurrent folds that run out unevenly on the stem, gives chanterelles a unique look. A rich, fruity aroma reminiscent of apricot (in Japan, it is known as "Anzutake, the apricot mushroom"), and dense, rubbery flesh, easy to recognize once handled, helps foragers learn how to recognize it quickly and makes it a safe mushroom in the Pacific Northwest. However, White Chanterelles have many more lookalikes and are only safe once you know your yellow chanterelles well. The four known chanterelle species in our region are easy to tell apart by cap color paired with underside color, as mycologist Michael Beug points out, see Table 1.

Chanterelles are widespread in the Northwest, and it seems every major forest type provides habitat for them, from species-rich old-growth forest to species-depauperate Douglas-fir "tree farms." Once you have found a chanterelle spot, you can return year after year

A basket of Pacific Golden Chanterelles

PROCESSING AND COOKING CHANTERELLES

The tissue of chanterelles persists for a very long time. A long-term chanterelle study in Oregon recorded one Golden Chanterelle fruiting body that lived for nearly two months! The firm flesh that makes handling and storing them so easy seems to extend their shelf life to that otherwise achieved only by indigestible conks. I have kept chanterelles stored in a paper bag (with a little bit of misting when they get too dry) fresh in my fridge for up to three weeks!

Many foragers, including my family, prefer cleaning their chanties as soon as possible once home. However, some people insist on storing their mushrooms as is and wait to clean them when they are ready to cook them, since cleaning can damage the mushrooms and reduce their fridge life.

Washing or spraying down chanterelles speeds up cleaning immensely, especially when working with sullied, rain-soaked mushrooms, and does not damage them. After washing, it is best to let them dry on a towel in a cool place for several hours before storing. When you want to fry them immediately, dry frying (also called dry sautéing) is best. Remember to save the mushroom juice to make your sauce after frying them. Another way to reduce moisture in mushrooms—for example, to improve consistency in fresh but soggy chanterelles—is a short parboil. However, in our household we have found that trying to rescue waterlogged late-season chanterelles by parboiling is not worth the time and energy use, especially taking into account how many pounds of chanties we usually bring home from the woods and the amount of hot water it takes to parboil big chanterelles.

For long-term storage, it is best to cook chanties before freezing them. When pressed for time or overwhelmed by abundance, spread chanterelles on a tray and oven bake them until they "sweat." Cool them before freezing, and make sure you include the tasty liquid as well! Label your frozen goods with the date. I have eaten seven-year-old frozen chanties from the back of my freezer without problem; they are still quite enjoyable, though not as good as those more recently frozen. Though dried chanterelles are sometimes offered in stores, most mushroom hunters prefer cooking and freezing over drying. Drying poses two challenges: chanterelles sometimes turn bitter (something that also frequently happens when freezing them uncooked), and even after rehydrating them all day long, they can retain an unpleasant core toughness. However, this toughness is attractive when making chanterelle jerky.

to enjoy their abundant fruiting. That is, until the chain saw strikes; chanterelles are mycorrhizal and depend on their living tree hosts.

It can take chanterelles many weeks to grow to full size, and during that time, they are continuously adding new cells. Sometimes chanterelles can

TABLE 1. THE FOUR DISTINCTIVE CHANTERELLE SPECIES OF THE PACIFIC NORTHWEST

CANTHARELLUS SPECIES	CAP COLOR	CAP UNDERSIDE	MAIN ASSOCIATED TREES
Golden Chanterelle C. formosus	Dull egg-yolk-yellow	Dull egg-yolk-yellow	Douglas-fir and hemlock, also pine and spruce
Cascade Chanterelle C. cascadensis	Bright egg-yolk-yellow	White	Douglas-fir, hemlock, and grand fir
Rainbow Chanterelle C. roseocanus	Rosy blush, fading to pale yellow	Very bright egg-yolk-yellow	Spruce, fir, and pine
White Chanterelle C. subalbidus	White	White	Douglas-fir, hemlock, pine, and madrone

produce odd growth forms, such as curled ridges on top of the cap. Under the cap, ridges or veins shape the hymenium, the fertile tissue that produces the spores. Actually, some chanterelles elsewhere have a nonfolded hymenium, for example, the Smooth Chanterelle (*C. lateritius*) common in the southeastern United States. Superficially, the veins can resemble the gills of agarics, but gills have a different cell structure.

PACIFIC OR GOLDEN CHANTERELLE
Cantharellus formosus

The color marking our Golden Chanterelle and other yellow chanterelles is derived from the yellow-orange pigment beta-carotene, which our bodies convert to the essential vitamin A. This carotenoid is also a powerful antioxidant that helps chanterelles suppress larval infection and animal feeding; it is a natural insecticide to prevent pesky flies from laying eggs so their maggot offspring

can feed on fungal tissue. It is assumed that the spiciness evident after thirty seconds or so when tasting a fresh, raw mushroom is part of this defense system. The German name *Pfifferling*, which translates to "pepperling" (very conveniently, the diminutive suffix *-ling* can turn anything into a mushroom in German), reflects this spicy flavor that cooks away. This antiparasitic power

Pacific or Golden Chanterelle, Cantharellus formosus

Pacific or Golden Chanterelle, Cantharellus formosus

is used in Lithuanian folk medicine to fight intestinal parasites. To use chanterelles to fight parasites, a person must ingest them dried (and powdered) or in an alcohol extract. In the field, worms occasionally manage to overpower the highly successful chanterelle immune system, but that usually occurs in old and soggy fruiting bodies.

IDENTIFICATION

- Golden to dull orange bald cap, (∅ 5–15, even 20, cm wide), sometimes with fine darker scales in a depressed center, cup-shaped to wavy.
- Decurrent (running down the stem) gill-like folds up to 2 mm deep, often forking, often with short secondary cross-veins; veins irregularly tapering out on stem.
- Stem color like cap or slightly paler; stem shape equal or narrowing toward base.
- Flesh firm and whitish, bruising slightly yellow and darkening to dull brown; flesh can be ripped lengthwise like string cheese.
- Fruity, apricot-like aroma (often after being banged around); mild to slowly developing spicy, raw flavor.

- Grows on ground with western hemlock, Douglas-fir, spruce, and lodgepole pine; endemic to the Pacific Northwest and northwestern California, widespread in the moist West up to Juneau, Alaska.
- Fruits from July through September into November or until a hard frost terminates the season.

NAME AND TAXONOMY: *Cantharellus* is derived from Greek *kantharos* "tankard" or "cup," and the diminutive *-ellus*, hence "small cup"; *formosus* Latin "beautiful, well-shaped." *Cantharellus formosus* Corner 1960, in old books listed as *C. cibarius* Fries 1821, a smaller European sister. Cantharellaceae, Cantharellales. *C. formosus* is actually a species complex with several species that have not yet been named.

CASCADE CHANTERELLE
Cantharellus cascadensis

The Cascade Chanterelle is the most recently described and least common of the three big Pacific Northwest chanterelles. It grows in a similar habitat as the Pacific Chanterelle but is more common in the southern Cascades and the Sierra. In shape it is closer to the White Chanterelle, but its cap color is closer to the Pacific Chanterelle. However, the cap of the Cascade Chanterelle is often brighter yellow than the cap of the Pacific Chanterelle and its stipe is club-shaped or widens at the base (just like the White Chantie), whereas the stem of the Pacific Chanterelle usually narrows or is at least equal at the stem base.

Cascade Chanterelle, Cantharellus cascadensis

Because it displays selected characteristics of both the Pacific and the White Chanterelle, commercial pickers called the Cascade Chanterelle the "hybrid"—even before it was recognized by mycologists as its own species. Another reliable characteristic of *Cantharellus cascadensis* is the thin cap margin. Noah Siegel and Christian Schwarz praise it for its fruity flavor in their excellent book *Mushrooms of the Redwood Coast*.

IDENTIFICATION

- Bright yellow, smooth cap (Ø 5–15 cm), cup-shaped overall, depressed center.
- Inrolled cap edge when young, later with wavy to crisped edges, always thin margin.
- Decurrent folds much paler than cap, whitish to pale pink, often changing to pale orange-yellow.
- Shortish stem often with wider base.
- Firm, white flesh stains orange-yellow and darkens to red-brown, flesh can be ripped lengthwise like string cheese.

- Odor indistinct to fruity.
- Occurs singly or in small clusters from late summer into early winter.
- Grows in deep humus with Douglas-fir, western hemlock, and true fir; found in coastal regions between the 48th parallel and Northern California, in the Cascades and Sierra, and also in the inland temperate rainforest.

NAME AND TAXONOMY: *Cascadensis* for Cascade Mountains. *Cantharellus cascadensis* O'Dell and R. Molina 2003.

RAINBOW CHANTERELLE
Cantharellus roseocanus

Many mushroom hunters collect the often smallish Rainbow Chantie without realizing that it is a different species when only the yellow segment of the rainbow is displayed. However, with the right weather conditions, the strong pinkish hue, especially obvious on young caps and caused

GOLDEN CHANTERELLE LOOK-ALIKES

Several species are common look-alikes to chanterelles and fool the novice mushroom hunter with their color, with their decurrent gills, and by fruiting at the same time and in the same habitats as true chanterelles. Frequently mistaken as chanterelles by newbies are the harmless False Chanterelle (*Hygrophoropsis aurantiaca*), edible Woolly Pinespike (*Chroogomphus tomentosus*), and the questionable Scaly Vase (*Turbinellus floccosus*). Mushrooms in the genera *Craterellus*, *Gomphus*, and *Polyozellus* also share a somewhat chanterelle-like appearance.

Most similar looking in the Pacific Northwest is the **False Chanterelle**, *Hygrophoropsis aurantiaca*, now recognized as a complex of several species. It might grow right in between your chanties, and it is hard to tell them apart from above. However, get closer and you will feel how lightweight it is compared to the solid chanterelle. Break a False Chanterelle and its flesh is fragile, not rubbery or stringy. It sports true gills—also forking and decurrent—under the cap: knife-blade thin, deeper than true chanterelles. If you still have been deceived, you will catch your mislabeling when frying the False Chanterelle, since its flesh browns more quickly than that of true chanterelles. If you happen to eat it, you will notice its lame flavor and flimsy texture. A few people may experience mild gastrointestinal distress; its toxicity was widely overstated in the past.

The edible but not incredible **Woolly Pinespike** (*Chroogomphus tomentosus*) grows in the same habitat, but a closer look will help you tell it apart by its woolly cap and true, decurrent gills (see Gilled Bolete Relatives in Gilled Mushrooms). It is also paler overall and will turn wine-red as it cooks, which

False Chanterelle, Hygrophoropsis aurantiaca

makes it easy to distinguish from chanterelles at this stage, if you missed the other signs while collecting and cleaning.

The most dangerous look-alikes to the highly edible chanterelles are poisonous **Jack-o'-Lantern Mushrooms**, which have true gills and always grow on wood. The **Western Jack-o'-Lantern** (*Omphalotus olivascens*) is absent from much of the Pacific Northwest, though there have been a few reports from southern Oregon and from Whidbey Island in Washington by Puget Sound

Chantycap, Cantharocybe gruberi

Mycological Society identifier Colin Meyer. However, we miss out big time, since as their name indicates, Jack-o'-Lanterns glow in the dark—no UV light required. Once your eyes adjust to the darkness, you can see their mind-glowing spooky green radiation. If that bioluminescence does not impress you, at least note that it's not a radioactive chanterelle!

Very rare, apparently restricted to drier conifer forests east of the Cascades and more common in southerly areas, is the gorgeous *Cantharocybe gruberi*, or **Chantycap**, a midsize to big (⌀ 5–20 cm) cap, fat-stemmed, and firm-fleshed member of the waxgill family, Hygrophoraceae. When it's young, its bright yellow cap and yellow gills can fool you, but when you look at the true, blade-like gills, it is easy to tell apart from a chanterelle. Often the cap color fades with sun exposure and age. Little is known about the edibility of this white- to pale-yellow-spored mushroom, but Alan Rockefeller, a citizen mycologist based in Oakland, California, regards it as solidly edible. In the West, it fruits in spring and summer, but in the Rockies of the Southwest, it also appears in fall.

by a thin ephemeral cell layer, is astounding and contrasts beautifully with the stunning golden, gill-like folds. It is even more striking when the cap shows pale white tones. And then there are some *roseocanus* who only display chanty orange—go figure!

Whatever their coloration, Rainbows are highly rated for their firmness and a more concentrated flavor than their bigger sisters. It is hard to fill a basket in the Cascades, but at the right time in the

right place, spruce habitats can be very productive.

- Bald, uneven plane cap (⌀ 2–12 cm), at first (pinkish) pale yellow fading to gray with yellow hues or golden.
- Cap with inrolled margin when young, differentiating to wavy, lobed with crisp edge.

Grows with Sitka spruce and shore pine on the coast and Engelmann spruce and true fir in the Cascades and Rockies.

NAME AND TAXONOMY: *Roseo-canus*, Latin "rose-colored" and "hoary or gray." *Cantharellus roseocanus* Redhead, Norvell, and Moncalvo 2012, previously *C. cibarius* var. *roseocanus* 1997.

WHITE CHANTERELLE
Cantharellus subalbidus

White Chanterelles are much better adapted to drier conditions than their golden sisters. Often they share the habitat of the gilled mushroom matsutake—sandy, fast-draining soils in conifer forests—and many of them have a bit of sand or volcanic ash on the base of their stems. Also like matsutake, their heads don't always reach above the duff; to find them, you sometimes have to check for mushrumps. Not surpris-

Rainbow Chanterelles, Cantharellus roseocanus

- Short, stocky stem with decurrent, bright apricot-yellow, forking folds with irregular veins.
- Not or hardly bruising when handled; if bruising, very slowly.
- Fruity apricot aroma and often spicy flavor.
- Occurs singly or in groups in soil, often in small clusters, June to October.

ingly, they are more common than Golden Chanterelles in drier regions east of the Cascades.

White Chanterelles are not a beginner's mushroom, because there are too many look-alikes. However, if you know Golden Chanterelles well, it is only a tiny leap to their white sisters. Greatly underrated by the mushroom industry, due to the absence of the iconic gold, though they stain yellow, they bring everything else to the table, including, some people insist, an even firmer flesh. I treasure them above *formosus*, but that could be due to the fact that I find them less frequently in my usual hunting grounds. They are, however, more common than Golden Chanterelles in the inland temperate rainforest regions.

IDENTIFICATION

- Cap (Ø 5–15 cm), at first flat or with downcurved margin, becoming irregularly lobed or wavy.

White Chanterelle, Cantharellus subalbidus

- White to cream colored; finely felty, going bald; broadly depressed to funnel-shaped with age.
- Stocky stem with deeply decurrent, cap-colored, forking folds with irregular cross-veins.
- Thick, firm, white flesh; stains dark yellow; flesh rips lengthwise like string cheese.
- Mild to spicy flavor and (usually) fruity, apricot-like aroma.
- Fruits solitarily or in groups and clusters from summer through fall.
- On the ground, singly or in groups, with Douglas-fir and/or western hemlock, from the Pacific coast (north of Monterey Bay to southern British Columbia) up to the Cascades and Rocky Mountains.

LOOK-ALIKES: The very common Short-stemmed Brittlegill (*Russula brevipes*) is similar but does not stain yellow, has true bladelike gills, and has a chalk-like consistency, not fibrous or stringy. In general, there are many white gilled mushrooms with gills running down the stem, but they all have bladelike gills, like funnelcaps (*Clitocybe* spp.), which include several toxic species. White, without decurrent gills, but very toxic are the all-white *Amanita smithiana* and *A. silvicola* (see The Amanitas in Gilled Mushrooms with White or Pale Spores).

NAME AND TAXONOMY: *Subalbidus* Latin "whitish;" *Cantharellus subalbidus* A. H. Smith and Morse 1947.

Winter Chanterelle, Craterellus "tubaeformis" *with close-up of underside of cap and a bounty in a basket*

WINTER CHANTERELLE
Craterellus "tubaeformis"

Everyone loves chanterelles, but many foragers are strangely oblivious to the Winter Chanterelle. What it lacks in size, it tries to compensate for in abundance—a lost challenge to the Golden Chanterelle. Its firm texture, bite-size-package, and rich aroma close to that of chanterelles render it a choice edible. The wide-ranging colors of caps and veins can be a bit confusing, but when handling these fragile mushrooms, you will quickly become familiar with them. However, harvesting Winter Chanterelles requires patience because you must either pinch them off by hand or trim them in bunches with scissors. Due to their small size and fragility, it is wise to collect them in their own container or bag.

The range of common names is helpful in describing them, and I am not referring to the discussion about calling them Yellow

Look-alikes: **LEFT:** *Goldgill Navelcap*, Chrysomphalina chrysophylla; **RIGHT:** *Pinewood Gingertail*, Xeromphalina campanella

Feet or Yellow Foots, which depends on how deep you are in the woods. "Yellow Foot" describes the bright yellow stipe, often a very distinct color from the riblike veins. "Funnel Chanterelle" points out the funnel-shaped center of the cap that deepens with age, and "Tuby" takes that depressed center a bit farther down the stipe until it is hollow. "Winter Chanterelle" informs us about its main fruiting season. This is especially true along the West Coast where *Craterellus tubaeformis* peaks in late fall and can fruit through winter; it survives and retains its composure through snow and frost! In the interior it is often the last mushroom to be gathered and is sometimes the first in spring. They are excellent when fried and gorgeous when sprinkled on a pizza.

IDENTIFICATION

- Smallish (∅ 3–8 cm) brown to dull orange wavy cap; central depression can grow into funnel shape; normally inrolled margin might flatten with age.
- Decurrent, forking, and interveined shallow folds under cap, colored pale whitish to grayish-brown or violet-gray or dull orange.

- Hollow, smooth, yellow to dull yellow-brown stem (3–8 cm tall, 0.5–1 cm wide), contrasting with the lighter, often flattened or grooved veins.
- Fragile but firm, thin flesh, whitish to pale yellow.
- Grows often in big groups and clusters potentially year-round, but especially in late fall and winter.
- Fruits from late-stage decaying conifer wood and in wet soil, bogs, and moss; mostly associated with western hemlock but also Douglas-fir and Sitka spruce. Widespread.

LOOK-ALIKES: Very similar, growing on dead wood, but also much rarer is the probably harmless Goldgill Navelcap (*Chrysomphalina chrysophylla*), which is also known as False Yellow Foot. The only clear difference is that the Navelcap has regular gills. Somewhat similar but smaller and more common is the inedible Golden Navel (*Chrysomphalina aurantiaca*). *Hygrophoropsis aurantiaca* and some wax-caps (*Hygrocybe* spp.) look similar. Not as close in looks, smaller, but also occurring in big groups on decaying conifer wood is the ubiquitous Pinewood Gingertail

Black Trumpet, Craterellus calicornucopioides

(*Xeromphalina campanella*), which is bitter but not poisonous. All these look-alikes have true gills.

NAME AND TAXONOMY: *Crater-ellus* Latin for "big mixing bowl," and *-ellus* diminutive for "small;" *tubae-formis* Latin "trumpet" (*tuba*) and "shaped" (*formis*). *Craterellus tubaeformis* (Fr.) Quel. 1888, also previously listed as *Cantharellus tubaeformis* or *C. infundibuliformis*; David Pilz suggested *C. neotubaeformis*, with prefix *neo*, Greek for "new," as in New World, but was unfortunately not validly published since genetic research shows our species needs its own name. Cantharellaceae, Cantharellales.

BLACK TRUMPET
Craterellus calicornucopioides

What a unique mushroom! Black Trumpets are special in several ways. Unfortunately,

they are rare in the Pacific Northwest, but their relative absence makes them even more special. Their favorite habitat is southwestern Oregon's coastal forests, with tanoaks intermixed, or coastal Northern California, with rare appearances in the Cascades. Trumpets sometimes also grow with Garry oaks, madrones, manzanitas, and some huckleberry species without tanoaks around. Yet so seldom are they encountered in the Puget Sound area that their presence seems solely mythical.

Spotting Black Trumpets can be a real challenge! At first, they are nearly invisible, aptly described as invisible black holes broadcasting between fallen leaves. Their darkness and shape inspired the name Trumpet of Death, a translation commonly used in most European languages. The easiest way to spot them is when they pop out of contrasting green moss. Once you spot one trumpet, you will realize you are standing in the midst of a spread of them, perhaps one reason they are also known as the Horn of Plenty. Fortunately, they are very easy to identify, since they are unique in shape and color and have only two challenging look-alikes (see below). Black Trumpets handle washing well, dry easily—which also concentrates their excellent flavor—and rehydrate well. Their thin but firm texture in combination with a rich and earthy, smoky flavor is excellent for cooking and adds fungal magic to any dish or sauce.

IDENTIFICATION

- Fruiting body is 5–15 cm tall, funnel- to trumpet-shaped, grayish-black to very dark brown or black, darkening when wet.
- Caps (∅ 3–10 cm) are thin edged with fine scales, inrolled when young, and wavy and often splitting when mature.
- Underside is smooth or slightly wrinkled, ash-gray to sometimes blue-gray when young.
- Stem is hollow, with inside and outside surfaces continuous from cap to solid, narrow base that is often fused to other trumpets.
- The thin, tough, but flexible flesh is colored more or less like cap, with a mild flavor.
- Grows scattered or in clusters on ground or very rotten wood.
- Fruits mainly in late fall or early winter, especially with tanoak, but also with oaks and woody members of the Ericaceae, like madrone and manzanita. In Washington it grows with Oregon white oak (*Quercus garryana*) but is rarely found.

LOOK-ALIKES: Both the much rarer and also choice Veined Black Trumpet (*Craterellus atrocinereus*) and the thicker-fleshed, edible Blue Chanterelle (*Polyozellus* spp.) look similar to Black Trumpets.

NAME AND TAXONOMY: *Cali* for California or Greek "beautiful," *cornucopioides* Latin for "horn" (*cornu*) "of plenty" (*copia*). *Craterellus calicornucopioides* D. Arora and J. L. Frank 2015, previously regarded as synonymous with Eurasian *C. cornucopioides* (L.) Pers. 1825. Cantharellaceae, Cantharellales.

PIG'S EAR
Gomphus clavatus

Nobody expects to uncover a whole hog belowground when they spot a Pig's Ear! Still, finding this easily overlooked mushroom—the mellow brown color of the caps is not showy—is very gratifying. Fruitings can be substantial, and the mushroom withstands drought well due to its very firm flesh. Often the lobes have been hollowed out by larvae long gone, which should not stop anyone from enjoying this unique mushroom that would not seem misplaced growing in a coral reef.

The most eye-catching element is the purple underside with its chanterelle-like network of ridges, and it is sometimes also known as the Purple Chanterelle, a name readily embraced by mushroom dealers, though *Gomphus clavatus* is more closely related to coral mushrooms (*Ramaria* spp.) than to chanties. The firm flesh makes Pig's Ears very versatile in the kitchen. When overmature, the flesh is mealy and can be astringent; some people react with gastric upset.

- Caps or lobes (Ø 5–15 cm), dingy light brown with light purplish tinge when young and toward often depressed center; flattening with maturity and developing wavy edges, often one sided; young growth club-like.
- Underside with wrinkly veins and ridges, usually purplish, sometimes pale brown, running nearly down to stem base.
- Stem can be branching (5–20 cm tall), solid if not hollowed by larvae.
- Flesh white, very firm, stringy.
- Grows scattered, but mostly in fused or compound clusters (sometimes making fairy rings) on ground or very rotten wood in conifer forests.
- Fruits mainly in late summer through fall.

Pig's Ears, Gomphus clavatus

LOOK-ALIKES: Nothing really above the high-tide line.

NAME AND TAXONOMY: *Gomphus* Greek "plug" or "wedge-shaped nail," and *clavatus* Latin meaning "club-shaped." "Pig's ear" is also used for a spring-fruiting morel relative (see Thick Cup in Ascomycetes). *Gomphus clavatus* (Pers.) Gray 1821. Gomphaceae, Gomphales.

SCALY VASE
Turbinellus floccosus group

The beautiful Scaly Vase often excites new-bies into thinking they have finally found a chanterelle. However, it turns out this mushroom, a.k.a. Woolly Chanterelle, is only a chanterelle in common name. Recent DNA studies reveal that it is closely related

to corals (*Ramaria spp.*) and Pig's Ears (*Gomphus clavatus*). Also, when it comes to edibility, there is plenty of confusion. Some people eat it regularly without problems, and Scaly Vases (possibly different species) can be found in markets in Mexico and East Asia. However, *Turbinellus* is absent in Europe. While wrongly maligned as containing an accumulative liver toxin, and hence widely shunned in the Pacific Northwest, for some people it can cause loose stools, the runs, or even wicked gastrointestinal symptoms, sometimes delayed by eight to fourteen hours.

Scaly Vase, Turbinellus floccosus *group; close-up of underside of cap*

IDENTIFICATION

- Vase-shaped, 8–20 cm tall, with reddish to orange-buff, big, coarse, scaly cap.
- Whitish to pale, deeply decurrent underside with forking and interconnecting veins.
- Stem at first solid becoming hollow and narrowing toward base, often rooted deep into topsoil.
- Flesh fibrous, white or pallid, color unchanging; mild flavor.
- Grows scattered or in clusters on ground under conifers in late summer or fall.

LOOK-ALIKES: Very similar or identical based on DNA work is *Gomphus bonarii*. Lighter colored, with cream to pale orange or brown cap tones, and often bigger is *Turbinellus kaufmanii*, which in terms of edibility shares the traits of Scaly Vase. There is a superficial resemblance to Golden and Cascade Chanterelles.

NAME AND TAXONOMY: *Turbin-ellus* Latin for "small spinning top," *floccosus* Latin "floccose, covered in woolly tufts."

Blue Chanterelle, Polyozellus atrolazulinus

In Yunnan in southwestern China, one of the most appreciated and expensive edible mushrooms is Ganba Jun, Thelephora ganbajun, a close relative of the Blue Chanterelle. Though quite leathery, it is sliced very thin and fried hard.

Turbinellus floccosus (Schwein.) Earle ex Giachini and Castellano 2011 = *Gomphus floccosus* Singer; *Gomphus bonarii* (Morse) Singer 1945. Gomphaceae, Gomphales.

BLUE CHANTERELLE
Polyozellus atrolazulinus

With unique hues of blue and purple, Blue Chanterelles are spectacular in their prime—a visual splendor that does not last, nor does it translate into culinary fireworks. This member of the Earth fan family only superficially resembles the chanterelle and is not closely related. Disappointed foragers claim that Blue Chanterelles might be best cooked in the mushroom dyers' pot, where they produce colors ranging from blue-grays and green-grays to mossy-greens.

While in the past we misapplied the name *Polyozellus multiplex*, which we now know grows in eastern North America and East Asia, recent genetic analysis has turned up instead three new species in the Pacific Northwest. *Polyozellus atrolazulinus* contains purple and violet colors (as does *P. purpureoniger*), while *P. marymargaretae* lacks these colors and is much bluer, especially when young. Blue Chanterelles contain poly-ozellin, a molecule that has generated research interest due to anticancer propensities as well as an ability to inhibit PEP (prolyl endopeptidase), an enzyme implicated in memory loss and senile dementia.

After picking and when old, all three species fade to black and seem, so far, in other ways indistinguishable—perhaps they are also indistinguishable by taste. Maybe one of them is more enjoyable than the very bland Blue Chanterelles I have tasted in the past. However, their texture and taste after turning them into jerky has been praised.

- Purplish or blue, fleshy mushroom (Ø 8–20 cm) with fan- or spoon-shaped, lopsided, sometimes branching caps fusing into a wide stem base.
- Underside (hymenium) covered in irregularly forking network of decurrent veins, forming a reticulate or almost poroid (pore-like) surface, sometimes nearly smooth over large areas; blue and purple when young, fading in age to gray and black.
- Flesh deep purple to bluish, soft but brittle.
- Grows in dense, fused clusters from soil in mycorrhizal association with conifers, especially spruce and fir.
- Fruits in late summer and fall.

LOOK-ALIKES: Only Black Trumpets (*Craterellus calicornucopioides*) come close, but they have much thinner flesh. Somewhat similarly shaped, but differing in color with their purple veins and brown caps, are Pig's Ears (*Gomphus clavatus*).

NAME AND TAXONOMY: *Poly* for "many" and *oz* meaning "branch," both Greek, plus *-ellus* Latin diminutive "small;" *atro-lazulinus* means "dark blue" in Latin. *Polyozellus atrolazulinus* Trudell and Koljalg 2017, previously known as *P. multiplex* (Underw.) Murrill 1910. Thelephoraceae, Thelephorales.

TOOTHED FUNGI

One of the smallest groups of mushrooms categorized by the shape of their hymenium are the toothed fungi, which exhibit slender, tapering extensions known as spines or teeth, most of them downward hanging. The phenomenon of a toothed hymenium under a protective cap to maximize surface area for spore production has developed separately several times in the fungal tree of life, a phenomenon known as "convergent evolution." From a culinary perspective, there are three genera of toothed fungi of relevance: hedgehogs (*Hydnum*), Hawk's Wings (*Sarcodon*), and Lion's Manes (*Hericium*), their spines exposed and not protected by a cap.

HEDGEHOG MUSHROOMS
Hydnum species

Some of the easiest, safest, choicest edible mushrooms worldwide are hedgehog mushrooms, named for the resemblance of their "teeth" under the cap to the spines of a hedgehog, the old world relative of shrews and moles with a porcupine hairdo. Many hedgehogs have pale whitish caps; some are a bit darker, with pale, light brown or dull orange tones. Their spines are usually lighter colored, white to cream. The

PROCESSING AND COOKING HEDGEHOGS

When cooked, hedgehogs have a nicely tender and meaty texture. Extended cooking can remove the bitterness of older specimens. A slight peppery note can be enjoyed when using raw, young hedgehogs in salads. The firm flesh is fragile: especially when collecting Bellybutton Hedgehogs, be careful to pick the whole mushroom and not crumble the poor thing into the duff. Also, they will appear more attractive in the kitchen if you separate them and protect them from basket bullies like boletes or chanterelles while transporting them. Pick them as cleanly as possible; having to pick dirt out of the spines is tedious. Hedgehogs store well in your fridge, often lasting longer than a week, but bigger ones might turn bitter with time.

length of the spines reflects the age of the cap, since the spines continue to grow as the cap matures. Hedgehogs have a very pleasant, mild flavor (hence the common name "Sweet Tooth/Teeth"), somewhat similar to the closely related chanterelles, but sometimes they turn bitter with age.

Like chanterelles, hedgehogs are not known to cause digestive distress, and also like chanterelles, hedgehogs fight off

Oregon Bellybutton Hedgehog, Hydnum oregonese

pesky larvae successfully. Many mushroom hunters have long interchangeably cooked hedgehogs and chanterelles, so similar the flavor and texture, whereas taxonomists used to classify them separately due to the difference in fertile tissue—spines versus ridges. Recent molecular studies confirm that preparing them in the same frying pan is not only culinarily acceptable but taxonomically correct, as they both belong to the Cantharellales order!

OREGON BELLYBUTTON HEDGEHOG

Hydnum oregonense

These little spiny "Sweet Teeth" are easily recognized by their small size and bellybutton center in a pale orange (never white) cap. They are late fruiters and in coastal regions fruit all winter. Often, they love to grow under evergreen huckleberry (*Vaccinium ovatum*). Until recently they were known as Bellybutton Hedgehogs, *Hydnum umbilicatum*. However, *umbilicatum* is now considered an East Coast species, and ours is genetically different enough to be newly described as *H. oregonense*. Similar to it is the Honey Hedgehog (*H. melitosarx*).

IDENTIFICATION

- Small (< 5 cm), slender mushroom with cream to pale orangish cap with central depression, resembling a belly button.
- Teeth 0.2–0.5 cm long, mostly non-decurrent, first whitish, later pale light brown.
- Stem whitish, slender (and sometimes long), bruising very pale brownish.
- Flesh firm but fragile, mild tasting—though possibly turning bitter when old.
- Grows on ground with conifers and evergreen huckleberry; fruits from late fall through mild winters, sometimes into spring.
- Distributed along the Pacific coast and in the Cascades, from Alaska to the redwoods.

HEDGEHOGS: TWO SPECIES BECOME SEVEN!

Formerly two species were reported in the Pacific Northwest: the small Bellybuton Hedgehog (*Hydnum umbilicatum*) and the bigger Spreading Hedgehog (*H. repandum*). A recent DNA study[31] by Tuula Niskanen and others (including Lorelei Norvell and Joe Ammirati, two well-known Pacific Northwest mycologists) added five more species: *H. olympicum*, *H. oregonense*, *H. melitosarx*, *H. melleopallidum*, and a Pacific Northwest version of the Finnish–Tibetan *H. jussii*. The latter is difficult to distinguish from the others and is so far known only from the deep interior or far north.

Unfortunately, many of the DNA-derived taxonomic insights have not yet been successfully applied to telling these new species apart in the woods, but this is no obstacle for the pot hunter, since they are all exquisitely delicious!

Honey Hedgehog, Hydnum melitosarx

NAME AND TAXONOMY: *Oregonense* latinized "from Oregon," *umbilicatum* Latin for navel, *melito-sarx* Latin "honey" and Greek "body/flesh." *Hydnum oregonense* Norvell, Liimat., and Niskanen 2018; formerly *H. umbilicatum* Peck 1901, *H. melitosarx* Ruots., Huhtinen, Olariaga, Niskanen, Liimat., and Ammirati 2018. Hydnaceae, Cantharellales.

HONEY HEDGEHOG
Hydnum melitosarx

Hydnum melitosarx is a medium-sized (cap Ø 2–6 cm), slender, often long-stemmed (3–7 cm) hedgehog mushroom. The (pale) orange-brown cap is at first convex but flattens out with an incurved, paler margin and rarely has a depression in the center. The pointed spines are usually non-decurrent, sometimes with scattered, small decurrent spines on the top of a whitish stem that turns orange-brownish where scratched. It is found in the Pacific Northwest, including Alaska, and in northern Eurasia.

WESTERN WOOD HEDGEHOG AND OLYMPIC HEDGEHOG
Hydnum washingtonianum and *H. olympicum*

These bigger hedgehogs have been reported all over the Pacific Northwest, and they are safe choice edibles. The combination of color, mild flavor, and especially the unique spines under the cap allows for identification in a loose group; all are edible. The taxonomic uncertainty does not allow an outline of clear distribution areas and fruiting season, but some morphological features are becoming clear. Based on the recent description, the Olympic Hedgehog

Western Wood Hedgehog, Hydnum washingtonianum

(*Hydnum olympicum*) has a longer stem (up to twice the cap diameter), while the Western Wood Hedgehog (*H. washingtonianum*) is stouter: often the stem is as long as the cap is wide. But fruiting bodies frequently ignore such concepts. To conclusively identify them, you must examine their spores under a microscope; when it comes to spore size, the Olympic Hedgehog is the champion.

There may still be other undescribed hedgehogs hiding in the woods. Hopefully in the coming years we will learn how to tell these new species apart in the field, and maybe there will also be subtle differences in their culinary qualities. Research with European *H. repandum* has shown mild antibiotic activity, potent antitumor activity in vitro, and in animal studies the capacity to lower blood cholesterol.

Orange Spine, Hydnellum aurantiacum, *good for the dyeing pot, but not for the cooking pot*

- Medium-sized mushrooms with a cream to pale orangish cap (Ø 5–15 cm), and often lighter cap edge.
- Teeth (0.2–0.7 cm long), pale white to cream color, mostly decurrent, running down the stem.
- Stem whitish, bruising very pale brownish: longish in *H. olympicum*, shorter in *H. washingtonianum*.
- Flesh firm but fragile; mild tasting but possibly turning bitter when old.
- Grows on ground, mostly with conifers; fruits in summer, fall, and winter.
- Distributed from Alaska down to California and east to the Rockies.

LOOK-ALIKES: The most commonly found soft-fleshed and toothed look-alikes to hedgehogs are Hawk's Wings and their relatives (see next entry). Also toothed, but tough fleshed, are inedible *Hydnellum* and *Phellodon* species. The Orange Spine (*Hydnellum aurantiacum*) comes close in color to hedgehogs, but it is easy to tell apart by its very tough flesh and, when young, by the contrast of white spines to orange stem—although this darkens and changes greatly with age (see photo). The closely related but extremely rare and inedible Aromatic Earthfan (*Sistotrema confluens*) is very small (down to 1–2 cm), pale white, and prone to enveloping needles and sticks in its often-fused fruiting bodies. Its stem is off-center and the cap is more fan-shaped.

NAME AND TAXONOMY: *Repandum* Latin for "spreading." Current species names are all Pacific Northwest geographic references. *Hydnum washingtonianum* Ellis and Everh. 1894. = *H. neorepandum* Niskanen

and Liimat. 2018; *H. olympicum* Niskanen, Liimat., and Ammirati 2018. Misapplied: *H. repandum* L. 1753 = *Dentinum repandum* (L.) Gray 1821. Hydnaceae, Cantharellales.

HAWK'S WING
Sarcodon imbricatus

Hawk's Wings are striking, easily recognized mushrooms that are beloved in some regions of the Northern Hemisphere while ignored in others, as a result perhaps of several factors. Apparently, there is some hidden diversity in *Sarcodon* that is slowly being researched and revealed; might similar-looking species have different flavor profiles, or might the local soil impact the flavor as *terroir* does wine? Often it is some form of bitterness that deters people from eating Hawk's Wings, a bitterness that intensifies with age but that can be removed by boiling the wings for ten to twenty minutes or frying them for an extended period, something David Arora learned between publishing *Mushrooms Demystified* (1986) and his 1991 pocket guide, since his appreciation evolved from "edible, but of poor quality" to "excellent, if sautéed for at least 20 minutes, otherwise it is apt to be bitter."

The aroma of Hawk's Wings is impressive; they hold their ground with their strong and earthy fungal flavor and are the perfect fungal ingredient for stews and casseroles. Each year at Telluride's fantastic Mushroom Festival in Colorado, Hawk's Wings are the key ingredient in the highly appreciated festival goulash. An unlucky few people discern a strange metallic flavor that for them kicks the "choice" out of this choice edible. In subalpine spruce forests, Hawk's Wings are some of the most abundant and easily collected mushrooms. They also make fragrant mushroom powder for spicing soups, sauces, and popcorn. When I collected mushrooms in the Alps as a child, we used Hawk's Wings only for making mushroom spice, a tradition I still keep.

IDENTIFICATION

- Medium to big (Ø 5–20 cm) centrally depressed mushroom, overall brown (darkening in age), with big dark brown to black raised chunky scales that reduce in size outward to a mostly inrolled cap edge.
- Spines soft, 0.2–1.5 cm long, white, grayish, or light brown.
- Central or off-center stem is wider at the base, staining, with a top hollowing in age.

ABOVE: *Collecting Hawk's Wing in eastern Tibet;*
BELOW: *Hawk's Wing, Sarcodon imbricatus, with close-up of underside*

- Flesh firm and brittle (not tough or woody), white to light brownish with a mild to bitter flavor.
- Grows in groups on the ground with conifers, especially Engelmann spruce, and sometimes hardwoods; fruits in late summer and fall.

LOOK-ALIKES: There are at least a dozen similar mushrooms in the Pacific Northwest, some of which are being moved from *Sarcodon* to *Hydnellum*, like the gorgeous purple Violet Tooth, *H. fuscoindicum*, and many others in all hues of brown. None seem poisonous, but some are way too bitter and tough to eat.

NAME AND TAXONOMY: *Sarcodon* Greek *sarco* "flesh" and *odon* "tooth"; *imbricatus* Latin "tiled" or "with overlapping tiles," as in shingled. *Sarcodon imbricatus* (L.: Fr.) P. Karst. 1881 = *Hydnum imbricatum* L. ex Fr.; *Sarcodon squamosus* (Schaeff.) Quel. 1886. Bankeraceae, Thelephorales.

Bear's Heads, Lion's Manes, and Bearded Teeth (Hericium *spp.*)

Some of our most stunning and distinctive mushrooms are members of the genus *Hericium*, with a range of imaginative common names. Mean people might label them as "idiot proof," but there is nothing bad about a beautiful, tasty mushroom that is easy to recognize. Though telling some species apart can be difficult when you don't pay attention to the host tree, that should not worry any fungal forager: they are all delicious! In addition, *Hericium* is the subject of great interest for its neuroregenerative potential.

The four species of *Hericium* in North America are all present in the Pacific Northwest. They start fruiting after sufficient rain in summer or wait for the return of precipitation in fall and fruit into late fall. They share color, texture, and flavor, though each has a slightly different structural appearance and each prefers a different host tree. Most common is Bear's Head, *Hericium abietis*—also called Conifer Bear's Head because it feeds on conifers. Often mistaken for Bear's Head is the Coral Tooth fungus, *H. coralloides,* which is more common in the interior, where it loves growing on black cottonwoods. Another true hardwood feeder is Lion's Mane, *H. erinaceus*. It is easily recognized by its long, well-combed, and evenly cut spines. Many people refer to all Hericiums as Lion's Manes. Feeding on all types of wood, but found mostly east of the Rockies, is the American Tooth, *H. americanum*. It tends to have orderly tufts of spines (1–4 cm long) hanging from the tips of a branched fruiting body.

All species of Lion's Manes are highly sought after for their delicious mild flavor and firm consistency. Chef Becky Selengut, author of *Shroom*, has declared them as close to a lobster or crab as a mushroom can get, and without any annoying fishiness (I would add). It is also traded under the French name *pom pom blanc*, as in "white cheerleader tufts." To cook Lion's Manes, it is best to separate their branches into manageable tufts. That's easy with the more highly branched fruiting bodies; the more compact ones are best sliced in line with the teeth. The core can be tough and might need to be cooked lon-

Bear's Head, Hericium abietis

ger or can be dried and ground. Chinese-farmed Lion's Manes are sold dried, but cooking and freezing freshly collected specimens preserves the delicate flavor better.

Lion's Mane contains brain neuron growth factor compounds that are water soluble and cross the blood-brain barrier. Two small human clinical trials on senile dementia have shown it to increase memory capacity for elderly people. Its neuroregenerative capacity has inspired Paul Stamets to suggest taking Lion's Mane with *Psilocybe* mushrooms to increase neurogenesis—the development of new brain neurons to speed up evolution of human consciousness. A double-blind, placebo-controlled trial on menopausal symptoms including depression, anxiety, and sleep problems showed positive results.

Hericium seems to reduce tremors caused by Parkinson's disease and other neuromuscular conditions. Apparently these qualities are not new to traditional Chinese medicine, where it is used for nerve conditions and general debility, and for digestive issues and gastric ulcers. It also has been shown to cause anticancer activity in several in vitro trials. Native people applied Lion's Mane powder as a styptic to stop bleeding of wounds.

BEAR'S HEAD
Hericium abietis

This choice edible mushroom could be described as a frozen waterfall, hence it is also known as the "Icicle Mushroom." When you find it on a dead conifer trunk, always check thoroughly on all sides of the host—often there are multiple fruitings—and check back the next year as well!

IDENTIFICATION

- Big to huge, multibranched mushroom, white to pale salmon.
- Teeth up to 1 cm long (sometimes to 2.5 cm), soft but brittle, always pointing down with maturity. Teeth grow along the branches or are clustered at the branch tips.
- Grows solitarily or in groups, especially on Douglas-fir and true fir. Host is usually dead, either a standing snag or on the ground.
- Fruits in summer to late fall; endemic to western North America.

LOOK-ALIKES: Bear's Head, Coral Tooth, and the other Hericiums are not closely related to coral mushrooms (*Ramaria* spp.)

and White Coral Fungus (*Clavulina coralloides*; for both, see Corals, Clubs, and the Like), whose branches point upward while the spines of *Hericium* point downward. Some people confuse white Bear's Head with white immature slime molds, but as soon as you touch the white stuff of slime molds (in an entirely different kingdom of organisms), you'll realize the difference.

NAME AND TAXONOMY: Latin *abietis* meaning "of fir" (*Abies*). *Hericium abietis* (Weir ex Hubert) K. A. Harrison 1964. Hericiaceae, Russulales.

CORAL TOOTH
Hericium coralloides

Very similar to Bear's Head, and just as delicious, but less common in the coastal region is Coral Tooth. It grows solitarily or

Coral Tooth, Hericium coralloides, *in Sicamous, British Columbia, growing on cottonwood*

Lion's Mane, Hericium erinaceus; *Lion's Mane from below*

in groups on hardwoods, which include woods as soft as cottonwood, willow, birch, and alder and as hard as maple and oak. Great places to encounter Coral Tooth are cottonwood stands along rivers in fall. Sometimes only a few tentacle-like branches form a more open, less densely packed fruiting body, especially when young. Also, the teeth tend to be the shortest of all Hericiums with a length of 0.3 to 1 cm, just growing at branch endings up to 3 cm. Fruiting occurs in summer to late fall; it's widespread in North America.

NAME AND TAXONOMY: *Corall-oides* Latin "coral-like." *Hericium coralloides* (Scop.) Pers. 1794 = *H. ramosum* (Merat) Letellier 1826.

LION'S MANE
Hericium erinaceus

Lion's Mane beats out all other mushrooms when it comes to the number of highly imaginative names: Bearded Tooth, Satyr's Beard, Old Man's Beard, Bearded

American Tooth, Hericium americanum

Hedgehog, Medusa Head, Pom Pom du Blanc, and Monkey Head (the latter translated from the Chinese *hou tou gu*). It is easily recognized by its well-combed and cleanly trimmed spines (using "teeth" in this context hurts!) or, more technically, by its less branched mass of long (2–5 cm), closely packed, hanging spines. When young, it can be compact enough to look like a brush. It is found only on hardwoods, especially on oak. It is being cultivated in great amounts in China, since 1959, and is often sold dried in Asian food stores.

NAME AND TAXONOMY: Both *Hericium* and *erinaceus* are references to the spiny European hedgehog, *Erinaceus europaeus*. *Hericium erinaceus* (Bull. Ex Fr.) Pers. 1825.

AMERICAN TOOTH
Hericium americanum

Feeding on all types of wood, but mostly east of the Rockies, is the American Tooth, *Hericium americanum*; it tends to have orderly tufts of hanging spines (1–4 cm), mainly from the tips of its branches.

NAME AND TAXONOMY: *Americanum* for distribution area. *Hericium americanum* Ginns 1984.

BOLETES

Before we start hailing the king, it is best to introduce boletes as a group. They are fleshy mushrooms with a typical mushroom gestalt of round cap and central stem and are easily recognized by a hymenium under the cap composed of spongelike tubes ending in pores. When they are young, the pore surface can be firm and dense, but when they are mature and producing spores, the pores become a soft sponge of open tubes; as a layer, the pores are easily peeled from the flesh of the cap.

Nearly all boletes are mycorrhizal and they all belong to the order Boletales. Very similar in many regards are Stalked polypores (*Albatrellus*, *Scutiger*, Russulales) and Kurotake (*Boletopsis grisea*, Thelephorales),

but pores of these bolete look-alikes do not turn soft. Stalked polypores are presented with polypores in the next chapter. Non-mycorrhizal polypores (order Polyporales), which often grow from wood, also have pored fertile tissue, but the tissue is much tougher, often perennial, and is very difficult to peel from the cap. The order Boletales also contains a few genera of often edible gilled mushrooms: *Phylloporus* (gilled boletes), *Gomphidius* (slime-spikes), *Chroogomphus* (pinespikes), *Hygrophoropsis* (False Chanterelle), and *Paxillus* (Rollrim). Some of these let you peel their gills away just like the sponge layer of a bolete. Most of these genera are described with the dark-spored gilled mushrooms.

Admirable Bolete, Aureoboletus mirabilis, *growing from a decaying conifer trunk*

TABLE 2. BOLETES AT A GLANCE

GENUS	CAP	STAINING	PORE COLOR
BOLETUS King boletes	Smooth to finely velvety	None	White stuffed, maturing greenish
BUTYRIBOLETUS Butter boletes	Pinkish-red to gray or brown	Intense bluing of pores	Yellow
LECCINUM Scaberstalks	Bald to finely hairy	None or sometimes blue and green in stem; flesh sometimes reddish and gray	Pale white to gray
SUILLUS Jacks	Scaly or sticky	Occasionally bluing, some browning	Yellow
XEROCOMELLUS Suede boletes	Suede-like, dry	Slight to intense bluing	Yellow
CHALCIPORUS Pepper boletes	Bald to finely hairy	Almost none	Brown and orange
CALOBOLETUS Bitter boletes	Bald to hairy; velvety	Pores and flesh bluing	Yellow
RUBROBOLETUS Red-pored boletes	Bald to finely hairy with red tones	All parts bluing	Orange and red

BOLETES: MUSHROOMS WITH SPONGY PORES

Distinguishing between your boletes is luckily not as hard as making sense of the diversity of gilled mushrooms, a much bigger and more dangerous group. The DNA revolution has brought us more clarity and a better handle on our boletes (though it has also brought us many new genera), helping differentiate the edibles from the few nonedibles. More than 90% of our region's boletes are edible—some need special treatment, while other boletes are the safest and tastiest edible mushrooms we know, edible even raw. So it is really worth the effort to learn the main genera (see Table 2). One great resource for learning your boletes is the set of *Boletes of Western North America* flash cards Gary Gilbert and I published.

In the Pacific Northwest, we have only a few black sheep in the bolete family, or rather red sheep, since the most toxic boletes are the usually rare, red-pored boletes (*Rubroboletus*). Several boletes are very bitter (*Caloboletus*) and inedible. Bluing tissue alone is not an indication of toxicity—and some boletes that stain blue are choice edibles, like butter boletes (*Butyriboletus*), or good edibles, like suede boletes (*Xerocomellus*). Yet some of the best boletes, like the King

TASTE	STEM	STATURE	EDIBILITY
Sweet and nutty	Reticulated, typically fat	Large and dense	*The best!*
Mild	Thick, yellow, and reticulated	Large and dense	Excellent
Mild	Scabers	Large and dense	Some very good, some bland, some cause gastric upset
Mild	Glandular dots or ring; often slender stem	Stout to slender	Wide range
Mild, sometimes lemony	Very fine red specks	Slender stems, fragile	Pretty good when young
Spicy	Slender and fragile	Smallish and fragile	Used as spice
Bitter	Slender, short	Large and dense	Inedible
Mild	Thick and reticulated	Large and dense	Poisonous

Bolete (*Boletus edulis*) and other members of the genus *Boletus*, do not stain blue. Bluing in most boletes is caused by variegatic and xerocomic acids being exposed to air by injury, triggering enzymatic oxidation.[32]

What also unites the boletes is that many of them get attacked by several *Hypomyces*, aptly called the Bolete Eaters (and often much worse names, especially when encountered on your porcini!). When a bolete is infected by the white and fluffy Bolete Eater, the victimized specimen's flavor and texture are ruined. (Whereas another *Hypomyces*, *H. lactifluorum*, turns a bland brittlegill into a tasty Lobster

Bolete Eater, Hypomyces chrysospermus, *maturing on a* Xerocomellus

PROCESSING AND COOKING PORCINI

King boletes, a.k.a. porcini, need to be processed rather quickly and carefully monitored if stored fresh. Firm young kings are delicious eaten raw and are served in this manner in some restaurants in Italy. At our home we love them best raw as a salad (see photo below), but that dish may cause a bellyache for some people. The king's quality diminishes every day it is stored in the fridge: young buttons last up to a week, while older kings may become corrupt in a few days. Storing them just a bit above freezing will preserve them the longest. If you collected more than you can eat, you have several choices. Firm kings can be frozen as is and are best used within several months. You can also cook your kings, be it alone or in a recipe, and freeze them for later use. Next most convenient is slicing your boletes and drying them—then they can last for years.

King boletes best develop their rich nutty flavor when fried and caramelized. I never overload my frying pan; I want all the pieces to sizzle in oil or butter in contact with the hot iron. Once the pieces are browned, I turn them over and may slice them even smaller. In my experience, kings need pre-frying to bring out their full flavor. Lower-quality boletes are always best dried. It concentrates their flavor and reduces mushiness when rehydrating them. Plenty of people insist that drying kings fully develops their aroma. Also, drying mature sponge layers is a great way to use them and turn them into an excellent spice. Some boletivores swear on using sponge powder as a veggie or steak rub!

For later use, boletes can be quickly rehydrated: half an hour is sufficient. Hot water speeds the process and ten to twenty minutes might suffice, but some people insist on soaking them overnight. Much of the flavor will be in the rehydration liquid. Use it for infusing your food with the king's aroma. Drying should not be limited to the old, flabby, and infirm; prime boletes store awesome dried too. They are great to serve up as gifts, especially in places where this deliciousness is unheard off, be it in the Amazon rainforest or at the in-laws' home in the Midwest. Less-than-prime kings also make a great broth.

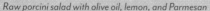

Raw porcini salad with olive oil, lemon, and Parmesan

Fresh Hypomyces chrysospermus *growing on an Admirable bolete,* Aureoboletus mirabilis

Mushroom.) The infected bolete turns into a foul mush while the parasite's surface brightens to golden as its spores mature (*chryso* is Greek for "golden-yellow"). Note there is a second, very similar Bolete Eater, *H. microspermus,* which can only be told apart by its smaller spores, hence *micro-spermus.*

King Boletes

Some of the best mushrooms under the sun are king boletes (*Boletus* spp.), loved for their sweet, nutty aroma and firm flesh! The Pacific Northwest is blessed with a range of king bolete species that fruit from spring into late fall, should the rains cooperate. The Cascades, interior mountains, and coastal spruce and shore pine forests are their main habitats. In May and June, the Spring King (*Boletus rex-veris*) keeps court in the eastern Cascades. When I first moved to the Pacific Northwest, I had a hard time fulfilling my desire for *Steinpilz,* the German name for king boletes, meaning "stone mushroom" for its firm flesh.

I was hoping to find them in the same western hemlock–Douglas-fir forests where I found my Golden Chanterelles; however, the residing king in the lowland forest is the Fiber King, which is not abundant.

Over the years, I realized the best bet for meeting the king is by wandering in spruce, true fir, and shore pine forests, in the mountains or along the coast. Timing is everything. Kings will not raise their heads in summer or fall until several weeks after a good soil-soaking rain. And then they all fruit at the same time. Come a week or two too late after they have started to grow, and all you find are larvae-ridden carcasses and yucky puddles of mush. It is heartbreaking to see royalty reduced to rubbish! Well, at least the spores got out, though without any assistance from you.

According to current revised taxonomy, king boletes are big, fleshy mushrooms with rounded caps that can be bald or velvety (many boletes go bald with age) and fat stems that are wider in the lower half and covered with netting (known as reticulation) on their upper half. They have white, mild, often sweet and nutty-tasting flesh that does not change color when cut, and a tube layer—the spongelike fertile tissue under the cap—that is white or pale yellow when young, softening in maturity to dull yellowish-greenish, olive, or brown.

Now, in nature, a few organisms always break the rules, and so it is with king boletes. The flesh of some kings may stain a tiny bit very slowly after peeling the sponge layer, or sometimes a faint bluing occurs when the mushroom is old and waterlogged, but these rule infractions are rare. Until recently, the genus *Boletus* contained a wide variety of boletes, including boletes that are small, fragile, thin stemmed, bluing or that changed colors in other ways, bitter or spicy tasting, and with pores in all kinds of colors. We still call many of these mushrooms "boletes," but many now have their own scientific genera (see Table 2). Only the most recent mushroom books have integrated these new names and taxonomic refinements.

Though boletes as a group are royalty when it comes to aroma and size, they have their weaknesses. They are easily invaded by larvae, and their flesh is relatively short-lived. A giant king is often just a week or ten days old, though smaller firm ones sometimes remain for two weeks in seemingly arrested growth, especially when the weather turns cold. Overmature boletes quickly turn into habitat for a range of insects, especially fungus gnats. Eggs laid by these tiny flies hatch and eat their way as larvae through the kings, fattening up and spoiling our feast. A few worms are fine by many boletivores, but most of us prefer that fungal protein clearly outweighs insect protein. We get very upset when it seems like the infested mushroom is trying to slither out of the frying pan! It's highly recommended that you check if your mushroom is worm-ridden in the woods, and if so, leave it behind.

Corrupted kings can still produce spores for a while and do so much better in the woods than in your compost bin, unless of course you are trying to inoculate the right trees by spreading the sporulating sponge around them. Any tubes I bring home that are too old to process, I spread around potential host trees in my neck of the suburban woods. Successful introduction is not guaranteed, but we have anecdotal evidence that it can work. It can take many years before mycelium bear fruit, perhaps outlasting the average duration of homeownership.

King boletes are very safe in the Pacific Northwest. There are no toxic fat-stemmed, white to yellow or olive, sponged look-alikes that have the same mild-nutty flavor, reticulation on the stem, and unchanging (not bluing) flesh. Inedible bitter boletes (*Caloboletus*) look similar from a distance, but they blue quickly and taste bitter.

Medicinal Benefits

Kings are loaded with beta-glucans; the stems contain up to 58%, while the caps might reach 17%. Beta-glucans are polysaccharides and soluble fibers found in the cell walls of most fungi. They may lower the risk of heart disease and prevent the body from absorbing choles-

King Bolete, Boletus edulis; detail of stem showing the typical reticulation, the netlike pattern

terol from food. Glucans show bioactive properties such as immune-modulating, antitumor, antiviral, and hepato-protective effects.[33] Also, *Boletus edulis* RNA enhances natural killer cell activity and thus possesses immunomodulatory potential.[34] Furthermore, porcini are a great source of selenium,[35] a trace element our body needs to produce antioxidant enzymes that prevent cell damage and thus certain cancers. However, there are no clinical studies yet where patients are being treated to delicious dishes of porcini while the control group has to settle for zucchini.

KING BOLETE
Boletus edulis

The DNA revolution has not challenged the throne of our King in the Northwest! His Excellence has aced repeated molecular studies and retained taxonomic loftiness, remaining the King Bolete, *Boletus edulis*. The Pacific Northwest DNA is extremely close to the European *B. edulis*, so this name is still applied. Don't tell the King that its Italian name *porcino* (plural *porcini*), a name lauded in culinary circles, means "piglet"; the same Italian swine reference is also the base of *Suillus* boletes, meaning "the little sow." However, a strong sense of hierarchy is evident in its Austrian name, *Herrenpilz*, understood either as "lord of the mushrooms" or "mushroom of the lords"; the latter explanation is usually supplied with the story that commoners were allowed to pick the mushroom on the local estate only for the local aristocrat, not for themselves. In Italy, porcini is defined as consisting of five species of *Boletus*, including *B. edulis*, but these days when buying Italian porcini, they probably contain Chinese *Boletus* as well!

In the Pacific Northwest, several species have been differentiated that are easily distinguished in the field. Once cooked, they share the delicious porcini flavor, firm texture, and generous portions. In the mountain conifer forests of the Rocky Mountains (especially Colorado, New Mexico, and

A mighty King Bolete, Boletus edulis

Arizona) and Mexico grows the similarly gorgeous, but more reddish-capped, Red King Bolete, *Boletus rubriceps.*

NAME AND TAXONOMY: *Boletus* from Greek *bolos* "lump" or "bulb." The King's common names elsewhere: Penny Bun (UK); Rodellón (small, round boulder, Spanish) or Panza (belly, Spanish); Steinpilz (stone mushroom, German); Eekhoorntjesbrood (squirrel bread, Dutch); Bilyy hryb (white mushroom, Ukrainian); Borowik (pine mushroom, Polish); Karljohanssvamp (named for King Karl XIV, Swedish); Mei Niuganjun (beautiful cow stomach fungus, Mandarin). *Boletus edulis* Bull. 1782; *Boletus edulis* var. *grandedulis* D. Arora and Simonini 2008. Boletaceae, Boletales.

CALIFORNIA KING
Boletus edulis var. grandedulis

A king even more majestic! The California King is larger and taller (up to 40 cm in height) than *Boletus edulis,* hence it is also commonly called Barstool Bolete. It is most easily told apart by the sponge that turns not olive but cinnamon-brown. The cap (Ø 10–50 cm) is often more reddish, too, and the stem a bit darker. *Grandedulis* fruits in wet summers in true fir–spruce forests of the Cascades (where they might outnumber regular *B. edulis*) and possibly also in the Rockies (at least there are big brown-pored kings). Another habitat is coastal pine forest, where it can be associated with bishop pine (*Pinus muricata*) in southwestern Oregon, Monterey pine (*P. radiata*) in California, and lodgepole or shore pine (*P. contorta*) elsewhere. *Grandedulis* also shows up in suburbia with planted pines, spruce, and oaks.

IDENTIFICATION

- Midsize to huge mushroom with warm brown, smooth cap (Ø 5–40 cm); surface greasy when wet.
- Fat, white stem; top part or the whole stem shows a fine whitish surface network called reticulation.
- Firm white sponge layer deepens and softens in age, turning pale yellow to deep olive.
- Flesh is firm and white with a sweet, nutty flavor; not bluing (though sometimes slightly browning).
- Grows solitarily or in groups with true fir (*Abies*) and spruce (*Picea*) in mountains and along coast, rarely with birch or other hardwoods in parks. Kings love to fruit at the forest edge.

QUEEN BOLETE
Boletus regineus

The gorgeous Queen prefers to abide in warmer climes of the Pacific Northwest. When young, Her Majesty has a powdered cap, a fine frosting over a nearly black head that with age lightens to a reddish brown, similar to the King. Often there is a spottiness to the color change. The stem is white or light brownish overlain by netting, whitish at first, then often darkening with age. The tubes change from white to yellow-green or olive. Queen Boletes fruit in fall and associate with hardwoods such as oaks, tanoaks, chinquapins, and madrones, as well as conifers such as pines. They occur in Western lowlands from Southern California into BC and in the warmer parts of the Cascades as well as in the Sierra.

California King, Boletus edulis *var.* grandedulis

Queen Bolete, Boletus regineus

NAME AND TAXONOMY: *Boletus regineus* D. Arora and Simonini 2008; previously the European name *B. aereus* Bull. 1789 was misapplied.

WHITE KING
Boletus barrowsii

The stately but often stouter White King is rare in its natural habitat in the eastern Cascades, with a distribution center in the Rocky Mountains of the Southwest, where it is mostly an associate of ponderosa pine but also mycorrhizal with other conifers such as Engelmann spruce. In recent decades, *Boletus barrowsii* has teamed up with a range of nonnative deciduous trees out West to pop up in parks and yards in the lowland western region north to southern Vancouver Island. We would not dare to label the wonderful White King an invasive species in the coastal areas and enrage the dedicated and grateful boletivores of this productive mushroom; rather, we extend a warm welcome to this absolutely delicious mushroom!

White King, Boletus barrowsii

- Midsize to big mushroom with pale white velvety cap (Ø 5–25 cm) with fine powder layer when young.
- Fat white stem (>3 cm) with whitish surface netting or reticulation, especially on upper part of stem.
- Sponge layer white and firm when young, softening and turning pale yellowish-olive in age.
- Flesh firm white; sweet, nutty flavor; turning brown, but not bluing.
- Grows solitarily, in groups or clusters with ponderosa pine, true fir (*Abies*), and spruce (*Picea*) in the eastern Cascades, Rockies, and Sierra, or with linden, elm, hornbeam, and oaks in landscaped, mostly coastal, areas.
- Fruits several weeks after sufficient summer rains and into fall; in lowlands sometimes in late spring.

NAME AND TAXONOMY: *Barrowsii* for Chuck Barrow from New Mexico, who collected the type specimens and was proud to be an "old, bold mushroom hunter." *Boletus barrowsii* Thiers and A. H. Sm. 1976.

SPRING KING
Boletus rex-veris

Relying on snowmelt-soaked soils for its fruiting, the firm and stately Spring King graces mountain conifer forests—you guessed it—in spring. For many years it passed as *Boletus edulis*. Each year the charismatic Spring King attracts subjects from far and wide to visit its remote realm (to the dismay of the subjects abiding close by). Spring Kings are plentiful after snowmelt and when spring rains extend soil humidity. In some areas, *B. rex-veris* tends to hide under a thick duff layer, and foragers will need to scope for mushrumps. Hot and dry spring weather seriously curtails the Spring King from spreading its splendor.

IDENTIFICATION

- Midsize to big mushroom with dry, reddish-brown to dull dark brown cap (Ø 10–30 cm) and a white bloom when young.
- Fat, whitish stem, often with curved base, turning tan or reddish-brown with age.
- Fine network reticulation on upper stem contrasts against paler background.

- Sponge layer first white and firm, softening in maturity to pale yellow, then yellow-olive.
- Flesh firm white; sweet, nutty flavor; no bluing.
- Grows solitarily, clustered, or in groups, with true fir (*Abies* spp.) and pines, such as ponderosa and lodgepole.
- Fruits in May and June in eastern Cascades, western Northern Rockies in the US, the Sierra, and dry slopes of southern BC.

NAME AND TAXONOMY: *Rex-veris* Latin "king of spring." *Boletus rex-veris* D. Arora and Simonini 2008.

FIBER KING
Boletus fibrillosus

The Fiber King (*Boletus fibrillosus*) calls the vast western Douglas-fir–hemlock forests, from the lowlands to the mountains, home.

Spring King, Boletus rex-veris

Fiber Kings, Boletus fibrillosus

It is the only king bolete that grows with Douglas-fir. This wonderful mushroom is surprisingly rare, taking into account how common Douglas-firs are, and its looks vary. Some of them have slender, curved stems like suede boletes, *Xerocomus*; others have the solid stems expected of stately *B. edulis.* Some are truly chocolate-brown, hence also endearingly known as the Chocolate King; others look like Dark Kings. But all share a more fibrous cap surface, hence "Fiber King." Luckily most of them have the delicious porcini aroma!

IDENTIFICATION

- Midsize to sometimes big mushroom with fibrous, dry, dark brown cap (Ø 5–15 cm).
- Sponge layer first creamy pale yellow and firm, in age softening and turning green-yellow.
- Slender to fat brown stem; reticulation contrasting against paler underground, often covers whole stem.
- Firm, white flesh; sweet, nutty flavor; no bluing.

- Grows solitarily or in small groups in Douglas-fir–hemlock forests.
- Fruits late summer to late fall in coastal areas up to southern British Columbia and in the Cascades and Sierra.

LOOK-ALIKES: Often mistaken for the edible Admirable Bolete (*Aureoboletus mirabilis*) and sometimes king boletes, especially the California King (*Boletus edulis* var. *grandedulis*). The also edible, rather rare Yellow-cracked Bolete, *Xerocomus subtomentosus,* is similar as well.

NAME AND TAXONOMY: *Fibrillosus* Latin for "fibrillose," covered in firm, fine hair. *Boletus fibrillosus* Thiers 1976.

OTHER BOLETES

There are many more boletes in other genera. Some, like scaberstalks (*Leccinum*) and jacks (*Suillus*), have been recognized for more than two hundred years now as separate from *Boletus.* Other genera were established more recently, fueled by the DNA revolution. For boletivores, once they overcame the shock of having to adapt to new scientific names, this progress has been very helpful, since the genera vary substantially in culinary quality. Some genera contain only toxic members (red-pored boletes, *Rubroboletus*) or inedible species (bitter boletes, *Caloboletus*), which are at the end of this section. Not included in this book are rare but edible Pacific Northwest boletes like the Dusky Bolete (*Porphyrellus porphyrosporus*), Smoothish-stemmed Bolete (*Hemileccinum subglabripes*), Powdery Sulfur Bolete (*Pulveroboletus ravenelii*), and several gastroid boletes (*Gastroboletus* spp.).

ADMIRABLE BOLETE
Aureoboletus mirabilis

Admired for its size, beauty, and willingness to fruit after sparse rains, the Admirable Bolete is common in coastal western hemlock forests, where it often emerges from late-stage rotten conifer wood. The decomposed wood works as a perfect sponge to retain every bit of water, allowing the "Admiral" to fly its flag when other root-associated mushrooms are still in aestivation. In edibility beloved by many, disdained by some, it is easily recognized by a bottom-heavy, club-shaped marbled stem that has most of its biomass in the lower half, much of it underground in rotten wood, a habitat used by few boletes.

IDENTIFICATION

· Midsize to big mushroom with velvety, dry, darkish (red-)brown to maroon cap (Ø 5–20 cm) with overhanging (yellow) margin.
· Club-shaped curved stem streaked with dark brown (like cap color); usually netted, often showing yellow mycelium on stem base.
· Sponge layer at first pale yellow turns lemon-yellow to greenish in age.
· Firm, white flesh; lemony to sweet and nutty flavor; not bluing.

Admirable Bolete, Aureoboletus mirabils, with close-up of underside of cap

Smith's Beauty Bolete, Pulchroboletus smithii

- Grows alone or in small groups from decaying wood in hemlock–Douglas-fir forests.
- Fruits summer and fall in coastal areas from Northern California to Alaska as well as in the Cascades and Sierra.

LOOK-ALIKES: Our *Xerocomus* and *Xerocomellus* boletes also have velvety caps and often curved stems, but they are smaller and lack the widened lower stem. Smith's Beauty Bolete (*Pulchroboletus smithii*) has a yellow stem base, is bluing, and lacks netting on its upper stem.

NAME AND TAXONOMY: *Aureo-boletus* Latin for "golden," a reference to the color of the sponge. *Aureoboletus mirabilis* (Murrill) Halling 2015, a.k.a. *Boletus mirabilis* Murrill 1912. Boletaceae, Boletales.

SMITH'S BEAUTY BOLETE
Pulchroboletus smithii

This bolete is beautifully colored and easily identified by both a red band on its upper stem (it is also known as the Red-banded Bolete) and red notes on the cap that overpower a pale yellowish-brown base (that fades to gray). This bolete is named in honor of the late American mycologist Alexander H. Smith. While Smith will be retained in the species name, the *Boletus* shall be lost soon (since it does not match the criteria of the current *Boletus* genus), and Smith's Beauty Bolete will be transferred to *Pulchroboletus*. Welcome to the DNA revolution. Not too many people seem to consume this bolete, which pops up early in summer in the Cascades after rains. At times when little else is fruiting, I have eaten it often, enjoying its mild, boletish flavor with no ill effects.

- Midsize to sometimes big mushroom with dry, velvety, brownish-yellow cap (Ø 5–15 cm) becoming red in maturity.
- Club-shaped stem, often curved, thinner at top; upper part usually red, lower yellow.
- Sponge layer yellow, maturing to olive-yellow with red at cap edge, bluing when touched.
- Flesh yellow, turning erratically blue when bruised.
- Grows alone or in small groups in conifer forests from summer to late fall.
- Distributed all over the Pacific Northwest from the coast to the Rockies, from Northern California to southern British Columbia.

LOOK-ALIKES: The bitter bolete *Caloboletus rubripes* has the stem coloration reversed, with red at the base, not the top. King boletes (*Boletus* spp.) are reticulated and have no red on their stems. The Admirable Bolete's cap is dark maroon and its stem a striated or marbled reddish-brown.

NAME AND TAXONOMY: *Pulchro-boletus* Latin "beautiful," and *smithii*, named in honor of Alexander H. Smith (1904–1986). *Boletus smithii* Thiers 1965; soon to be *Pulchroboletus smithii*, from its close relationship to the European *P. roseoalbidus* and southeastern North American *P. rubricitrinus*. Boletaceae, Boletales.

ROSY AUTUMN BUTTER BOLETE
Butyriboletus autumniregius

Rosy Autumn Butter Bolete is quite a sight! Unfortunately, this big, colorful bolete is limited to only the southernmost Pacific Northwest. The combination of mild

Rosy Autumn Butter Bolete, Butyriboletus autumniregius

flavor, pink cap tones, and yellow stipe and pores—as well as the pores' intense bluing, which magically disappears in the frying pan—makes this mushroom easily recognizable as one of the butter boletes. The firm, abundant, and pleasantly mild-tasting flesh makes butter boletes highly desirable edibles. The only problem if you live in Washington, upper Oregon, or Idaho is that butter boletes flourish more in California than the temperate Pacific Northwest.

The much grayer Mountain Butter Bolete (*Butyriboletus abieticola*) is found farther up north. Recent molecular studies reveal that we have a range of very similar butter boletes that can be told apart by fruiting season—the Spring Butter Bolete (*B. primiregius*) and Autumn Butter Bolete (*B. autumniregius*)—and by habitat, in the case of the oak-associated Oak Butter Bolete (*B. querciregius*).

IDENTIFICATION

- Midsize to big bolete with a cap (Ø 8–30 cm), usually pinkish-red to

Mountain Butter Bolete, Butyriboletus abieticola

wine-red overall; cap's dry, bald surface can break into fine red scales on yellow cap base.
- Sponge layer with tiny (pale) yellow pores, maturing to olive-yellow or greenish, bluing when touched.
- Stem club-shaped, stem skin stretching with age, butter-yellow with dark red stains toward base, upper part with fine netting, often bruising blue.
- Flesh firm, dense, and pale yellow, unchanging when very young, but with maturity bluing slowly and erratically.
- Grows solitarily or in small groups with Douglas-fir (and possibly hardwoods) in fall.
- Distributed in lowland or mid-elevation mountain forests in southern Oregon (and California).

LOOK-ALIKES: Very similar, but fruiting in spring with true fir, is the Spring Butter Bolete—*Butyriboletus primiregius*. Smith's Beauty Bolete (*Pulchroboletus smithii*) also has red tones in the cap, lacks reticulation on the stem, and stains blue. It tastes mild. Some bitter boletes, like *Caloboletus rubripes*, are similarly big and also stain blue quickly but lack the red tones in the cap and taste bitter.

The Mountain Butter Bolete (*Butyriboletus abieticola;* see photo at left) shares the great mild flavor and bluing yellow pores but lacks the intense pink tones in a grayish cap with fine, distinct scales. It can be encountered in the Cascades up into southern British Columbia and likes to fruit in summer and early fall. As its name *abieticola* implies, this butter bolete is a true fir associate, especially subalpine fir (*Abies lasciocarpa*), but it seems also to connect to mountain hemlock.

NAME AND TAXONOMY: *Butyri-boletus* from Greek *boutyron* for "butter." Species epithets are all combinations with -*regius*, in refence to the previously misapplied European species of *Boletus regius*: Latin *autumni-regius* for "autumn-king," *primiregius* for "spring-king," and *querci-regius* for "oak-king." *Butyriboletus autumniregius*, *B. primiregius*, *B. querciregius*, and *B. abieticola* D. Arora and J. L. Frank 2014; misapplied European species: *Boletus regius* Krombh 1832 and *Boletus appendiculatus* Schaeff. 1774, both now moved to *Butyriboletus*. Boletaceae, Boletales.

SUEDE BOLETES
Xerocomellus species

These small, slender-stemmed but beautiful boletes are not easy to tell apart. Suede boletes are eye-catching with contrasting colors of light to very dark brown velvety caps, bright yellow pores, and yellow to red stems. They blue to varying degrees. Historically, most of them passed as Zeller's Bolete (*Boletus zelleri*) and Red-cracking Bolete (*B. chrysenteron*). Only very recently has DNA analysis clarified the diversity of this group of enjoyable boletes,[36] all edible when collected young but short-lived, turning quickly softish and wormy. The genus name *Xerocomellus* is the diminutive of *Xerocomus*, derived from Greek *xeros* for "dry" and *kome* for "hair," and refers to the dry, velvety surface of the cap. The -*ellus* ending is a Latin diminutive suggesting that *Xerocomellus* boletes are smaller than the also edible *Xerocomus* suede boletes.

Xerocomus is represented in the Pacific Northwest by at least three species, all probably edible; taxonomy is still in flux. Suede-like caps (Ø 5–15 cm) range from yellowish-brown to dull olive-brown, becoming darker brown with age. The stem is usually paler, sometimes reticulated. The pores are (sometimes brightly) yellow, and the flesh may slowly blue. The most commonly applied species name is *Xerocomus subtomentosus*, the Yellow-cracked Bolete.

DEEP PURPLE BOLETE
Xerocomellus atropurpureus

These often Douglas-fir associates are commonplace in the wet western Pacific Northwest and can fruit in big numbers, though with their dark caps, they are well camouflaged among the sword fern and Oregon grape understory. In the past, they were known as Zeller's Bolete, but fruiting bodies sampled for DNA reveal that Zeller's Bolete (*Xerocomellus zelleri*) is relatively rare and possibly restricted to humid coastal areas in the Pacific Northwest. More frequently encountered is the newly described

Deep Purple Bolete, Xerocomellus atropurpureus

Deep Purple Bolete (*X. atropurpureus*), which tends to have a bumpy, wrinkled cap, its surface smooth to finely hairy. The cap is dark blackish-purple to dark wine-red or olivaceous-black. In contrast, Zeller's Bolete's cap is less bumpy, lacks purple hues, and has a more pronounced pale cap edge. Telling them apart visually is a taxonomic challenge that pot hunters do not have to master to safely enjoy either species.

IDENTIFICATION

- Small to midsize mushroom with smooth to finely hairy, dark brown to purple-tinged, coarsely wrinkled cap (Ø 4–12 cm).
- Slender stem, often streaked, with wider yellow base densely covered in fine red spots on yellowish background.
- Sponge layer pale yellow maturing to dingy yellow, rarely bruising blue.

- Flesh firm, marbled yellow, tasting mild to lemony, unchanging or bluing slightly in older specimens.
- Fruits singly, in groups, or in troops in conifer forests from late summer into late fall (but also in spring).
- Distributed along the coast to the Cascades and into the Sierra, from mid-California up to southern British Columbia, including the inland temperate rainforest.

LOOK-ALIKES: Quite similar is a range of smallish velvety-capped boletes, all edible, such as *Xerocomellus diffractus* (see below), *X. zelleri*, *X. rainisiae*, *X. salicicola*, *X. amylosporus*, and *X. mendocinensis*—all with brown to dark brown caps—as well as edible suede boletes (*Xerocomus* spp.) and the usually bigger, marble-stemmed Admirable Bolete (*Aureoboletus admirabilis*). Red velvety caps are typical for the edible Garden Bolete, *Hortiboletus coccyginus*. Harmless Pepper Boletes (*Chalciporus piperatus*) have

Deep Purple Bolete, Xerocomellus atropurpureus

Red-cracking Bolete, Xerocomellus diffractus

dull orange to brown pores and caps and are spicy tasting.

NAME AND TAXONOMY: *Atro-purpureus* Latin for "dark purple." *Xerocomellus atro-purpureus* J. L. Frank, N. Siegel, and C. F. Schwarz 2020; previously included in *Boletus zelleri* Murrill 1912, now *Xerocomellus zelleri* (Murrill) Klofac 2008. Boletaceae, Boletales.

RED-CRACKING BOLETE
Xerocomellus diffractus

The Red-cracking Bolete is one of our most common smaller boletes. Luckily, its name describes its most distinguishing trait— that's what a common name should be all about! When this bolete is very young and fresh, there's little cracking, but scratching the cap reveals the prominent red layer under the darker velvety skin. With age, the skin cracks and reveals this red layer. Though no porcini, *Xerocomellus diffractus*

is still a pretty good meal when collected young and firm.

IDENTIFICATION

- Small to midsize mushroom with velvety, dry, olive-gray to leather-brown cap (Ø 3–12 cm).
- Cap ages lighter brown, sometimes reddish-brown, cracking to show whitish to pinkish-red flesh.
- Slender stem with wider base, densely covered with fine red spots on yellowish background.
- Sponge layer pale yellow, maturing to dingy yellow; slowly bruising blue.
- Firm flesh, mild flavor, whitish to pale yellow in cap, pale yellow in stem, irregularly and slowly bluing.
- Fruits singly or in groups in conifer and hardwood forests from summer into late fall, sometimes also in spring.
- Distributed all over the Pacific Northwest and in the Rockies, from Arizona up to southern British Columbia.

LOOK-ALIKES: See comments above for Deep Purple Bolete, *Xerocomellus atropurpureus*.

NAME AND TAXONOMY: *Diffractus* Latin for "broken, cracked." *Xerocomellus diffractus* N. Siegel, C.F. Schwarz, and J. L. Frank 2020; in the past European *Boletus chrysenteron* Bull. 1789 was misapplied.

PEPPER BOLETE
Chalciporus piperatus

Pepper Boletes seem often a bit too small to enjoy as a meal by themselves, and some people have unpleasant reactions; however, their peppery flavor provided by the alkaloid chalciporon makes a nice condiment to spice up any dish and excite any mushroom nerd. When dried and ground, they are easy to use. Cooking often reduces the spiciness and risks of gastric upset. Pepper Boletes tap into the mycelium of Fly Agarics (*Amanita muscaria*). Are they just parasites, or do they spice up their bud-

Pepper Bolete, Chalciporus piperatus

dies' underground life? Who knows! Also unknown is how many species we have in the Northwest, but it is surely a species complex, and future DNA work will tell. For now, any blue-staining Pepper Bolete is *Chalciporus piperatoides* and is just as tasty or inedible as you wish it to be.

IDENTIFICATION

- Dull yellowish-brown to reddish-brown small mushroom with a cap (Ø 3–12 cm), dry to slightly sticky, bald or finely hairy.
- Pale yellow to copper to cinnamon-brown pores that do not blue.
- Slender stem colored like cap, with bright yellow base.
- Peppery tasting, fragile, yellow to yellow-brown flesh that does not turn blue.
- Fruits singly or in groups with conifers and hardwoods in summer and fall, often in the company of *Amanita muscaria*.
- Distributed all over the Pacific Northwest and in the Rockies.

LOOK-ALIKES: The copper-colored to brown pores are unique; the Pepper Boletes have no close look-alikes.

NAME AND TAXONOMY: *Chalciporus* from Greek *chalco* "copper" and Latin *porus* "pore," a reference to the color of the pores; *piperatus* Latin for "peppery." *Chalciporus piperatus* (Bull.) Bataille 1908;

formerly *Boletus piperatus* Bull. *Chalciporus piperatoides* (Smith and Thiers) Baroni and Both 1991. Boletaceae, Boletales.

Scaberstalks

Scaberstalks (*Leccinum* spp.) are easily recognized as their own group of boletes by their typical "scabers"—soft, scale-like tufts of hair, standing out darkly from a ringless stem. However, narrowing an identification down to a particular species is another thing, unless it is a suburban Birch Scaberstalk, *Leccinum scabrum*. MycoMatch lists sixteen species in the genus *Leccinum* in our region. We are still waiting for a decisive DNA-based revision of scaberstalks in the Pacific Northwest to gain clarity.

A helpful factor for identification is their mycorrhizal partner tree. *Leccinum* works with a wide range of deciduous trees (including aspen, cottonwood, birch, and willow) and conifers (mostly spruce and pine, but also true firs), as well as with madrones and manzanitas in the rhododendron family (Ericaceae). Most species are quite particular regarding their mycorrhizal partners. In the past, blue or green color in the stem base (which can develop during cool, humid weather) was a key criterion for species delineation, but DNA work for European species has shown that this criterion is not reliable for separating species,[37] contributing to valid species in Europe being reduced from 36 to 14. It's so nice when molecular work reduces complexity!

Several of our scaberstalks have been implicated in intestinal upsets with flu-like symptoms, probably due to xerocomic acid, a yellow-and-orange pigment, that can be neutralized by cooking for fifteen to twenty minutes. Apparently xerocomic acid is less of an issue in Leccinums like birch boletes, whose caps are not red or orange colored and whose flesh displays no yellow or red hues. All kinds of scaberstalks are beloved by many people; they are collected in huge amounts and are available in wild mushroom markets all around the Northern Hemisphere. In the Pacific Northwest, besides the Birch Scaberstalk, several, often huge red-toned Scaberstalks excite foragers and offer an impressive haul. They are processed just like king boletes and can be used as a substitute, though they are of lower quality. Scaberstalks, especially the orange and red-brown ones, should not be eaten raw and need to be approached with caution. Plenty of people have regretted eating them.

BIRCH SCABERSTALK
Leccinum scabrum

Even many tree-illiterate people can recognize a birch from her white, paperlike bark. Birch Scaberstalk foraging will convince any novice forager that knowing your trees makes a big difference. *Leccinum scabrum*, also known as Birch Bolete, first arrived on the roots of introduced *Betula pendula*. Now this ectomycorrhizal European transplant is widespread. A smooth to finely felty brown cap and black stem scabers make this bolete easy to recognize and one of the safest edible mushrooms. Thanks to irrigation in urban yards, this bolete often fruits in late summer and early fall when its wild brethren are still hoping for the rains to return.

Birch Scaberstalk, Leccinum scabrum

- Midsize to big brown mushroom with dry to slightly sticky, bald or finely hairy, pale brown to reddish-brown cap (Ø 5–20 cm).
- Pale white pores turn grayish or dark brown with age, sometimes staining yellowish but no bluing.
- Longish, whitish stem with brown to black scabers, sometimes bluing in stem base.
- White flesh that may turn brownish.
- Grows singly or in groups with birch trees from early summer through fall.
- Distributed all over the Pacific Northwest and in the Rockies, usually in human settlements.

LOOK-ALIKES: Other edible and native scaberstalks with brown caps include (up north) the often-splotched-capped Alaska Birch Scaberstalk (*Leccinum alaskanum*) and farther south, *L. cyaneobasileucum.*

NAME AND TAXONOMY: *Leccinum* is derived from a local Italian name. *Scabrum* Latin for "rough, small scales." *Leccinum scabrum* (Bull.) Gray 1821. Boletaceae, Boletales.

 ASPEN SCABERSTALK
Leccinum insigne

Whenever you find yourself among aspen in summer or fall, you might be surrounded by this very noticeable and impressive big scaberstalk. In some regions, like around Telluride, Colorado, foragers ignore the Aspen Scaberstalk, since too many people encounter unpleasant digestive upsets, while in other locations, collectors are excited to find this handsome and productive bolete. The flesh turns black when cooked. It is unclear if the GI upsets are specific to a small number of people who are more sensitive, or if the mushrooms were simply not cooked long enough to neutralize the xerocomic acid, which is involved in coloring the cap and flesh. Peeling the cap may reduce the risk of poisoning, but we lack enough information to do more than speculate.

Identifying Aspen or Red-capped Scaberstalks (see below) to specific species is a challenge, and researchers hope that molecular studies will soon clarify our species concepts. Most helpful would be clear answers about which species associate with what trees and how conclusive color changes in the flesh are.

Aspen Scaberstalk, Leccinum insigne

However, for most pot hunters, this may not be a question they lose sleep over, as long as they cook their orange or reddish scaberstalks well enough.

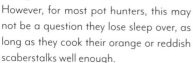

IDENTIFICATION

- Midsize to big mushroom, cap (Ø 5–20 cm) orange-brown to reddish-brown, paling with age, bald or very finely hairy.
- Flaps of cap tissue hanging over the edge.
- White to olive-buff to yellowish sponge layer, sometimes stains yellowish but does not blue.
- White stem with brownish scabers that age to blackish; stem base sometimes turns blue.
- Whitish flesh when cut darkens to purplish-gray, then blackish, without going through a reddish phase.
- Fruits singly or in groups with aspen from early summer through fall.

- Distributed all over the Pacific Northwest and in the Rockies.

⚠ RED-CAPPED SCABERSTALK
Leccinum aurantiacum group

The Red-capped Scaberstalk is very similar to the Aspen Bolete (*L. insigne*) and is best told apart by its flesh, which turns red before it dulls to gray after exposure. The solid, white stem is covered first in white scabers; with age, it turns orange-brown or even blackish. It was originally described with aspen, but there are claims of association with pine as well. *Leccinum aurantiacum* is a European name for a hardwood-associated species, and it very likely will be given another name, possibly *L. discolor*; DNA analysis will reveal its identity in time.

In addition, the name *L. aurantiacum* has also sometimes been used for large (dark) red-brown-capped scaberstalks associ-

firm flesh and good fungal flavor after cooking it well.

LOOK-ALIKES: A range of edible scaberstalk species feature orange to orange-brown to reddish-brown caps, and it makes little sense to list all these species until we have a DNA-based revision for the Pacific Northwest. Other orange- or red-capped boletes (none listed below with the scabers typical for *Leccinum*) include the toxic red-pored boletes (*Rubroboletus*), set apart by red and orange pores; the edible, but much smaller, more fragile Garden Boletes (*Hortiboletus*) with yellow pores; and the choice edible butter boletes (*Butyriboletus*), with bright yellow stems and pores that quickly blue.

NAME AND TAXONOMY: *Insigne* meaning "marked, distinguished," *aurantiacum*: "orange;" *discolor*: "differing color"—all Latin. *Leccinum aurantiacum* Gray 1821; *L. insigne* and *L. discolor* A. H. Smith, Thiers, and Watling 1966.

TOP: *Red-capped Scaberstalk,* Leccinum aurantiacum group; BOTTOM: *Conifer Scaberstalk,*
Leccinum "vulpinum"

ated with conifers in the Pacific Northwest. When exposed, the stem base blues and the whitish flesh grays, often passing through a reddish phase. This Conifer Scaberstalk still lacks its own scientific name, but molecular analysis has placed it close to the European conifer associate, Foxy Scaberstalk (*L. vulpinum*). I have eaten Conifer Scaberstalks collected in the Cascades and enjoyed the

MANZANITA SCABERSTALK
Leccinum manzanitae

The reddish-brown Manzanita Scaberstalk is best differentiated by its association with madrones and manzanita growing in coastal areas in fall. Often sticky when young, the dome-shaped cap has flattened hairs. Tough off-white pores soften when

Manzanita Scaberstalk, Leccinum manzanitae; close-up of cap flaps and stem scabers of L. vulpinum

mature to olive-grayish, bruising brownish. Initially pale stem scabers darken to deep brown or black with age. The stem base can be greenish-blue, and the flesh in the upper stem and cap slowly turns pinkish. There are reports of people having adverse effects, but thorough cooking should render this often-large mushroom a safe edible.

NAME AND TAXONOMY: *Manzanitae* Latin for "growing with manzanita." *Leccinum manzanitae* Thiers 1971.

Nonedible and Toxic Boletes

Two groups of boletes will kill your desire to feast on boletes. Luckily, they are fairly easily recognized as nonfood. The bitter boletes reveal their inedibility quickly in a taste-and-spit test; the red-pored boletes warn you by flashing an orange or red sponge.

Red-pored Boletes

Foragers in Europe and East Asia enjoy a range of red-pored boletes, but here in the Pacific Northwest, we have few species and limited experience with their toxicity status. The toxicity of red-pored boletes with clear stem reticulation (such as *Rubroboletus pulcherrimus*, *R. haematinus*, and *R. eastwoodiae*) is most infamous—the exact toxins are not yet clear. In general, all red-pored or intensely orange-pored boletes (*Rubroboletus*, *Neoboletus*, and *Suillellus*) are regarded as toxic, or at least highly suspect, in the PNW. In the case of Scarletina Bolete,

sidered for protection as rare or threatened species. However, fungophobia manifests in North America in the lack of such protective "red lists" or regulations—or is it just the spirit of freedom that is invoked until resource exhaustion is evident to most? Given the many safe, edible boletes, there is no need to experiment with red-pored boletes.

Red-pored Beauty Bolete, Rubroboletus pulcherrimus, *has been blamed in the death of a man in Spokane, Washington; with close-up of underside of cap and stem*

Neoboletus "luridiformis," and Liver Bolete, *Suillellus amygdalinus* (not closely related to jacks, *Suillus*), cooking might render them edible for most people. I carefully tested both without getting sick but have talked to people who got sick from *Suillellus*. However, I was unimpressed by their flavor and texture.

The rarity of red-pored boletes is another valid reason not to eat them. Many of these boletes should be con-

Bitter Boletes

The extreme bitterness of *Caloboletus* species disqualifies them from nutritional services, unless you are one of those rare people who genetically cannot taste bitter: please report! Bitter boletes are responsible for uncountable severe cases of heartbreak, when the illusion for budding bolete hunters of finally having hit upon the porcini pot of gold is shattered. The quick bluing would make seasoned foragers instantly suspicious. With experience, foragers begin to recognize bitter boletes by their dull brown, grayish cap color, without needing to endure terrible taste testing. Let the bitter bolete sporulate in place, fulfilling its destiny.

JACKS, *SUILLUS*

The jacks, or *Suillus* species, are highly appreciated, choice edibles elsewhere around the world, but mushroom foragers in the Pacific Northwest mostly

ignore this abundant group of boletes. Partially this situation can be explained by the abundance of seemingly superior edibles in the Pacific Northwest, but it is also caused by a lack of knowledge in how to bring out the best in jacks. Knowing them better broadens your foraging options: suburbanites do not need to drive an hour or two to find jacks. You can hit the fungal jackpot all over town. Pick young, firm *Suillus* for their texture, more mature *Suillus* for their aroma. Drying jacks deepens their rich, woodsy flavor and balances textural challenges.

As a genus, *Suillus* is fairly easy to recognize, but with more than thirty species known in the region, identifying the specific species can be tricky. Collectively, their caps are colored yellowish or brown, often with some reddish-brown tones. Their tube layers are usually yellow, some of them with striking, radially aligned, angular pores. The stems often have a ring or ring zone; if the ring is missing, there will be dots on the stem. These stem characteristics help differentiate them from most sponge mushrooms. All *Suillus* are ectomycorrhizal, and in the Pacific Northwest, they are all associated with conifers. Every pine tree, whether on the coast, in the mountains, or in a parking lot, will have a cohort of *Suillus* growing from the ground beneath it in fall. Other favorite tree partners of jacks are larch and Douglas-fir. Larch is associated with *S.*

TOP: *Redfoot Bitter Bolete*, Caloboletus rubripes, *is recognizable by bitterness, bluing, and its red stem base.* **BOTTOM**: *Some Conifer Bitter Boletes*, Caloboletus conifericola, *lack red stem coloration.*

ampliporus, *S. clintonianus*, *S. elbensis*, and *S. ochraceoroseus* (only the latter is too bitter to enjoy), and Douglas-fir teams up with *S. lakei*, *S. caerulescens*, and *S. ponderosus*.

Many *Suillus* have a slippery cap, hence "slippery jacks" is an often-used colloquial name for the whole genus. It was originally coined referring to the now globally distributed *S. luteus*, very slippery and one of our most common pine associates (one out of over a dozen). Yet some *Suillus* caps are mostly dry or

PROCESSING AND COOKING JACKS

The main culinary challenges with jacks are their flabby texture and slimy caps. If you are frying one of the viscid jacks, like *Suillus luteus*, *S. acidus*, *S. ponderosus*, *S. clintonianus*, or *S. elbensis*, it is best to first remove the sticky cap tissue, which peels easily, along with plenty of stuck debris. However, if you want to use your jacks for soup or sauce, the glutinous pileipellis, a.k.a. cap skin, can be a welcome thickening agent. It is not digested well by all people, but I have never had an issue. Furthermore, there are reports of GI upset, probably caused by an acid that may be neutralized by at least twenty minutes of cooking.

There are several tricks to improve the texture of jacks. Removing the sponge will reduce the mush factor, especially when your jack is mature. Young pores are firm and there is no reason to remove them: often in boletes, the tubes are the most aromatic tissue. Dry frying *Suillus* firms the flesh more than when it is fried in oil, which tends to close the outer cells and lock the water inside the tissue. Dry frying reduces water content, firms up the tissue, and concentrates flavor. You still will not have the sweet, nutty flavor of kings or queens; you are dealing with jacks! However, a rich, woodsy fungal flavor is always in the cards.

Drying jacks and pulverizing them produces a high-quality mushroom powder that can be used in sauces and soups, or on popcorn and more. In his seminal mushroom cookbook, retired chef Jack Czarnecki of the famous mushroom restaurant Joel Palmer House in Dayton, Oregon, praises *S. luteus* for their unique flavor. Current chef Chris Czarnecki, fourth-generation myco-chef, offers several dishes that use slippery jacks, such as a delicious mushroom tart and a rich wild mushroom soup, as well as a risotto combined with porcini.

scaly, like the very common Western Painted Jack, also called the Matte Jack, *S. lakei*. Dry or viscid caps in descriptions are not absolutes: humidity and life stage also influence their relative stickiness. Several of the jacks that have fibrils (very fine hairs) when young may lose these with age or in the rain, when the smooth subsurface is revealed and becomes stickier when moist.

The DNA revolution had surprisingly little impact on the genus *Suillus*, and it was not split up into many new genera like the other boletes; in fact, the genus *Fuscoboletinus* was reintegrated into *Suillus*. Some species, like *S. brevipes*, are turning out to be a species complex,

meaning a grouping of several similar species. Molecular studies reveal that *Suillus* are more distantly related to *Boletus* than other genera in the Boletaceae family. They have received their own family, Suillaceae, which is more closely related to gilled members of the Boletales, including slimespikes (*Gomphidius*) and pinespikes (*Chroogomphus*)—fascinating!

In general, most *Suillus* species are regarded as edible, or at least harmless enough to make such a sweeping statement. Some are avoided for their bitterness, like the larch-associated *S. ochraceoroseus*. Others are frowned upon for being too thin and soft fleshed for good eating. There have also been anecdotal reports that slimy-capped jacks speed up digestion and induce flatulence or diarrhea. Other people report gastrointestinal upset, probably caused by an acid that can be neutralized by thorough cooking.

PURPLE-VEILED SLIPPERY JACK
Suillus luteus

Who would have imagined that a lowly mushroom with a mocking name like slippery jack could be a winner in the era of globalization! Plantations of fast-growing pines have spread slippery jacks from the Northern Hemisphere to all southern continents; for example, in Argentina and Chile, people fell in love with this mushroom. Great amounts of slippery jacks return as powder to the north, where they are used to infuse their umami flavor in dried soup mixes and processed food; they are supposedly frequently admixed to porcini powder without declaration. Purple-veiled Slippery Jack is one of the most common and most delicious jacks, while also being relatively easy to identify due to the purple veil and super sticky cap. In addition, it is abundant in landscaped areas as long as pines are present.

IDENTIFICATION

- Midsize (to big) mushroom with a smooth, slimy to sticky peelable cap (Ø 5–15 cm) colored red-brown to yellow-brown, with cap tissue hanging over the edge.
- Pale yellow aging to olive-yellow sponge tubes, sometimes decurrent, not bruising.
- Thick, white stem (maturing to pale yellow) with yellowish to brown

Purple-veiled Slippery Jack, Suillus luteus

glandular dots. Base often covered by brownish sheathlike remnants of partial veil.

- Felty white partial veil with purplish-gray to brown tones on the underside.
- White to pale yellow flesh (nonstaining), mild flavor.
- Fruits singly or in groups or clusters with conifers, especially pine; mainly in fall, but also other seasons.
- Distributed all over the Pacific Northwest and in the Rockies and beyond.

LOOK-ALIKES: Edible Slimy Fat Jack, *Suillus ponderosus*, is very similar but tends to be bigger. It is massive and associates with Douglas-fir. While its stem base turns green, the upper stem can be reticulated. The sticky ring around the yellow veil of the Slimy Fat Jack is not purple as is typical for *S. luteus*.

NAME AND TAXONOMY: *Suillus* Latin diminutive for *sus*, "the sow;" *luteus* Latin for "lemon-yellow," in reference to the pores. *Suillus luteus* (L.) Roussel 1821. Suillaceae, Boletales.

SHORT SLIPPERY JACK
Suillus brevipes

One of the most common two-needle pine buddies, the Short Slippery Jack is fairly easy to identify due to the absence of stem dots (when it is young), lack of partial veil, and hence no overhanging tissue from the slimy cap. As the name implies, this short-stemmed *Suillus* plays it safe close to the ground.

IDENTIFICATION

- Midsize bolete with bald, sticky, peelable cap (Ø 5–12 cm), colored red-brown to brown, sometimes fading to tan in age.
- Sponge white when young, maturing to pale yellow and dull yellow, not bluing.

Short Slippery Jack, Suillus brevipes

Fat Jack, Suillus caerulescens

- Short stem (Ø < 3 cm), white to pale yellow. Glandular dots missing when young, but inconspicuous when old.
- No veil or veil remnants.
- Soft white to pale yellow flesh, non-staining. Mild flavor.
- Grows alone in groups or clusters with two-to-three-needle pines (including lodgepole pine) from summer into winter.
- Distributed all over the Pacific Northwest, the Rockies, and the rest of North America.

LOOK-ALIKES: The similar Weeping Jack, *Suillus weaverae,* has a stem with clear dots when young and a cap that is usually more mottled. The Veiled Short Slippery Jack, *S. pseudobrevipes,* has a yellow-brown cap with an overhanging margin and a clear veil.

NAME AND TAXONOMY: *Brevi-pes* Latin "short foot." *Suillus brevipes* (Peck) Kuntze 1898.

FAT JACK
Suillus caerulescens

Fat Jack is one of the larger and firmer *Suillus* that does not fight off insects and foragers with a slimy cap. When encountered young enough, it has a nice, firm texture. It sports a pale cinnamon-yellowish cap, brown-staining pores, and blue-staining stem, especially in the stem base, and grows mostly with Douglas-fir.

IDENTIFICATION

- Midsize to big mushroom with sticky to dry cap (Ø 5–15 cm), pale cinnamon to yellow-brown and lighter on the edge, with fine, pressed-down fibrils when young, but mostly bald in age, with tissue overhanging the edge.
- Pale yellow sponge maturing to golden and bruising dingy ochre-brown, with angular to irregular pores.
- Stem yellow above partial veil, sometimes weakly netted without dots, dull

Matte Jack, Suillus lakei

ochre-brown below, staining brown when bruised, bluing in the tapering stem base.

- Fibrillose band-like veil, white with dull ochre-brown tones on the underside.
- Mild to slightly sour-tasting flesh; whitish when young, maturing to yellow, bruising brown.
- Grows solitarily or in groups with conifers, especially Douglas-fir, from late summer into early winter.
- Distributed all over the West, from central British Columbia down to Mexico.

LOOK-ALIKES: The Slimy Fat Jack, *Suillus ponderosus*, is very similar; see details under Purple-veiled Slippery Jack, *S. luteus*.

NAME AND TAXONOMY: *Caerulescens* from Latin *caeruleus* "sky-blue," with the ending *-escens* expressing the process of bluing. *Suillus caerulescens* A. H. Sm. and Thiers 1964.

MATTE JACK
Suillus lakei

Its striking cap of brownish fibrils or scales in age on a yellowish background, brown staining of the sponge when bruised, and growth with Douglas-fir make Matte Jack one of the easy-to-identify jacks. The bluing of the stem is most pronounced in young specimens; older Matte Jacks may not blue. It is mycorrhizal with the ubiquitous Douglas-fir all over the West. In addition, it seems to fruit whenever the soil is sufficiently moist but not frozen. When it comes to edibility, the Western Painted Jack (as the Matte Jack is also known), is solid. When used appropriately, it can transform a freeze-dried dehydrated backpacking meal into a wild-crafted mushroom dinner.

It could also be dry fried and spiced taco-style as a meat alternative in a quesadilla.

- Midsize cap (Ø 5–15 cm) with brown to reddish-brown fibrils or scales on yellow base, often with flaps of the cap tissue hanging over the edge.
- Yellow to yellow-brown, sometimes decurrent sponge layer with longish stretched-out pores, staining brown.
- Stem smooth and yellow above a thin, whitish to yellow, banded ring veil that might disappear with age; lower stem streaked brown, sometimes turning blue or green when scratched; old bases usually brown.
- Thick, firm, yellowish flesh, slowly changing to pinkish- or red-brown when exposed, not bluing in cap; somewhat sour flavor.
- Grows singly or in groups, mostly with Douglas-fir but also spruce; fruits sometimes in spring, but mostly summer through fall into winter.
- Distributed all over the Pacific Northwest and in the Rockies, from British Columbia down to Mexico.

LOOK-ALIKES: The Slimy Fat Jack, *Suillus ponderosus*, is very similar but tends to be more massive. The Hollow-stalked Jack, *S. ampliporus*, has a hollow stem and is strictly associated with larch.

NAME AND TAXONOMY: *Lakei* for mycologist E. R. Lake, who collected the type in 1907 in Corvallis, Oregon. *Suillus lakei* A. H. Sm. and Thiers 1964.

TAMARACK JACK
Suillus clintonianus

The beautifully colored Tamarack Jack, *Suillus clintonianus*, grows with larch. Until recently, it was mislabeled as *S. grevillei*, a

Tamarack Jack, Suillus clintonianus

Olive Jack, Suillus acidus

name still valid for its European cousin. It is recognized by its sticky cap, pronounced sticky ring, and pale orange-yellow flesh that bruises pinkish-brown. Also found under larch is S. *ampliporus,* the Hollow-stalked Tamarack Jack—a Matte Jack look-alike with a hollow stem, formerly known as S. *cavipes*—and the Gray Larch Jack, S. *elbensis* (formerly S. *viscidus*), the palest of all jacks with white to gray pores. Several webcaps (*Cortinarius* spp.), before they open their caps and show their gills, are strikingly similar to young *Suillus clintonianus.*

OLIVE JACK
Suillus acidus

Suillus acidus, is associated with five-needle pines such as western white pine (*Pinus monticola*), sugar pine (*P. lambertiana*), and whitebark pine (*P. albicaulis*). It is recognized by its thin and darkly dotted stem with banded sticky veil remnants and a sticky, pale brownish cap with an olive tinge—hence Olive Jack (*acidus* is Latin for "olive"). The flesh is nonstaining and the flavor pleasantly lemony. When dry fried, it is surprisingly firm, with a subtle bolete-like flavor.

POLYPORES AND ALLIES

Some of the most common, easy-to-spot mushrooms are wood-digesting polypores or bracket fungi, also commonly called conks, since many of them are big and perennial, and thus visible year-round. Like boletes, they produce their spores in tubes, but unlike boletes, their spore-bearing tissue cannot be easily peeled off; it is tough and normally fused with the cap tissue. Most polypores have very firm "woody" tissue, disqualifying them as food for the table. Also, many of them are intensely bitter. Stalked polypores, including edible *Albatrellus* and allies, have firm pores—not a spongelike hymenium—but they are ground-dwelling mycorrhizal fungi, and their cap and stem gestalt distinguish them well from most other polypores.

STALKED POLYPORES: *ALBATRELLUS* ET AL.

A handful of firm-fleshed ground-fruiting mycorrhizal mushrooms, all formerly classified in the genus *Albatrellus*, have firm flesh and pores similar to polypores, and they were perceived as polypores for most of their taxonomic existence. The DNA revolution has flushed them into the Russulales order, which includes edibles from the genera *Russula*, *Lactarius*, *Artomyces*, and *Hericium*. In the process it was suggested that *Albatrellus* be split into several genera, but that has not been fully accepted yet.

SHEEP POLYPORE AND HAZEL POLYPORE
Albatrellus ovinus and *A. avellaneus*

Neither sheep nor hazel, these two *Albatrellus* species are still enjoyable, long-lasting edibles, with a firm, meaty consistency and a sweet flavor. They turn a beautiful golden color when cooked.[38] It is next to impossible to tell them apart decisively without checking spore size: Hazel Polypore, *A. avellaneus*, has bigger spores. Sheep Polypore, *A. ovinus* is more consistently sheepish-white compared to Hazel Polypore, with a cap ranging from cream-white to brown. Both species are mostly associated with spruce, growing with Sitka spruce along the coast and Engelmann spruce in the Cascades and Rockies.

Hazel Polypore, Albatrellus avellaneus

- Whitish to yellowish, dry, dull cap (Ø 4–15 cm) with orangish or pinkish areas, surface finely scaly, sometimes slightly reddish-brown, possibly darkening in age.
- Cap margin rolled in when young, cap becoming wavy as it expands.
- White pore surface with tiny tubes, firmly attached to cap tissue, turning yellowish with age or possibly yellowish-orange when bruised; white spores.
- Solid white stem, often fused, upper part roughened by decurrent pores, yellowing with age and developing rusty spots, browning at base.
- White, firm flesh staining yellow when cut.
- Odor mild to faintly anise; flavor mild to slightly bitter.
- Grows in clusters or groups (rarely alone) in soil, especially with spruce but also with hemlock.

- Fruits in fall, often found in younger coastal forests.

LOOK-ALIKES: A third edible and very similar species is *Albatrellus subrubescens*. It grows mostly in montane pine forests and can be told apart from Sheep and Hazel Polypores by its amyloid spores. There are several other probably edible *Albatrellus* and *Albatrellopsis* species in the Pacific Northwest, some of them still undescribed. Also similar are very firm, white, pored Kurotake, *Boletopsis grisea,* but they have light gray to black caps; see below.

NAME AND TAXONOMY: *Albatrellus* from French *albâtre*, "whitish," as in *alabaster*; *ovinus* Latin "pertaining to sheep," and *avellaneus* Latin "pertaining to hazel," both references to color. *Albatrellus avellaneus* Pouzar 1972, *Albatrellus ovinus* (Schaeff.: Fr.) Murrill 1903. Albatrellaceae, Russulales.

Goat's Foot, Scutiger pes-caprae

GOAT'S FOOT
Scutiger pes-caprae

Probably if *Scutiger pes-caprae* were known as Goat's Shank or God's Food instead and were more abundant, this unique and impressive mushroom would get the culinary attention it deserves. It is a good edible when found young enough, but since the firm fruiting body can last for many weeks, often it is found too old to impress. Though renowned in Europe for its excellent nutty flavor, it is also protected in many countries due to its rarity. It is rare, too, in the Northwest, where it dwells in conifer forests.

Very similar in most aspects, but with flesh that turns greenish and often a lighter yellow-brown cap, is the also enjoyable Greening Goat's Foot, *S. ellisii.* Both Goat's Feet are recognized by their firm flesh, off-center stems, and white decurrent pores, and their brown caps covered in dense hair forming tufts and scales. When encountered young, the firm flesh is appreciated by foragers lucky enough to find these rather rare mushrooms in their prime. When they are overmature, their odor and flavor can turn unpleasant, with a rancid note. It's best practice not to pick such old mushrooms so they can keep spreading their spores. Also, when it comes to such rare mushrooms, please don't take them all—always leave some growing in the ground.

Greening Goat's Foot, Scutiger ellisii

- Yellowish to brownish, dry, very hairy to scaly, tufted cap (Ø 8–25 cm), round to fan-shaped with inrolled margin at first, flattening or becoming wavy with age.
- Pores are white to cream, round to angular and decurrent, producing white spores, darkening in age, sometimes greening.
- Stem is brown and off-center if not laterally attached.
- Firm, mild-tasting, whitish flesh (when green in parts or slowly staining greenish, it is most likely *S. ellisii*).
- Grows singly, in groups, or clustered in soil under conifers.
- Fruits from late summer through fall from southern British Columbia down to Northern California and in the Rockies.

LOOK-ALIKES: The common and inedible but handsome Bitter Iodine Polypore, *Jahnoporus hirtus*, has a velvety gray to brown cap that is asking for petting. The rare and often bigger Lavender Polypore, *Polyporoletus sylvestris*, has a dark purplish-gray, fading to cinnamon-buff to olivaceous-tawny, non-scaly cap with purplish-gray pores that turn olive with age. The Western Polypore, *Bondarzewia occidentalis*, has a less scaly, velvety cap with concentric rings of different hues of brown (lighter to the edge) and white to cream pores that can ooze milk (it is related to the milkcaps). It often grows in huge multistemmed clusters that together form a round assembly. Several *Bondarzewia* species are edible, unless too spicy and old, when they turn bitter and tough. On the East Coast, Berkeley's Polypore, *B. berkeleyi*, is frequently eaten, and in Bhutan, the Mountain Polypore, *B. montana*, can be found in markets.

NAME AND TAXONOMY: *Scutiger* Latin for "shieldlike," *pes-caprae* Latin for "foot of goat." *Ellisii* named for J. B. Ellis. *Scutiger pes-caprae* (Pers.) Bondartsev and Singer (1941) = *Albatrellus pes-caprae* (Pers.) Pouzar 1966; *Albatrellus ellisii* (Berk.) Pouzar 1966, synonymous *Scutiger ellisii* (Berk.) Murrill 1903.

BLUE-CAPPED POLYPORE
Albatrellopsis flettii

Once you have found this matchless mushroom, with its often intense blue tones, including sky blue or royal blue, you will not forget it. The cap can be less flamboyant, pale blue or gray-blue, and often there are tan or dingy orangish areas on the fruiting body, especially with age. Its firm flesh tastes mild and is edible, especially when young. Recent molecular study results suggest that American *Albatrellopsis flettii* is identical to European *A. confluens*, which never is found blue. Future research will show, by using several other DNA regions for species delineation, whether these two mushrooms are actually the same species or contradict the current confusing impression. Furthermore, if they are the same species, what causes the bluing? Is there possibly a secondary fungus involved?

- Bluish with tan or rusty spots, dry, velvety to bald cap (Ø 5–20 cm), centrally depressed with inrolled margin when young, wavy with maturity.
- White, very fine pores that lose brightness with age.

Blue-capped Polypore, Albatrellopsis flettii

- White to pale bluish-gray stem with deeply decurrent pores, aging brownish or reddish.
- Very firm to rubbery whitish flesh, mild flavor.
- Grows scattered or in groups, often fused in clusters on soil under conifers.
- Fruits in coniferous forest from the Alaska panhandle down to Northern California and in the Rockies.

LOOK-ALIKES: Similar but lacking the blue tones is the edible, but often bitter Fused Polypore, *Albatrellopsis confluens*. Also blue or grayish is the closely related, probably edible Blue-pored Polypore, *Neoalbatrellus subcaeruleoporus*, distinguished by its bluish pores and much smaller size (cap ⌀ 1.5–3 cm).

NAME AND TAXONOMY: *Albatrell-opsis flettii*: *opsis* Greek "like, similar to," *Albatrellus flettii* for J. B. Flett, who found the holotype in Bremerton, Washington. *Albatrellopsis flettii* (Morse ex Pouzar) Audet 2010; *Albatrellus flettii* Morse ex Pouzar 1972. Albatrellaceae, Russulales.

KUROTAKE
Boletopsis grisea

Resembling somewhat a stout king bolete (*Boletus* spp.), with a well-defined cap and stem and shallow pores (not tubes), this ectomycorrhizal mushroom is also called False Bolete. It is more similar to stalked polypores (*Albatrellus*) in having very firm flesh and tiny white pores. However, Kurotake are usually bitter, and their caps tend to darken with maturity to gray or black. In Japan, Kurotake, meaning "black mushroom," are traditionally consumed after an overnight bath in cold water and parboiling to eliminate their bitter flavor. Eastern British Columbia naturalist Tyson Ehlers reports that brining and sautéing them until crispy renders them quite good, but he states: "Some find it still too bitter; sort of like if you have a taste for double-salted black licorice, you might like *Boletopsis*."

Kurotake are closely related to neither boletes nor stalked polypores (*Albatrellus*), but they are in the order of earth fans (Thelephorales), which includes Hawk's Wing

A youngish Kurotake, Boletopsis grisea

(*Sarcodon imbricatus*) and Blue Chanterelles (*Polyozellus* spp.).

IDENTIFICATION

- Whitish to gray, smooth or finely hairy cap (Ø 5–20 cm), centrally depressed with margin rolled in when young and becoming wavy with maturity, sometimes darkening to black.
- Small, roundish pores are bright white turning grayish (sometimes with red hue) with age.
- Thick, whitish to pale gray stem with slightly to deeply decurrent, shallow pores.
- Thick, white flesh, slowly staining reddish-gray, becoming brittle in cap.
- Bitter flavor, but also sometimes mild.
- Grows solitarily or in groups in soil under conifers, especially pine.
- Uncommon but widespread from the Alaska panhandle down to Northern California and into the Northern Rockies.

LOOK-ALIKES: Very similar and also edible, but often darker, is *Boletopsis leucomelaena*, which prefers to associate with spruce. Otherwise, the members of the Albatrellaceae family—*Albatrellus* (see above), *Albatrellopsis*, and *Scutiger*—can be similar. The ground-growing polypore Bitter Iodine Polypore (*Jahnoporus hirtus*) has a gray-brown to occasionally purple-brown velvety cap. And as the name implies, there is a similarity to boletes, but a bolete's pore tissue can be easily peeled from the cap.

NAME AND TAXONOMY: *Boletopsis* Greek "similar to *Boletus*," *grisea* Latin for "gray." *Boletopsis grisea* (Peck) Bondartsev and Singer 1941. Bankeraceae, Thelephorales.

POLYPORES

Polypores have a long history of human use, especially as medicinals (like Reishi, *Ganoderma* species, and Agarikon, *Lariciformis officinalis*), and as crucial devices to sustain a glowing ember when traveling in times before there were matches (including Tindor Conk, *Fomes fomentarius,* and False Tinder Conk, *Phellinus igniarius*). We have a few great edible polypores in the Northwest, including Chicken of the Woods and Cauliflower Mushroom, though unfortunately almost no Hen of the Woods (Maitake, *Grifolia frondosa*), an eastern American favorite that feeds on oaks.

CHICKEN OF THE WOODS
Laetiporus conifericola

Chicken of the Woods, Laetiporus conifericola

Mushroom hunting does not get easier than spotting Chicken of the Woods. Record-setting specimens of this huge annual polypore reach fifty pounds, screaming for attention in flaming yellow and orange colors; hence it is also known as Conifer Sulfur Shelf. As with all mushrooms, timing is everything, and most encounters with *Laetiporus conifericola* do not yield delicious food. The mushroom must be caught early enough, basically while emerging and while still growing at the edges. Arrive too late and the consistency of the mushroom is closer to fiberboard shelf than tender chicken. Remember the location, and return earlier next year. Talk about early! Chicken of the Woods starts growing after summer rains that inspire few other mushrooms in the Pacific Northwest to pop their heads aboveground.

When very young, especially when still dripping with guttation, the whole baby is delicious; grown a bit more, the fresh edge is edible (all that should be harvested at this point), being soft and pliable, with a texture

like chicken breast, though more succulent. When the tissue gets fibrous, there is still protein, but the delight, as expressed in the scientific name (*laetus* means "delight" in Latin), is immensely reduced and might turn into intestinal distress! The toughness of mature *Laetiporus* is evident when encountering the styrofoam-like white carcasses of last year's fruiting piled on the base of a dead tree. As it ages, Chicken of the Woods turns sour: according to mycologist Paul Stamets, that change is caused by bacteria.

It can be harmful, too, when eaten raw or undercooked, making it hard to digest; also, some people can suffer allergic reactions. Sadly, an Oregon woman even died from anaphylactic shock after eating sulfur shelf![39] Some mushroom suppliers only sell them precooked to restaurants, since inexperienced chefs too often undercook them, which does not sit well with patrons' digestive systems.

Always check the consistency—it needs to be tender, not tough. If you are in doubt about its consistency, try soaking it overnight in an acidic liquid like buttermilk or lemon water. As the name Chicken of the Woods suggests, you can use your tender mushroom just like chicken. Slice it and cook it first like chicken breast and use it in a stir-fry or a cream sauce. It is also awesome breaded and deep-fried. To store it, slice it and then dry it. Slices can be rehydrated later or powdered for soups or sauces. It also makes great mushroom jerky. In my experience, it is best to cook Chicken of the Woods first and then freeze it.

IDENTIFICATION

- Big to huge yellow-and-orange polypore growing fan-shaped fruiting bodies.

- Smooth to suede-like upper surface with a lighter, yellow edge and tiny, yellowish pores below.
- Firm flesh, tasting somewhat mild or acidic, turning sour and astringent.
- Grows in clusters (shelving or in rosettes) on mature, dead conifers.
- Fruits after summer rains and into fall; widespread, from Alaska to California and the Northern Rockies.

LOOK-ALIKES: It is easy to confuse it with *Laetiporus gilbertsonii*, another delicious, often less orange, more yellow sulfur shelf encountered in the Pacific Northwest and California on hardwoods. There are at least half a dozen species of *Laetiporus* in North America, and all are choice edibles when harvested early enough. The inedible, shiny Sponge Polypore, *Pycnoporellus fulgens*, exhibits similar (though browner) colors, but it is much smaller and its pores are wider and neither bright orange nor yellow.

NAMES AND TAXONOMY: *Laetus* "delighted and bright," *porus* for "pores" as in polypore, and *conifericola* "belonging to conifers" (all Latin). *Laetiporus conifericola* Burdsall and Banik 2001, *L. gilbertsonii* Burds. 2001; in old books both are called *L. sulphureus* (Bull.) Murrill 1920, a species common east of the Rockies. Fomitopsidaceae, Polyporales.

CAULIFLOWER MUSHROOM
Sparassis radicata

Everyone loves to find this easy-to-recognize, uniquely shaped mushroom! It is best described as a pile of wide egg noodles or head of a cauliflower. It can grow to an impressive size, and you can

Cauliflower Mushroom, Sparassis radicata, *looking like a pile of egg noodles*

share your twenty-pounder with a lot of friends and family. Cleaning as you go, taking the mushroom apart and washing it, works great; it is a sturdy mushroom. If there is any drawback, it is the firmness of the cauliflower flesh—no surprise, as it is a polypore.

There is no mush in this mushroom; it is all texture! It absorbs all kinds of flavors well. Ethnomycologist Elinoar Shavit loves to parboil her *Sparassis radicata* and dip it in Chinese *jiaozi* sauce with a touch of wasabi, or blanch it in orange juice with a tiny bit of cinnamon before adding it to fruit salad. It tastes great roasted in the oven marinated with teriyaki sauce (see Teriyaki Roasted Cauliflower Mushroom in Part Three). If you do not like your mushrooms crunchy, just keep simmering it. Some people swear by simmering the Cauliflower Mushroom in a broth for up to an hour before further processing; I never have. The firm structure offers an interesting contrast in an omelet or at the core of a deep-fried battered tempura.

Cauliflower Mushroom found in Oregon

Unfortunately, we do not find Cauliflowers often enough. It favors big, old live trees in old-growth forest, and very little old growth remains. Luckily, it can occur in older secondary forests too, sometimes still feeding off the root network of old stumps. And I am curious when "Hanabiratake," as it is known in Japan, where it is cultivated successfully and available in stores, will be for sale here.

- Unique big to huge, pale white to creamy-yellow mushroom comprised of tightly packed and contorted flat branches.
- Attached by a single extended base to roots of old conifers.
- Fruits in fall and early winter from British Columbia down to California.
- Parasitic on old conifers, most common in old-growth forest.

MEDICINAL: *Sparassis* species increase collagen and improve skin conditions. Trials have shown beneficial effects for diabetes and cancer treatment, reduction of blood lipids, and anti-inflammatory, antioxidative, and antifungal activities.

LOOK-ALIKES: The False Sparassis (*Daleomyces phillipsi* = *Peziza proteana* var. *sparassoides*) is a very rare ascomycete growing usually in burned areas.

NAME AND TAXONOMY: *Sparassis* Greek from *sparassein*, "ripped in pieces," for its many branches, and *radicata* Latin "rooted," for the rootlike connection to its host tree root. *S. radicata* Weir 1917; misapplied name: *S. crispa* (Wulfen) Fries. Sparassidaceae, Polyporales.

TURKEY TAIL
Trametes versicolor

Turkey Tails are striking mushrooms that even attract the toadstool trampler's benevolent attention! Their contrastingly colored concentric rings in a diversity of colors uniquely mark the leathery, velvety caps that grow in overlapping clusters on dead wood. Tiny pores dot their white undersides. Most famous as a medicine, *Trametes versicolor* has a rich fungal flavor, reminiscent of cream of mushroom soup, that is easily turned into a delicious and healthy broth—quite different from the innate bitterness of other medicinal polypores. The fruiting bodies can also be powdered and used as tea, their medicinal benefits unlocked after boiling them for twenty minutes.

When collecting Turkey Tails, be aware that they often are full of invisible fly eggs that will turn into devouring larvae, even when dried. If you want to store them dried,

Turkey Tail, Trametes versicolor

Close-up and pore detail of Turkey Tail, Trametes versicolor

it is best to freeze them first for twenty-four hours to shatter the eggs, and then dry the mushrooms! (See Making a Tasty Medicinal Mushroom Extract.) A surprisingly tasty way to enjoy fresh Turkey Tails is as "fungum"! Just pick a fresh piece in the woods and treat it like chewing gum. Once you are done chewing, you can swallow your "fungum" and get the benefits of chitin and fiber as well!

IDENTIFICATION

- Pliable, leathery cap (Ø 3–10 cm) marked by concentric zones or bands of contrasting colors: white, gray, beige, brown, light red, and purple; sometimes stained greenish from algae.
- Zone surfaces alternate between velvety and silky smooth.
- Underside snow-white (browning slightly in age) and covered with tiny pores, discernable to the naked eye or with a hand lens.
- Absent or undeveloped stem.
- Grows shelflike, but also tiered or in rosettes on dead hardwood, especially alders, oaks, and fruit trees, rarely on dead conifers.
- Common and widespread, fruits year-round, but especially winter into late spring.

MAKING A TASTY MEDICINAL MUSHROOM EXTRACT

Clean any attached woody debris from your Turkey Tails with strong scissors before slicing them thinly or grinding them in a blender. Immerse your ground Turkeys in water (10 ounces of water for 1 ounce fresh mushrooms, 20 ounces water for 1 ounce dried mushrooms). Boil gently for twenty minutes, or use a pressure cooker for more intense extraction, and then simmer till the broth is reduced by a quarter. Filter the liquid using cheesecloth or a fine sieve.

Now, to extend the use of your water extract, pour it into ice cube trays and freeze it. Once the cubes are frozen, put them in a ziplock bag and label them. Now, you have easy-to-use broth waiting in your freezer. Turkey Tail cubes are great to add to any soups and sauces when you're looking for umami flavor plus medicinal benefits.

MEDICINAL: Turkey Tails are some of the most thoroughly researched medicinal fungi. The Chinese have long used them to treat infections and inflammation of the respiratory, urinary, and digestive tracts. They are reported to be anti-inflammatory, immuno-modulating, analgesic, antiviral for hepatitis and human papillomavirus, and useful for regulating cholesterol. Modern medicine has focused on cancer-fighting capacities, especially using Turkey Tails for immune support during chemotherapy, which clinical trials have shown improves patient well-being and overall outcomes.

Turkey Tails are full of beta-glucans, huge sugar molecules (polysaccharides) that stimulate the human immune system. In 1977, polysaccharide-K (PSK) from Turkey Tails was approved in Japan as an anticancer drug.[40] This section does not cover all the information you need to self-medicate, but rather may it inspire you to research medicinal mushroom use and make the case for integrating them in treatments and for preventive use to avoid serious health problems.

LOOK-ALIKES: False Turkey Tails, *Stereum* species including *Stereum hirsutum*, also have concentric rings but tend to be thinner. They are yellow-orange to reddish-brown and have fine white hair on their caps, while Turkey Tail hairs share the colors of the cap rings. The clearest difference is that the yellowish underside of False Turkey Tails lacks pores, even under a hand lens. All *Trametes* have pores. Hairy Turkey Tail, *T. hirsuta*, is less brightly colored and hairier, and it tends to be larger. Ochre Turkey Tail, *T. ochracea*, is paler, not pliable, and bland.

NAME AND TAXONOMY: *Trametes* from Latin *trama*, "weft," a reference to the thin woven cap tissue; *versi-color* Latin for the diversely colored zones. *Trametes versicolor* (L.: Fr.) Pilat 1936 = *Coriolus versicolor* (L.: Fr.) Quel. Polyporaceae, Polyporales.

BEEFSTEAK POLYPORE
Fistulina hepatica

The fruiting body of the Beefsteak Polypore looks like a piece of raw meat; it is also known as Ox-tongue Fungus or Poor-man's Beefsteak. It even oozes red "blood" when young.

Staying true to the theme, it is best eaten thinly sliced raw when young to enjoy the lemony flavor of its watery but firm texture! Best to think carpaccio style, maybe with a little extra *balsamico*. David Arora recommends Beefsteak Polypore jerky, slicing them into ⅜-inch (1 cm) strips and marinating them overnight in peppered salt water—yummy! Older Beefsteaks are tougher and possibly more acidic; they might need an overnight soak to get that darn age out of the bones. Unfortunately for us, this fungal oddity—no polypore, but closely related to the tiny, tough but edible and medicinal Split Gill, *Schizophyllum*

Split Gill, Schizophyllum commune

Beefsteak Polypore, Fistulina hepatica; *close-up of unique individual tubes*

commune—is found rarely in much of the Pacific Northwest since it loves oaks and chinquapins, in short supply across much of the region.

IDENTIFICATION

- Rough-surfaced cap (Ø 8–30 cm) with fine yellow spots growing in kidney, tongue, or fan shapes (often lobed); when older, subsurface jellylike layer drips over margin.
- Cap coloration changes with age from pinkish-beige when young to blood-red and reddish-brown or dark red when old.
- Pores underneath are white to yellowish, comprised of singular, closely packed tiny tubes that turn brown when bruised or aged.
- Stem is absent or short and laterally attached, fused into cap but tougher and colored like cap.
- Thick, soft, "marbled" flesh, oozes dark reddish juice when young; sour flavor.

- Grows solitarily or in small clusters on lower trunk of chinquapins, sweet chestnuts, tanoaks, and oaks (all in the beech family, Fagaceae).
- Fruits in fall and early winter in western Oregon and Washington, rare and mostly restricted to oak habitat.

LOOK-ALIKES: Though the Beefsteak Polypore is unique (and is not a true polypore), there are a few reddish or pinkish polypores. The highly medicinal and very bitter Oregon Reishi, *Ganoderma oregonense,* has an orange-red to red-brown cap, but the surface is shiny as if lacquered and it lacks "blood." The also very bitter, pinkish-pored Rosy Polypore, *Rhodofomes cajanderi,* is much tougher, but when young it can excrete a red liquid in a process known as guttation.

NAME AND TAXONOMY: *Fistulina* Latin "little tube," and *hepatica* Latin "liver," for its liverlike consistency. *Fistulina hepatica* (Schaeff.) With. 1792. Fistulinaceae, Agaricales.

PUFFBALLS AND STINKHORNS

Before molecular genetics seriously reshaped the fungal tree of life, one peculiar branch was known as the Gasteromycetes (from *gaster*, Greek for "stomach"), which included such illustrious mushrooms as puffballs, stinkhorns, earthballs, and birds' nest fungi. These fungi produce their spores inside their fruiting body. Now we know this grouping is unnatural, but we still call this assembly "gasteroid fungi." However, it is helpful for the purposes of a field guide to know about some of these shared characteristics. Interestingly, DNA revealed that puffballs are part of the Agaricaceae family, which is home to common gilled mushrooms like the Prince (*Agaricus augustus*), shaggy parasols (*Chlorophyllum* spp.), and Shaggy Mane (*Coprinus comatus*).

Puffballs and stinkhorns are intriguing organisms and fairly easy to recognize. Usually everyone lucky enough to have spent some time in nature will be familiar with puffballs. Many people cannot help themselves, and simply have to step on the balls to puff them, thus actually helping them disperse their spores.

The warning you encounter everywhere when speaking about edibility is not to confuse them with what I

Puffing Lycoperdon *balls, or puffball, in eastern Tibet*

call bluffballs! Bluffballs are roundish, white, and young, but hardly innocent; these mushrooms belong to the genus *Amanita*, which has many nasty toxic and even deadly members. However, the bluff is easily called by slashing the bluffball in half and exposing the unhatched *Amanita* mushroom with cap, gill, and stem structure enveloped by the universal veil. If your cutting exposes white flesh, often of marshmallow-like consistency, that will first turn greenish and slimy before maturing into dark, powdery spore dust, you are looking at a genuine puffball. Some puffballs have a visibly different tissue below the spore mass at their base.

Bluffballs of Amanita aprica: *Note the gill and stem structure in the sliced white tissue. White-capped Amanitas do not show a colorful cap skin when sliced.*

Other tricksters halfway in the ground are earthballs, *Scleroderma*. They are tougher, close to a potato in consistency, and unless very young harbor a dark, purple-blackish center that tells you "We are poisonous!" Unidentified toxins will upset your digestive system and might cause nausea, vomiting, tingling, and spasming, or hardened muscles.

Luckily, avoiding poisonous look-alikes to the edible puffballs is not rocket science. Puffballs are heavy metal accumulators, so be careful where you pick them (see Heavy Metal Concentrators in Part One) and do not overindulge.

PROCESSING AND COOKING PUFFBALLS

Due to their spongy consistency and often subtle flavor, puffballs are predestined for marinating (see Lebanese Puffball Steaks in Part Three) and frying in batter, the latter a beloved approach in the giant-puffball heartlands of the Midwestern US. This fungal protein bomb can be sliced into slabs, strips, or cubes—and treated like tofu or eggplant. They can also be used as a pizza base (see Giant Puffball Pizza in Part Three), enjoyed grilled or fried like steak with butter and pepper, but it is advisable to add the salt toward the end of the frying, so that the slab does not dehydrate. Puffball slices fried in butter and garlic until crispy make great soup or salad croutons. Gem-studded Puffballs are very rich in protein; Turkish research reports that 100 grams of dried *Lycoperdon perlatum* contains 44.9 grams of protein, 42 grams of carbohydrates, and 10.6 grams of fat[41]—impressive!

Western Giant Puffball, Calvatia booniana

MEDICINAL USE

Humans have used and appreciated a wide variety of puffballs since time immemorial. Even in otherwise fungophobic cultures, puffballs are known as safe and seasonally abundant edibles. However, more interesting is the global use of puffballs as medicinals, especially to staunch bleeding or to dress wounds. The spores of puffballs are antiseptic, and being the same size as red blood cells, they slow blood flow and help coagulate blood. Many Native American tribes ate puffballs, sometimes called prairie mushrooms, but also used their spores as a styptic for treating injuries.[42] The Blackfeet know them as *ka-ka-taos* (or "fallen stars"), and they decorated the base of their tipis with painted puffballs.[43] Indigenous people also used puffballs to heal the navels of newborns: the Lakota name for puffball is *hokshi chekpa* (or "baby's navel").[44]

Mother Nature packages the medicinal spores conveniently in a small "leather satchel" so it can be kept for later use. Actually, "leather satchel" was how archaeologists in fungophobic Great Britain mislabeled puffballs found in Bronze Age graves until very recently. I was told in Bhutan and eastern Tibet that puffball spores are used to relieve diaper rashes. In a less healing and more destructive fashion, innovative US scientists used their highly combustible spores as backup ignition systems for the first atomic bombs in the twentieth century.

WESTERN GIANT PUFFBALL
Calvatia booniana

This humongous fungus in the shape of a spherical white cloud titillates the imagination of every new pot hunter. This is an enjoyable and interesting edible mushroom for many, but its sheer size and immaculate appearance raise expectations that its marshmallow texture and bland aroma are challenged to fulfill. You have found it too late when the aroma has turned unpleasant and the inside is yellow or greenish. Some people experience laxative effects from Giant Puffballs. When mature, the billions of spores are brown and embedded in a fiber mesh, and the cottony surface and firm rind break into polygons. The whole mature mesh or parts of it get blown through the landscape like a tumbleweed to spread its spores. Older field guides sometimes call it *Calvatia gigantea*, the current valid name east of the Rockies and in Eurasia. Record specimens can grow to be more than fifty pounds.

Sculptured Puffball, Calvatia sculpta

LOOK-ALIKES: Other big puffballs, bluffballs, river rocks, stranded buoys, dinosaur eggs, and dessert meringues.

NAME AND TAXONOMY: *Calvatia* from Latin *calva* "skull shape," and *booniana* for William J. Boone, first president of College of Idaho and a mushroom friend. *Calvatia booniana* A. H. Sm. 1964; previously included in *C. gigantea.* Agaricaceae, Agaricales.

IDENTIFICATION

- Huge (Ø 20–60 cm), round, white, stemless fruiting body found in pastures, among sagebrush, and juniper, or in grassy open areas.
- Grows mostly in late spring through summer.
- Occurs in drier parts of the Pacific Northwest, such as rain shadow areas and wide valleys east of the Cascades, and all over the Rocky Mountain region.

SCULPTURED PUFFBALL
Calvatia sculpta

You will stop in your tracks when you encounter this fungal sculpture in late spring or summer as you are hiking in or near mountain conifer forests. Facing this puffball at eye height, you'll see the hairdo of a comic book hero! The outer rind of this midsize white, roundish puffball cracks into tall pyramidal warts, often coiling up into pointed tips. The immature interior is white and firm, and when mature, the interior turns yellowish-white and slimy, becoming dark brown to purple-brown when mature.

IDENTIFICATION

- White, roundish mushroom (Ø 10–20 cm) with sculpted top wider than base, comprised of sterile tissue.

TOP: *Shadow Puffball*, Lycoperdon umbrinum; BOTTOM: *Gem-studded Puffball*, Lycoperdon perlatum

- Extended, longish, sometimes coiling "warts" that become pyramidal.
- Growing alone or in groups in the mountains—often under ponderosa pine—from the eastern Cascades into the Sierra.
- Fruits from April into August.

LOOK-ALIKES: Also edible *Calbovista subsculpta* has much shorter warts and can resemble a pineapple.

NAME AND TAXONOMY: *Cal-bovista* is a contraction of *Calvatia*—Latin *calva* "skull shape"—and the scientific name of a puffball, *Bovista*, latinized old High German for "fox fart;" *sculpta* Latin "sculpted." *Calvatia sculpta* (Harkn.) Lloyd 1904.

GEM-STUDDED PUFFBALL
Lycoperdon perlatum

This widely distributed common puffball is easily recognized by its upside-down pear shape with a topping of small spines or "gems," and a narrower supporting stem base that includes sterile flesh. The spines can wash off with age. The inside of the rounded top is at first firm white, but it soon softens to marshmallow-like tissue and matures into spore mass, transitioning from slimy yellow to olive to a powdery dark brown. This stage and its cloudlike eruption from a top hole or pore surely inspire its classic scientific name *Lycoperdon* which translates from Greek as "wolf fart."

Pear Puffball, Apioperdon pyriforme

Some people report enjoying these puffballs raw, halved into salads; the soft tissue absorbs dressing well. I mostly eat them when out in the woods, adding a few raw slices into my sandwich for a refreshing fungal flavor. Once the puffball matures beyond the fully white phase, its days as human food are long gone.

The much darker Shadow Puffball, *Lycoperdon umbrinum*, is also edible as long as its interior is all white (*umbrinum*, Latin for "dark brown").

- Golf ball–size pear-shaped fruiting body, white to light brown, with slight nipple on top.
- Thin skin covered with small (1–2 mm), geometrically arranged cone-shaped warts that can wash or rub off in age; in maturity, the top breaks open with a central pore.
- Visible sterile area in stem base, mycelial strands attached.
- Grows in groups in fall in duff, moss, needle litter, humus, and disturbed areas along roads.
- Common and globally distributed.

LOOK-ALIKES: A range of similar puffball species are often hard to ID without microscopy. As long as your fruiting body has a marshmallow-like, white core, they are all edible. But beware of bluffballs! Always slice your puffballs in half vertically to make sure there is no cap and stem structure hidden inside.

NAME AND TAXONOMY: *Lyco-perdon* from Greek *lykos* "wolf" and *porde* "fart," a reference to spore dispersal; *perlatum* Latin "widespread." *Lycoperdon perlatum* Pers. (1794); *Lycoperdon umbrinum* Person 1801. Agaricaceae (formerly Lycoperdaceae), Agaricales.

PEAR PUFFBALL
Apioperdon pyriforme

Similar to the Gem-studded Puffball, but deprived of all but a few crown jewels, is the white to pale brown Pear Puffball, *Apioperdon* (*Lycoperdon*) *pyriforme*, which is named for its pear-shaped fruiting body. (*Apion* is Greek for "pear," and *pyriforme* is Latin for "pear-shaped.") Its round head grows up to 5 cm wide. At first it is smooth, with a few small scattered spines on top, but the outer skin soon becomes finely cracked to form small patches that may peel off. Fruiting can start in summer, but mostly occurs in fall.

The Pear Puffball is also known as the Stump Puffball because it always grows on decaying wood, sometimes in big groups. Where it grows in soil, there are white, threadlike rhizomorphs connecting it to pieces of wood. It is edible when young (some people like to remove the peel), but sometimes an unpleasant coal gas odor dashes the culinary excitement.

NAME AND TAXONOMY: *Apioperdon pyriforme* (Schaeff.) Vizzini 2017, formerly *Lycoperdon pyriforme* Schaeff. 1774.

PURPLE EGG STINKHORN
Phallus hadriani

Every mushroom hunter remembers the first time they found a stinkhorn; the combination of stench and suggestive shape leaves a strong, lasting impression. Once the egg is mature, the fruiting body grows to full size in a few hours. The Purple Egg Stinkhorn

is known in Great Britain as the Dune or Sand Stinkhorn. It is most reliably found in coastal dunes in spring and fall. However, it can grow in all kinds of other habitats as well, like lawns, gardens, compost piles, and wood chip beds. In these sites the Common Stinkhorn (*Phallus impudicus*) can be encountered as well. It can be told apart by its whitish, saclike volva and the fact that it often grows a bit taller than *Phallus hadriani*. Unless initiated, no one will try to engage in stinkhorny culinary adventures.

However, *Phallus* is enjoyed in many cultures! In East Asia, the stinky cap covering and dirty volva are plucked away, leaving the stem to be used as a beloved ingredient in soups. It offers an interesting structure, and the mild fungal flavor readily absorbs other flavors. The closely related pantropical Veiled (or, as I like to call it, Cross-dressing) Stinkhorn (*Phallus indusiatus*) is widely cultivated in warm regions of China. In France and Germany, the immature eggs of stinkhorns are eaten

Purple Egg Stinkhorns, **Phallus hadriani**

raw or pickled; in Germany it is known as *Hexenei*, for "witches' egg." The universal veil and thin jelly layer below it are easily peeled off, and the remaining stem and the firm immature gleba—as the dark layer of spore mass is known—can be enjoyed raw. The magic here is that the stem has a harmless flavor, whereas the spore mass has an interesting radish-like flavor while offering a firm texture.

The stinky trick happens when the gleba is exposed to the air and develops its fragrance, which attracts flies that love to feast on the slimy stuff and spread its spores far and wide! In European folk medicine, the stinkhorn was used to treat rheumatism, gout, breast cancer, and epilepsy. Modern research shows that stinkhorns enhance immune system activity and have anti-inflammatory and in vitro antitumor activity.

IDENTIFICATION

- Fruiting body when young is egg-shaped (up to 6 cm high), covered by pinkish to purple (sometimes whitish) universal veil with gelatinous layer beneath.
- With stem elongation, the universal veil ruptures and the slimy swollen head (up to 4 cm wide) emerges, sometimes capped by a piece of veil.
- Head surface first covered with dark olive, foul-smelling spore mass; once

devoured by insects, the white honeycombed structure is revealed.
- Hollow, fragile, cylindrical stem (10–20 cm tall and up to 3 cm wide), white or sometimes pinkish in lower part.
- Purple saclike volva at stem base from which purple or white rhizomorphs run.
- Grows singly or in groups or clusters; saprobic in lawns, compost, wood chips, and rich soil, as well as in sand dunes along the coast; fruits in spring, summer, and fall.
- Occurring from southern British Columbia into Mexico and all the way back East.

LOOK-ALIKES: The Common Stinkhorn, *Phallus impudicus*, is very similar—just as fascinating, disgusting, and edible—but has a white egg shell. Puffballs as well as bluffballs (young poisonous *Amanita* eggs) also have an egg-like form. Without their spore mass, stinkhorns look similar to morels.

NAME AND TAXONOMY: *Phallus* Latin for "erect penis," *hadriani* named for sixteenth-century Dutch botanist Hadrianus Junius, who wrote a short book about the Dune Stinkhorn and named it Phallus; *impudicus* is Latin meaning "shameless." *Phallus impudicus* L. 1753, *P. hadriani* Vent. 1798. Phallaceae, Phallales, class Agaricomycetes.

CORALS, CLUBS, AND THE LIKE

Some of our coolest and most colorful forest mushrooms are the coral fungi. Their long-noted similarity to underwater organisms, like sea sponges and ocean corals, is astounding. Interestingly, the Latin *fungus* is derived from the Greek *spongos*, for "sponge," and in several other European languages, mushrooms are also known as sponges: *Schwamm* in German dialects or *svamp* in Swedish and Danish! It is easy to recognize a coral in general, with its multitude of rounded, always upward-pointing branches. Coral-shaped mushrooms have evolved in several fungal orders: White Coral Fungus and Ash Coral (*Clavulina* spp.) are part of the Cantharellales, and Crown Coral (*Artomyces*) belongs to Russulales. However, most common and diverse in the Pacific Northwest are *Ramaria* species (Gomphales). We have over eighty *Ramaria* species and varieties,[45] and many of them are quite difficult to identify. Often, their (at first) distinctive colors can fade quickly to dingy brown after picking or in age. Microscopy is necessary to accurately ID many of them.

CORAL MUSHROOMS: *RAMARIA* SPECIES

The challenge of identifying corals with certainty undermines appreciation of them as good edibles. Corals have a nice firm structure and a good fungal flavor and thus can be used in many ways—be it cooked simply or folded into omelets, added to sauces, or baked as "coral cakes." David Arora says corals made the best mushroom jerky he has ever had. Luckily, none of the corals are known to be dangerously poisonous—you don't have to worry about organ failure!—but an upset digestive system is in the cards when your identification is sloppy or should you have an allergic reaction. In truth, GI poisonings are much rarer than one would assume listening to Pacific Northwest foragers or leafing through guidebooks! In many cultures, corals are popular edibles, even in some fungophobic cultures. For example, most contemporary Colombians aren't interested in their wild edible fungi, even though Andean oak forests are full of chanterelles—but in one picturesque colonial small town, Villa de Leyva, I saw corals for sale!

Yellow Spring Coral, Ramaria rasilispora

In general, all *Ramaria* corals with a gelatinous core are on the do-not-eat list, as are some without gelatinous cores like the Yellow-tipped Coral, *R. formosa*, which has caused many unpleasant GI upsets. Excluding gelatinous core and yellow corals leaves a wide range of brown and red corals. For example, the bright red *R. araiospora* and *R. stuntzii* are edible, though the latter can be bitter. Nontoxic corals for the most part taste good and have an excellent structure for cooking. There are daring pot hunters out there netting all kinds of corals and eating them seemingly without major trouble.

A good rule of thumb is to avoid yellow corals in fall; most of the Yellow Spring Corals seem harmless. I hope with growing interest in edible mushrooms, and by spreading knowledge, we will overcome the prevalent ignorance regarding identification and edibility of low-risk corals.

YELLOW SPRING CORAL
Ramaria rasilispora

You might have been depressed not scoring any morels or Spring Kings during a droughty spring season, but perhaps you were lucky enough to encounter this fleshy coral and collect it. Its firm texture and good fungal flavor will save your dinner! This is definitely a mushroom that stands up to washing. Though identifying corals is usually challenging, there are only a few spring fruiting corals. And apparently, only two big yellow corals are found in spring in morel mountain forest habitat (see Look-alikes); both are edible when not bitter.

- Big (up to 20 cm tall), single-stemmed, stout coral, with white (not browning) base.
- Upper branches and tips are pale to apricot-yellow.
- Flavor of firm, white flesh is mild, not bitter.
- Grows in soil in mountain conifer forests, often not fully emerging from duff.
- Fruits in spring and occurs widely in Pacific Northwest and into California.

LOOK-ALIKES: Very similar and also spring fruiting, but often more intensely yellow, is the Bigfoot Coral, *Ramaria magnipes*. It is suspected to be eaten frequently as well, with few reports of bitterness after cooking. The fall-fruiting Yellow-tipped Coral, *R. formosa*, is quite similar, but its stem bruises brown when handled or old, it has light yellow-orange branches with light yellow tips, and it is frequently implicated in gastrointestinal problems. Christian Schwarz and Noah Siegel suspect *R. formosa* is a catchall name for many different yellow coral species.

NAME AND TAXONOMY: *Ramaria* from Latin *ramus* "branch," *rasili-spora* Latin for "smooth-spored," and *formosa* Latin "beautiful." *Ramaria formosa* (Pers. per Fr.) Quel. 1888, *R. rasilispora* Marr and D. E. Stuntz 1973. Gomphaceae, Gomphales.

PINK-TIPPED CORAL
Ramaria botrytis

In fall, when coral diversity peaks, the Pink-tipped Coral is the most commonly eaten coral, standing apart from her sisters by her pink or wine-colored to purple tips, while the white stem and lower branches are shared with many other corals. Identification of *Ramaria botrytis* is helped by the fact that the white stem stains yellowish to light

Pink-tipped Coral, Ramaria botrytis

LEFT: *White Coral Fungus,* Clavulina coralloides; **RIGHT**: Clavulina sp. *(still undescribed)*

brown. Also, the branch pattern is cauliflower-like, with dense terminal branching on a few thick, lower branches.

IDENTIFICATION

· Single, massive white stem with cauliflower branching pattern.
· Upper tips pinkish to wine colored when fresh.
· Firm, white flesh, no jelly core, not bitter; flesh bruises yellowish-brown.
· Grows on ground and fruits in fall in conifer forests, especially with spruce and true fir.

LOOK-ALIKES: There are a few other reddish-tinged corals in the region. Another choice edible, the purple-red-tipped Perma Pink Coral, *Ramaria rubripermanens*, fruits in late spring, summer, and into fall under white pine, Douglas-fir, and grand fir. It can grow up to 20 cm high and has a large to massive white base. Tasty Pale Pinky Coral, *R. rubrievanescens*, loses its pinkish branch tip color with maturation or after being picked and fades to white to pinkish-beige. It grows to 10 cm in height, fruits in spring and fall under conifers, and is also a choice edible. The much rarer Red-brown Coral, *Ramaria rubribrunnescens*, has a base and lower branches stained dark reddish-brown, often grows with hemlock, and fruits in fall (edibility unknown); when old, it releases a sweet anise-like odor.

NAME AND TAXONOMY: *Botrytis* Greek "grape-shaped," *rubri-* from Latin *ruber* "red," *permanens* "permanent, lasting," *evanescens* "vanishing," and *brunnescens* "browning," all Latin. *Ramaria botrytis* (Pers. ex Fries) Ricken 1918; *Ramaria rubripermanens, R. rubrievanescens,* and *R. rubribrunnescens,* all Marr and D.E. Stuntz 1973.

WHITE CORAL FUNGUS
Clavulina coralloides

White Coral Fungus and her sisters are some of the most common non-*Ramaria* corals. From fall and into winter, these

LEFT: *Ash Coral* (C. cinerea); RIGHT: *A coral fungus infected by* Helminthosphaeria clavariarum

mycorrhizal mushrooms sprout all over the ground in Douglas-fir forests and can easily save the day if someone else has snatched up all "your" chanties or hedgehogs. Ignored by most mushroom hunters, these small whitish corals have a surprisingly appealing flavor and texture, which is less surprising when looking at their provenance: Clavulinas are closely related to chanterelles and hedgehogs. They make great Coral Cakes (see Part Three for recipes).

The Pacific Northwest has several highly variable, apparently all edible species. *Clavulina coralloides* has thin, whitish branches with busy tips often becoming tinged yellowish or ochre with age. Young *Clavulina cinerea* has also thin white branches, but they turn gray in age, hence Ash or Grey Coral. *Clavulina reae* and *C. rugosa* (*rugosa* Latin "wrinkled") both have noticeably thicker branches. Wrinkled Coral is often much less branched and does not gray with age, while *C. reae* turns evenly gray. Danny Miller's online DNA research informs us that our clavulinas will need several new species names, including for the depicted, still unnamed, clustered wrinkled coral fungus.

Matters are complicated by *Helminthosphaeria clavariarum*, a parasitic Ascomycete that grows from the stem base upward on *Clavulina*, causing graying before turning dark bluish-gray, as well as contorted growth. No one recommends eating infected specimens, but since infection is common and no one has reported becoming sick from it, it is likely not toxic.

IDENTIFICATION

- Smallish coral (5–10 cm tall) consisting of multiple, often irregularly forking erect branches arising from a short base and ending in profusely and delicately branched tips (less branching in *C. rugosa* and *C. reae*).
- *Clavulina coralloides* is white, yellowing with age; Ash Corals are white and turn grayish with age.
- Firm but brittle, mild-tasting flesh.
- Grows singly, scattered, and in large groups on the ground in conifer and mixed forests.
- Fruits in fall and winter; widespread.

LOOK-ALIKES: There are at least two very rare and probably harmless small

Peppery Crown Coral, Artomyces piperatus

White Coral look-alikes. Slender Coral, *Ramaria gracilis*, has pliable branches, and *Ramariopsis kunzei* is bright white with a clean look and smooth, round branches without crested tips. Several *Lentaria* species are similar, but their branches are pliable, not brittle, and they have white rhizomorphs and grow on wood, not in soil.

NAME AND TAXONOMY: *Clav-ulina* Latin "small club," *corall-oides* Latin "coral" with Greek suffix *-oides* "like," and *cinerea* Latin "ashlike." *Clavulina coralloides* and *C. cinerea* (Fries) J. Schroet. 1888. Clavulinaceae, Cantharellales.

PEPPERY CROWN CORAL
Artomyces piperatus

The rare Peppery Crown Coral (*Artomyces piperatus*) is easy to tell apart from other corals. It has unique crown-like branch endings, and as its scientific name suggests, it is spicy like pepper. In addition, it grows on dead wood, while most Ramarias, being mycorrhizal, grow on forest ground, though they are also found on late-stage decaying tree trunks. The yellowish-brown to cinnamon-brown, branched fruiting body is paler higher up on the crown. The distribution area of the bigger and whiter Crown-tipped Coral Fungus, *Artomyces pyxidatus*, starts east of the Rocky Mountain Trench and in the Wasatch Range and reaches all the way to the Eastern Seaboard. Cuisines featuring spicy food make great use of Crown Corals!

NAME AND TAXONOMY: *Artomyces piperatus* (Kauffman) Julich 1981 = *Clavicorona piperata* (Kauffman) Leathers and A. H. Sm. 1967. Auriscalpiaceae, Russulales.

CLUB CORAL
Clavariadelphus truncatus

As a large club coral, *Clavariadelphus truncatus* is unmatched! The strange flattened top and sometimes veined sides of this bright golden club are reminiscent of chanterelles, but they are not closely related. The most intriguing quality is a sweet flavor, which can be subtle or intense. Tasting the perfect sweetness is unforgettable! Club Corals are good edibles and can be used in a variety of desserts or as a sweetener in other dishes. Only when they are old do they turn bitter. Club Corals contain clavaric acid, which reduces the rate of tumor development in lab mice, suggesting that this acid may have therapeutic value for treating certain cancers.[46]

IDENTIFICATION

- Club-shaped, flat-topped fruiting bodies (5–15 cm tall, ∅ up to 8 cm at top).
- Pale yellow to golden or orange-brown, but also pinkish-brown to ochre; top usually brighter; fruiting body smooth, wrinkled, or veined.
- Solid white to ochre flesh, becomes soft and spongy upward as club head enlarges.
- Sweet to bittersweet flavor.
- Grows scattered or in groups on soil and in duff under conifers, especially mountain spruce forests.
- Fruits in fall; widespread and common in the Pacific Northwest.

LOOK-ALIKES: Also edible but not as sweet is Pale Candy Coral, *Clavariadelphus*

Club Coral, Clavariadelphus truncatus

Purple Fairy Club, Alloclavaria purpurea

pallidoincarnatus, a slender species with gold only on the flat top; the sides are grayish-orange to ochre-beige. Furthermore, there are a range of smaller, much paler, yellow or whitish club corals, such as the Strap Coral, *C. ligula,* and its doppelgänger *C. sachalinensis.* They are not eaten.

NAME AND TAXONOMY: *Clavariadelphus:* Latin *clava* "club," Greek *adelphos* "brother," as in closely related; *truncatus* Latin "truncated," for the flat or cutoff top. *Clavariadelphus truncatus* (Quel.) Donk 1933. Clavariadelphaceae, Gomphales.

PURPLE FAIRY CLUB
Alloclavaria purpurea

Foragers often miss this beautiful and unique purple mushroom in the woods. But when found, it generates a lot of *oohs* and *aahs* and many pictures are taken. Few bother to pick the Purple Fairy Club for din-ner. I collected them several times in spruce forests and do not recall frying them up. Probably I did not keep them separate from the other mushrooms, and so their fragile fruiting bodies broke and sullied, rendering them unappealing. Apparently, we are looking at a novelty, emergency food.

IDENTIFICATION

- Slender, club-shaped, unbranched, fragile, purplish and brownish fruiting bodies (10–15 cm tall).
- Grows in tufts or extensive patches on ground or in moss near conifer wood in spruce and fir forests.
- Fruits in spring through early fall.

NAME AND TAXONOMY: *Allo-clavaria: allo* Greek "other," *clavaria* Latin "club-shaped;" *purpurea* Latin "purple." *Alloclavaria purpurea* (Fries) Dentinger and D. J. McLaughlin 2007, *Clavaria purpurea* Fr. 1821. Probably Rickenellaceae, Hymenochetales.

GILLED MUSHROOMS

Gilled mushrooms are an immense and diverse group. Look-alikes abound. This group also contains most of our region's deadly and toxic mushrooms. Very few are included in this guide's Fourteen Fantastic Fungi, which are safe, easy-to-identify, edible fungi (see sidebar on p. 32 in Part One). Gills as a spore-bearing structure have developed many times in several branches of the fungal kingdom, though the biggest group is contained in the order Agaricales, which represents more than a third of the edibles in this book. Brittlegills and milkcaps, members of the Russulales, are also beloved by foragers and featured here. Facing such diversity and many toxic members, you must pay special attention to fine structural details unimportant or nonexistent in other forms of mushrooms. Learning to distinguish between gill attachment and shape, rings and partial veils, universal veils, stem shape, and so on requires learning new, often difficult terminology. Refer to the Glossary in the back of the book and the detailed descriptions of species. Ignoring descriptive points when identifying your foraged food can have fatal consequences!

The best way to present gilled mushrooms is to sort them by spore color, an early taxonomic approach that has mostly stood the test of time. On the downside, judging spore color is often tricky on the spot, especially for the beginner; spore prints can take two to twenty hours, which feels like an eternity to a hungry forager. But spore color is still a powerful first step to identification.

Spore colors fall into four basic color groups: "white and pale" (including white to yellowish), "pink" (ranging from pinkish to salmon), "warm brown" (encompassing yellow-brown to rusty-brown), and "cold dark brown to black" (truly purple-brown to purple-black to black). These categories cover most gilled fungi, though there are mavericks, like the Green-spored Shaggy Parasol, *Chlorophyllum molybdites* (a poisonous species included with its edible white-spored shaggy brethren), and one of the oyster mushrooms, *Pleurotus ostreatus*, with its distinctly lilac-gray spores (described in this field guide with the other whitish-spored oysters).

SPORE PRINTING

Spore prints are easy to make and tremendously useful when identifying gilled mushrooms. The easiest and quickest method is to look for a "natural" spore print when picking. Mushrooms growing very close with their caps overlapping often deposit their spores onto their neighbors' caps. Sometimes that print might be on a leaf or wood beneath the cap. However, if Mother Nature is not providing splendid assistance in the field, take a cap, remove the stem, and place the cap gill-side down on a clean surface. Or poke a hole in a piece of paper, insert the stem, and get the paper as close to the gills as possible. Be aware that the paper color makes a difference; white prints are hard to see on white paper. Newspaper with its newsprint will show any spore color.

Now comes the hard part—you have to wait! Sometimes you can get a spore deposit in an hour, but usually it takes four to twelve hours. Your find from the afternoon hunt will not be readily identifiable for dinner. You can start a print earlier, in the forest, by using aluminum foil, and some people say that vibrations during the drive home may speed up the spore drop too.

Mushrooms use tiny drops of water to shoot their spores. They use the forces of water's surface tension, optimized by a hydrophilic solution of mannitol and hexose sugars, to propel spores. What is known as "Buller's drop" catapults the tiny spore with incredible force—several thousand times the force of gravity—to travel less than a millimeter from the side of the gill into the space between the gills. So, protect your specimen from drying out. An upside-down bowl over the cap keeps moisture in and reduces air movement, creating a clearer print. If the mushroom seems dry, you can increase humidity by spritzing it with water. Very wet mushrooms will soak your paper. Sometimes mushrooms do not cooperate if they are immature or already too dry.

Chocolate brown print of an Agaricus

GILLED MUSHROOMS WITH WHITE OR PALE SPORES

The biggest group of gilled mushrooms produces white or pale spore prints, the latter including cream or yellow as in many brittlegills (*Russula*), and white with a light gray-purple tinge as in some oysters (*Pleurotus*) and light pinkish as in the Blewit. While *Russula* and *Lactarius* are evolutionarily distinct and belong to the order Russulales, they are often recognized by their brittleness, which is caused by a roundish cell structure.

Most other gilled mushrooms belong to the vast order Agaricales that contains a lot of tasty fungal fruits in gilled genera, including *Amanita, Armillaria, Chlorophyllum, Clitocybe, Floccularia, Hygrophorus, Lyophyllum, Marasmius, Pleurotus, Tricholoma*, and many more. Also, white-spored mushrooms include plenty of inedible and poisonous mushrooms in many of these genera, including some deadly mushrooms—especially the dangerous Death Cap or *Amanita phalloides* (and allied destroyers) and the Fatal Dapperling or *Lepiota subincarnata*—and some nastily toxic to potentially deadly funnelcaps, like *Clitocybe rivulosa*, commonly called the Sweater or Fool's Funnel.

Many light-spored mushrooms grow on the ground. Others grow on wood. They are described here, with the groups growing on ground first, followed by those whose habitat is decaying wood.

Brittlegills, Russula

This group of white- to yellow-spored gilled mushrooms is fairly easily recognized by their firm but brittle flesh that breaks like chalk when fresh, often with a snapping sound. Closely related Milkcaps also break chalklike, but ooze a milklike liquid when injured. Our brittlegills and milkcaps are all ectomycorrhizal forest mushrooms, so it helps to know your trees so that you can find and identify them. Tasting their flesh is important for identification; both genera have not only mild but also spicy, peppery, and sometimes bitter-tasting species. Overall, however, they contain few known seriously poisonous species.

Brittlegills (*Russula* spp.) are some of the most abundant mushrooms in temperate conifer forests and fairly easy to recognize as a genus by their gestalt: a solid central unadorned stem and a rounded, smooth cap that can appear in any color of the rainbow, though it is most often white, beige, brown, or red. Identifying them at the species level is a whole different ball game. Brittlegills are the chameleons of the fungal kingdom, and cap color is often not indicative of their identity. Luckily in the Pacific Northwest, we have several easy-to-recognize species and none known to be seriously toxic.

In the past, edibility has been mostly tied to flavor, and any spicy species was regarded as inedible if not potentially poisonous. Though there are probably brittlegill species to avoid, we are learning from Eastern European mushroom lovers that spiciness can be removed by frying and especially by parboiling and using an overnight cold-water bath. Even the Sickener (a.k.a. Vomiting Brittlegill, *Russula emetica*), which upsets plenty of people's GI systems, is rendered edible in Eastern Europe. Consider, however, that spicy species

SHATTERING BRITTLEGILLS

The notorious *Russula subnigricans*, a pale-red-staining brittlegill, occurs in East Asia—and so far reportedly nowhere else. It has been responsible for several deaths in Japan from rhabdomyolysis, which triggers the breakdown of muscle tissue, resulting in kidney failure. Based on the Japanese experience, though without equivalent calamities elsewhere, careful foragers refuse to collect red-staining *Russula* species, in which the reddened flesh will later turn black, such as our region's Bloody Blackening Brittlegill, *R. nigricans*. However, there are no known poisoning cases here from *R. nigricans*, and more adventurous foragers appreciate this big and very firm brittlegill, which is easily identified by its blackening and distant gills. The closest relative of *R. subnigricans* is found in California, *R. cantharellicola*, where it is known as the Chanterelle Eater, since it always grows with chanterelles; no reports of poisonings have been documented.

Another group of brittlegills known as the fragrant brittlegills (*R. fragrantissima* group) feature powerful scents, with some being reminiscent of maraschino cherries and almonds. They are widely regarded as inedible and could potentially cause gastrointestinal upset. However, in Ukraine, the closely related Fetid Brittlegill (*R. foetens*) is eaten after being rendered edible through water baths. Go figure. Some green Russulas used to be eaten and considered quite safe on the West Coast. However, the invasion of the often-greenish-capped Death Cap, *Amanita phalloides*, has introduced a deadly look-alike. In the Pacific Northwest, the Death Cap is more or less still confined to urban areas, but in the Bay Area, it is very common and abundant!

have been tested much less for their edibility than mild-tasting Russulas.

Russula expert Danny Miller's suggestion for the adventurous forager is: "The denser and fresher the *Russula*, the more it snaps audibly like chalk, and that could be a good 'quality of edibility' test." He explains, "Some people eat all of them hot or not after detox, some people check for acrid ones and avoid them, but they still collect the fragile ones of poor texture. However, if you just eat the ones that are mild and dense enough to snap audibly, you should get a pretty high-quality mix without having to know the species. This way you would also avoid specimens softened by bacteria or old age."

SHRIMP BRITTLEGILL
Russula xerampelina

The Pacific Northwest's flagship edible brittlegill, the Shrimp Brittlegill, is common, abundant, and persistent in both forest and fridge. One of the bigger Russulas (which narrows down *Russula* look-alikes substantially), it offers the crunchiness typical of the Russulas, with a great fungal flavor! If present, the fishy odor is easiest to detect in the tissue of the stem base or where the stem meets the gills. Sometimes I cannot smell it in the woods, but two days later in my kitchen, it screams shrimp. Often the shrimpiness cooks away. And luckily, it is one of the Russulas that is positively identifiable down

Shrimp Brittlegill, Russula xerampelina

to the species—or at least to the species complex—without bringing a mobile DNA lab to the forest! The combination of purplish-red in the cap, a pink or red hue on a stem that stains brown when handled, and yellowish gills gets you very close to an ID. Add to that a shrimpy odor and mild flavor—not spicy. If chemically equipped with a drop of ferrous sulfate ($FeSO_4$) will turn the flesh of only our three to four species of Shrimp Brittlegills green, but no other *Russula*—a rare, conclusive chemical test.

However, there are a range of other cap colors as well, as Danny Miller reports in his excellent online Pacific Northwest Mushroom Pictorial Key: "In the space of a single hour at an identification table, I saw yellow, orange, brown, green, red, and purple ones, although they are 'usually' dark purple." Once you feel secure identifying the mainstream purple version, you can branch out to netting other Shrimpies (see Look-alikes), but in the beginning, there are enough purplish ones to enjoy and identify

with confidence. Be prepared, however, for *R. xerampelina* to soon receive a different name, since it is a European species and ours does not have its own name yet.

Though the brittle nature of Russulas requires careful handling after picking, it also contributes to their longevity. Brittlegills kept cool last for over a week unless they are worm infested. Their firm structure works well in stir-frys and makes them an interesting mushroom for brining.

IDENTIFICATION

- Midsize to big mushrooms with bald, slightly sticky caps (∅ diameter 5–15 cm, or bigger) when wet; depressed centers and inrolled margins into middle age.
- Cap color can vary widely: purple, purple-brown, or red tones are most common, or at least some purple in cap, but also brown, orange, yellow, and green.

- Cup cuticle can be peeled from the margin halfway toward the center (it takes a couple of tries to learn this technique of "skinning" Russulas).
- Crowded, broad, very brittle gills are adnate to adnexed; creamy-white becoming dull yellowish with age; spores yellowish.
- White or pale buff stem, often pink tinged, sometimes all pink; bruising brown where handled.
- Flesh firm and brittle, all parts snapping like chalk; mild tasting, often with fishy odor.
- Fruits alone or in groups on the ground with conifers in summer to late fall.
- Distributed in our whole region and the Northern Hemisphere in general.

LOOK-ALIKES: Where to start? As mentioned, the variability of cap colors allows for unlimited confusion, but when you stick with Russulas that match the traits described above, the following species or varieties also fit the shrimpy bill. A greenish cap with a dark brown center used to be *R. xerampelina* var. *elaeodes* and is now known as *R. favrei*. A pale pink or yellow-brown

Rainbow Brittlegill, Russula olivacea

shrimp mushroom with little stem flushing used to be *R. xerampelina* var. *isabelliniceps* and is now called *R. viridofusca* (Latin for "green gray-brown"). Also, the enjoyable Rainbow Brittlegill, *R. olivacea*, is very close, sometimes huge, and very colorful, but it lacks the fish scent. While regarded as a choice edible in the western US and British Columbia, in Europe there are rare reports of GI upset. However, our Rainbow Brittlegill is genetically distinctive and in need of its own scientific name. Furthermore, there are dozens of other brittlegills that match the above cap colors, but they are spicy or their stems do not stain quite as brown.

NAME AND TAXONOMY: *Russula* from Latin *russus* for "reddish," as in Italian *vino rosso,* and *-ula* as diminutive, hence "the little reddish one." *Xer-ampelina* Greek for "dry" (*xeros*) and "grapevine" (*ampelos*), referring to the dark red to purple color of dry grapevine leaves reflected in the cap. *Russula xerampelina* (Secr.) Fr. 1838. Russulaceae, Russulales.

SHORT-STEMMED BRITTLEGILL
Russula brevipes group

Short-stemmed Brittlegill's claim to fame is playing host to parasitic *Hypomyces lactifluorum*, which turns the bland but very firm *Russula brevipes* into the highly appreciated Lobster Mushroom (see Truffles and Lobsters in Ascomycetes). Short-stemmed Brittlegills that escape this hostile fungal takeover often have to endure being cursed in the woods, since they refuse to reveal themselves as a matsutake or White Chanterelle. Up close, the true identity of this big white *Russula* is easy to discern from

Short-stemmed Brittlegill, Russula brevipes *group*

its deeply sunken center, funnel-shaped cap, and crowded, brittle gills. And then there is that short, stubby stipe that is unable to push the mushroom through the duff layer, leaving the cap covered in some inches of needles and sticks.

Too many people just root this most common of the big Russulas up and leave it to rot in the woods. Young David Arora provided unflattering rhymes, bottoming out at "better trampled than sampled," but he, too, has come around to appreciating them for their texturally pleasing firmness. Short-stemmed Brittlegill can be an enjoyable edible in its own right and not just as the victim of *Hypomyces lactifluorum*. It can be pickled and preserved to light up dark winter nights, served best with vodka, as many Eastern Europeans will assure you. A great new way to enjoy this ubiquitous and abundant mushroom was recently revealed by wild mushroom chef Chad Hyatt who uses it as a protein base for an awesome chickpea-free hummus, allowing the olive oil, lemon juice, tahini, and garlic to shine!

Culinary progress has outpaced taxonomy at this point. DNA reveals that this East Coast name is misapplied out West and needs to be replaced. It becomes complicated, though, since we have probably half a dozen white brittlegill species hiding huddled together under that big *brevipes* cap.

IDENTIFICATION

- Big, dry, dull white cap (Ø 8–20 cm), dimpled when young with an inrolled margin and maturing into an upright vase often with yellowish or rusty patches at center.
- Crowded, fragile, white gills that are adnate or decurrent, stain yellow-brown with age, and produce white to cream-colored spores.
- Hard, short white stem, becoming hollowed with age.
- Crisp white flesh does not blacken or redden when bruised (but slowly browns when injured or with age).
- Mild flavor, or slowly revealing a slightly peppery flavor.
- Grows solitarily or in groups, often producing "mushrumps" covered by duff and topsoil.

- Fruits in conifer and mixed forests from fall into early winter; widespread and common.

LOOK-ALIKES: A wide range of large, white gilled mushrooms might be mistaken by novices, ranging from White Chanterelle, matsutake, and big cats to highly toxic white Amanitas like *A. smithiana*. None, however, have brittle flesh like the Russulas or the vase shape when mature. Once you are 100% sure of your identification of the genus *Russula*, the big, dry, vase-shaped cap, short stem, and flesh that browns with age will point you to the *R. brevipes* group. *Russula brevipes* var. *acrior* has a narrow greenish band at the top of the stipe where the gills attach and is always spicy, hence *acrior*, Latin for "spicier." The harmless Cascade Brittlegill (*R. cascadensis*) is very similar, but it is always spicy and has a cap diameter of usually less than 10 cm.

NAME AND TAXONOMY: *Brevi-pes* Latin for "short-foot." *Russula brevipes* Peck 1890.

GREEN BRITTLEGILL
Russula aeruginea

Green mushrooms are quite rare, and one of the greenest and shiniest brittlegills is *Russula aeruginea*, the cap color often a good match for oxidized copper. However, like many other brittlegills, its cap colors vary, and it also can have a yellow-green or olive-brown cap. We are fortunate that this color-challenged genus offers very low risk if you mistakenly eat another mild *Russula* species.

Simply out of luck are foragers who might pick and fry up the similarly colored Death Cap, *Amanita phalloides*, which is easy to tell apart by extracting the whole stem. Without the stem's base, a forager may miss the obvious cuplike volva (characteristic of deadly Amanitas) underground, possibly sentencing their human body to that same location a few awful days later. Luckily, the Death Cap is basically absent from Pacific Northwest forests, though it is spreading via nonnative landscape trees in urban areas.

Green Brittlegill, Russula aeruginea

- Caps (Ø 3–10 cm) olive-to-grass green (rarely olive-brown or yellow-green), smooth and slightly sticky when wet and shiny when dry, centrally depressed; the cap margin is striate, lined from the gills bellow.
- Cap skin can be peeled from the margin nearly two-thirds of the distance toward the center.
- Gills crowded and at first white, maturing to yellow, forking near the stem and producing pale yellow spores.
- Whitish stem may stain pale brown or reddish at base; stem is stuffed, becoming hollow with age.
- Flesh firm, brittle, white; flavor mild or slightly spicy.
- Grows singly or in groups on the ground, often with spruce or birch.
- Distributed widely in the Pacific Northwest and throughout the Northern Hemisphere.

LOOK-ALIKES: There are several green or greenish brittlegills. Most similar and quite common is the also edible *Russula graminea*, with a sticky cap, often grass-green (*graminea* in Latin) or olive-green with pale yellowish and brownish tones. *Russula occidentalis*, the Western Brittlegill, has a sticky cap and is sometimes all green or all purple, but more frequently purple with greenish-yellow. The flesh of this edible brittlegill common to Douglas-fir–hemlock forests turns slowly reddish and then gray or grayish-black. Its gills are dense and produce ochre spores. And I may as well say it again: there is the unforgiving green-capped Death Cap, *Amanita phalloides*, which has a very different stem base and a ring on the stem.

NAME AND TAXONOMY: *Aeruginea* is Latin for the color of oxidized copper. *Russula aeruginea* Lindblad apud Fr. 1863.

Milkcaps, Lactarius

Milkcaps (*Lactarius* spp.) share the firm but brittle texture of their relatives, the Russulas, but are characterized by a milky liquid they exude; hence they are called *Lactarius*, *lac* as in *galaxy* (the Milky Way) and *lactation*. Unless the mushroom is old or dry, the milk is easily detected by slicing the gills or the upper stem; it should ooze out. Sometimes there is only a subtle appearance of milk at the broken tissue. Most milks are white, but some are colorful; some white milks will stain the flesh where injured. The color of the milk and possible color changes from oxidation after exposure to air are very important criteria for identification. Many *Lactarius* can be detected by "remote sensing"—by looks alone. Their caps display a ring pattern, known as concentric zonation, where lighter and darker rings alternate. However, when this trait is missing and you are wondering what the heck this mushroom could be, it is always good to slice the gills and see if milk droplets appear, revealing it to be a *Lactarius*. Very few mushrooms ooze milk besides milkcaps. Most common in the Pacific Northwest is the Bleeder, the dainty and stunning red-bleeding *Mycena haematopus*.

Lactarius come in all sizes—the cap can grow to nearly a foot in diameter—but mostly the caps are 5 to 15 cm wide, and the stems are rarely no longer than the cap's width and often even shorter.

Gills are fragile and usually closely spaced, producing white to yellowish spores. All milkcaps are root associated, so being able to identify the trees growing in the vicinity of a mushroom will help in identifying them. Their firm structure and crisp texture make them attractive edibles, and they bring nice diversity to the table. There is not any known serious toxicity in milkcaps, but many of them can be intensely spicy or bitter. In Eastern Europe, people use a combination of parboiling for ten minutes with an extended cold-water bath of several hours to leach out or neutralize these unfavorable flavors while maintaining the mushroom's firm texture. Often I have run into Eastern Europeans picking *Lactarius* and *Russula* species, such as the habanero-pepper-spicy Redhot Milkcap (*L. rufus*) or the Olympic Milkcap (*L. olympianus*), neither of which are recommended as edibles in most mushroom books published west of Poland. In Eastern European culinary traditions, milkcaps are some of the most popular mushrooms for pickling. One great culinary use of the mild and popular Saffron Milkcap is to fry them in butter or oil with a pinch of salt and pepper, emphasizing their crunch. Be aware, however, that generous Saffron Milkcap intake may stain your urine red for a day, just as red beets do.

Bleeders, Mycena haematopus

SAFFRON MILKCAPS
Lactarius deliciosus group

Saffron Milkcaps and allies are easy to identify due to their orange or red milk that often changes fungal flesh to green. The unique color of their milk and the color change makes them the only terrestrial gilled mushroom that is included in the Fourteen Fantastic Fungi in Part One. Their crunchy texture with good fungal flavor makes them attractive edibles. In Spain, people are crazy about *níscalo* or *robellón*, in East Asia they love Hatsutake, *L. hatsudake*. However, the species epithet *deliciosus* builds up high expectations that Saffron Milkcaps do not always meet. Some of our Saffron Milkcaps in the Pacific Northwest can have a bitter flavor that takes the "fun" out of this fungus.

Similarly, in Central Europe, while the true *L. deliciosus* there receives high praise, very similar-looking, closely related

Saffron Milkcaps, Lactarius deliciosus *group*

milkcaps are appreciated by some people while ignored by others. DNA reveals that the European *L. deliciosus* does not appear to grow on the West Coast, and we still lack a valid name for most of our Western Saffron Milkcap species. Still, the lack of taxonomic clarity in the *Lactarius* section *Deliciosi* should not scare off foragers. In the future, perhaps we will have a clearer picture, which should include pointers to the most desirable and delectable species.

The safest to identify are the orangish-colored species whose flesh readily turns green, like the *L. deliciosus* group and *L. aurantiosordidus.* Also easy to identify is the tasty, red-bleeding *L. rubrilacteus.* The best of the bunch, in my experience, is the Summer Saffron Milkcap, *L. aestivus,* the least oxidizing (greening) of all of them.

- Colors of the bald, often mottled, dry cap (Ø 5–15 cm) range from dull orange to carrot-orange or pale orange-brown to gray with an orange tinge; cap turns green with age.
- Cap often has rings of these colors, often with a depressed center and inrolled margin when young that can turn upward with age.
- Close, adnate to slightly decurrent, fragile gills that are bright to dull orange, sometimes yellowish or orange-buff, and greenish where wounded; spores cream to pale yellow.
- Stem colored like cap, often with a white sheen when young that turns green with age.

- The bright orange but scanty milk turns green, sometimes after first turning deep red.
- Flesh is thick, brittle, and grainy, light brownish-yellow to orange, discoloring to orange-red or purple-red, and after injury or in age slowly staining greenish.
- Flavor mild to slightly bitter.
- Fruits on ground in fall, scattered or in groups with conifers; widespread.

Summer Saffron Milkcap, Lactarius aestivus

LOOK-ALIKES: Only recently extracted from the group, *Lactarius aurantiosordidus* can be told apart by its unchanging orange milk and growth with Sitka spruce. It also tends to be smaller and has a dingy pale orange cap with green stains and is edible too. The reddish-orange-bodied *L. rubrilacteus* bleeds red. Another bright orange milk-cap, edible and choice (and hardly green staining), is the Summer Saffron Milkcap (*L. aestivus*)—see below.

NAME AND TAXONOMY: *Lactarius* for "milk producing," *deliciosus* for "delicious," and *aurantio-sordidus* for "golden-dirty"—all Latin. *Lactarius deliciosus* (Fr.) Gray 1821, *Lactarius aurantiosordidus* Nuytinck and S. L. Mill. 2006. Russulaceae, Russulales.

SUMMER SAFFRON MILKCAP
Lactarius aestivus

Summer Saffron Milkcap, *Lactarius aestivus*, is the brightest and best of the Pacific Northwest *Lactarius* section *Deliciosi*. This new species is recognized by its bright to

pale orange cap color, zonate (ringed) cap pattern, and scant neon-orange milk that only rarely and, if so, slowly discolors greenish. When I encountered my first Summer Saffron Milkcap in the woods, I initially thought, *Nice, Rainbow Chanterelle!* The orange color is a good match, but upon closer inspection, that ID error is quickly revealed, eliminating any disappointment.

Aestivus means "summer" in Latin; in 2014 University of Washington mycologist Joe Ammirati and Dutch mycologist Jorinde Nuytinck with the Naturalis Biodiversity Center describe it as fruiting in summer.[47] However, his type specimens were collected in mid-September. This milkcap is often one of the last mushrooms standing in the true fir–hemlock forests in the Cascades before snow covers other mushrooms.

RED BLEEDING MILKCAP
Lactarius rubrilacteus

The red milk that runs when the mushroom is fresh and moist (forming dark red stains when it is not) and the greening of the fruiting body make the Red Bleeding Milkcap unique. It is good eating when young and crispy—as it ages, the grainy structure crushes the crunch factor and texture thrill.

IDENTIFICATION

- Fruiting body orange-brown to reddish-brown or tan.
- Bald, dry cap (∅ 4–14 cm) often ringed with these colors and paler zones, with an inrolled margin when young and depressed center with age.
- Fragile gills close, adnate to slightly decurrent, often forking near the stem; pale pink to buff-orange and staining red-brown or purplish-red with age.
- Stem colored like cap or paler and sometimes with large, shallow, sunken spots (scrobiculate).
- Dark red milk that is never orange; oxidized milk leads to green staining of all parts with bruising or in age.
- Flesh centrally thick, thin at margin, brittle and grainy; whitish, but showing orange-buff to reddish color from milk, turning green from injury after several hours.
- Fruits on ground in fall, scattered or in groups with conifers, especially Douglas-fir and two- and three-needle pines.

NAME AND TAXONOMY: *Rubrilacteus* Latin for "red milk." *Lactarius rubrilacteus* Hesler and A. H. Sm. 1979.

CANDY CAPS
Lactarius rubidus

Candy Caps excite everyone—and rightly so! The alluring odor of maple syrup, sometimes reminiscent of butterscotch, fenugreek, or burnt sugar, is unique in the often odoriferous fungal kingdom. The aromatic compound is quabalactone III, which hydrolyzes into the all-powerful sotolon, one of maple syrup's core aromas.[48] However, these sweet "fun-gals" don't make it too easy for the shy beginner: the scent is not a good field marker, as their alluring aroma is revealed only when they are heated or dried. They are small—often hiding in plain sight—and they share the forest with many look-alikes. Plenty of small orange milkcaps could be Candy Caps, but most are not. You first need to get familiar with the nature of milkcaps, their brittle consistency, and their lactation

Red Bleeding Milkcap, Lactarius rubrilacteus

Candy Caps, Lactarius rubidus. *Note the whey-like milk on the gills.*

habit. Mushrooms can be very unforgiving if you "miss-take" them!

The first field check is to feel the surface of the cap. Candies have a rough, wrinkly cap center. Sometimes the whole surface is nubbly, and when dry it is dull, not shiny. If the cap is completely smooth or sticky, there's no need to hurt the little thing; just let it go on popping out its spores. If you feel a nonsticky, nubbly, orange peel–like cap, it's time to pick her up, and now you need to bleed the poor thing with a merciless nick of the knife or by breaking off a piece. Most orangish milkcaps will ooze white milk: that means you have an imposter! If the latex is instead watery and wheylike, you are getting much closer. Though some people claim to detect a faint burned sugar odor, you will learn to know them best by sight and feel. However, there is an immediate on-site shortcut to get a positive ID without taking them home to dry overnight! I always carry a lighter with me for this purpose. Burn the edge of the cap. If it is a Candy Cap, you will smell the unmistakable maple syrup aroma, and then you know that you have found a *Lactarius rubidus*—awesome!

Nevertheless, some of my collections have exhibited a very weak maple syrup aroma that was nearly overpowered by a bitter aftertaste; consequently, I have thrown out several jars of Puget Sound Candy Caps through the years. But when using a delectable collection in my "butter cookies" around Christmas, I am always pleased. Still, the Candies I have collected on California's North Coast have a much stronger aroma and flavor.

When collecting Candy Caps, it is smart to separate them from other, more solid mushrooms like chanterelles. They break easily and can turn into dregs among bigger basket bullies. David Arora suggests drying them slowly in the open air, which supports the chemical process that brings on the maple syrup aroma. Cooking them also triggers this alchemical wonder!

NAME AND TAXONOMY: *Lactarius rubidus* (Hesler and A. H. Sm.) Methven 2013; previously *Lactarius fragilis* var. *rubidus* Hesler and A. H. Sm. 1979.

IDENTIFICATION

- Smallish, rusty-brown to burnt-orange mushrooms with centrally, very wrinkled caps (∅ 4–12 cm) that are never sticky.
- Straight attached (adnate) to decurrent gills, light pale orange, which can darken in age to match cap color; white to pale yellow spores.
- Fragile stem, same color as cap, smooth outside, stuffed when young and often hollow with age.
- Watery white, wheylike to clear milk that can be absent or scant.
- Flesh thin, brittle, and unchanging, pale orange-brown.
- Fruits (late) fall in Douglas-fir–hemlock forests, on or around late-stage decaying wood and often with extra leaf litter (e.g., alder) in the mix.

LOOK-ALIKES: Similar milkcaps are legion. Very common but with opaque white milk are *Lactarius luculentus* and several other more or less harmless milkcap species, some with spicy or bitter flavors. They can grow interspersed with Candy Caps. Keep checking for that watery milk marking your prize. For a newbie collector, the Deadly Skullcap (*Galerina marginata*) is a possible look-alike. Keep in mind milk, very brittle tissue, white spores, and hollow stems: Galerinas share none of these distinctive milkcap features and Deadly Skullcap grows on wood.

VELVETY MILKCAP
Lactarius fallax

So far, the striking Velvety Milkcap has been appreciated as an edible by only a few initiated foragers. Its Eurasian cousin, Chocolate Milkcap (*Lactarius lignyotus*), has been valued as a culinary mushroom for ages. Alas, it is found infrequently. The nearly black cap and scant fruiting hide Velvety Milkcaps from detection. The challenge of finding them contrasts with how easy they are to identify. The beauty of the dry, dark velvety cap matched by the stem surface and contrasting with the whitish gills and white milk is distinctive. There are two reported varieties, one with dark marginate gills (the gill margins are as dark as the cap, though the gills are white). *L. fallax* offers crunchy texture and a good fungal flavor. According to German mushroom expert Wolfgang Bachmeier, who maintains the awesome website 123pilzsuche. de, Velvety Milkcap can be enjoyed raw.

IDENTIFICATION

- Evenly dark brown to nearly black, dry and velvety cap (∅ 3–9 cm), often wrinkled toward the center.
- Straight attached (adnate) to decurrent crowded gills, white to cream colored, producing a stunning contrast when a few white gills are running down the dark top of stem; gills staining reddish when injured; yellowish spores.

- Fragile stem, colored like cap or a bit lighter brown, dry and slightly velvety.
- White milk stains wounded tissue dull reddish.
- Rather thin, brittle flesh, staining dull reddish; flavor mild to slightly peppery.
- Fruits in fall in conifer forests, often with true fir and often on or around late-stage decaying wood.
- Occurs from Alaska down to Northern California and east into the Rockies.

LOOK-ALIKES: Several other milkcaps have dark brown caps that are not dry, but slimy and sticky, such as the peppery Slimy Milkcaps, *Lactarius mucidus* and *L. pseudomucidus*.

NAME AND TAXONOMY: *Fallax* Latin for "misleading, deceptive" (probably regarding taxonomy). *Lactarius fallax* A. H. Sm. and Hesler 1962.

The Amanitas

Fascinating, famous, and fittingly feared fungi are found in the genus *Amanita*. In general, we advise people to stay well away, since most mushroom deaths in North America and Europe are caused by several *Amanita* species with evocative names such as Death Cap, Death Angel, and Destroying Angel. There are other Amanitas that might not kill you but will make you feel miserable, like the Western Brown Panther (*A. pantherinoides*) and Fly Agaric (*A. muscaria*), the most famous and emblematic toadstool. The red-capped, white-flecked *muscaria*, a Germanic good luck symbol, shape-shifts to show up in fairy tales, a multitude of cherished kitsch articles, and even as a sidekick to the hero in a classic video game.

Velvety Milkcap, Lactarius fallax

FEASTING IN FEAR ON AMANITAS

The Amanitas include some very tasty, tempting, truly choice edibles, the most treasured of which, the European Caesar Mushroom, has close relatives in California but is rare in the Pacific Northwest. And then there are the grisettes, edible and fairly clearly identifiable Amanitas with several species in our region. Recently David Arora has calculated *Amanita augusta* to be edible as well. But because of the highly poisonous Amanitas, ingesting *any Amanita* should be left to expert mushroom hunters; it is never worth putting your own life or health—and especially your friends' and family's lives—on the line. When you are first starting out as a forager, never eat an *Amanita* based solely on your own ID. Find someone who has survived foraging for Amanitas to initiate you. In any case, the fear generated by this genus will likely overpower the enjoyment unless you can be 100% sure of your mushroom. Interestingly, some toxic Amanitas, like the Fly Agaric, can be rendered edible. Others, like the Destroying Angel, can never be rendered safe.

Amanitas are white-spored, white-gilled ectomycorrhizal mushrooms recognized by their universal veil, a membrane that envelops the whole mushroom when young but rips into pieces when the mushroom's stem stretches and opens its cap. The universal veil might leave behind many small spots on the cap or a single big cap patch that can easily detach. The lower half of the universal veil might leave a clear membranous or felty cup or obscure pieces of tissue at the base of the stem where the universal veil attached. Volvas are rare in the mushroom world; best known are the pink-spored *Volvariella* species, like the edible Silky Rosegill, *V. bombycina*, and the Paddy Straw Mushroom, *V. volvacea*, commonly used in Thai cuisine and sometimes popping up in Pacific Northwest greenhouses, rich soils, or compost piles.

Many (but not all) Amanitas also have a partial veil that leaves a ring on the stem—either a membranous ring (Fly Agaric and Death Cap) or a flocculose ring (*Amanita smithiana*). Grisettes as a group have no ring.

 WESTERN GRISETTES
Amanita pachycolea group and *A. constricta*

We have good edible Amanitas in the Pacific Northwest, but the enjoyment of Western Grisettes is curtailed not only by all the pitch-black sheep in the *Amanita* "family" but by taxonomic challenges within the grisettes. Luckily, the latter can be viewed as an academic problem. But no one should approach the culinary enjoyment of any *Amanita* without sound knowledge of the genus. There are plenty of expert mushroom hunters who refuse to cross this line.

Grisettes are often tall, graceful mushrooms with a clear volva in a sac shape that is often found hidden in the soil or duff. The shape of the volva is a key element in distinguishing among our Northwest grisette

Basket of Fly Agarics, Amanita muscaria *group*

species. As the French name indicates, many are grayish mushrooms, or rather in the Pacific Northwest, grayish-brown.

Grisettes have a ribbed margin on the cap, first visible as fine lines on the cap's edge that deepen into grooves aligned in a radiating fashion. And they have naked stems and no ring. Ringed mushrooms can lose their rings, for example, when the cap opens and the ring does not detach properly from the cap edge or gets pulled off the stipe entirely. However, the grisette's partial veil is integrated in the surface of its stem, and when the stem grows, it rips the "veil," leaving fine scales on the stem.

The exact species identity is still frustratingly unclear. The late University of Washington mycology professor Daniel Stuntz first described *Amanita pachycolea*, most frequently a tall and impressive mushroom, in 1982, but there seem to be several species going by that name. It is still uncertain which species occur in the Pacific Northwest. For now, the grisettes are treated as a group until we know more.

IDENTIFICATION

- Cap (Ø 8–25 cm) dark brown when young, brown to grayish-brown or paler when mature, often with a darker halolike ring (retained as a broad umbo) on the inside of crisp marginal radial lines, a.k.a. striations, that deepen into grooves as the cap opens.
- Cap texture smooth, sticky when wet, usually with one or more white to pale gray felty patches of universal veil, lost easily to rains or handling.
- Gills close, finely attached at first, then becoming free; white to whitish-gray with dark brown edges when young.
- Tall (up to 20 cm), slender stem without a bulbous base, light gray, covered with fine gray-brown to brown scales on a pale background; stuffed core that becomes hollow with age.
- No ring on the stem.
- Whitish volva (often with rusty stains), fragile, felty, and saclike; connected to

stem only at base, sometimes with an hourglass shape.
- Soft, white flesh; mild flavor.
- Grows on ground alone, scattered, or in small groups associated with conifers, especially spruce, but also with hardwoods.
- Fruits in fall all over the Pacific Northwest and beyond.

LOOK-ALIKES: Grisettes can be confused with a wide range of brown-capped mushrooms. Anyone hunting them must know how to confidently recognize Amanitas, especially the toxic ones. The probably toxic gray-veiled Amanita, *A. porphyria*, has a definite ring; a grayish marbled stem with a basal bulb; and a cap, often with a purplish cast, that features gray or whitish wartlike remnants of a universal veil. Harmless *A. augusta* has a brown cap, bulbous base, and yellow universal and partial veils that fade in age to gray. The greenish-, or white-capped Death

TOP: *Western Grisettes:* A. constricta; **MIDDLE:** *A budding Grisette;* **BOTTOM:** Amanita pachycolea *group*

MORE ABOUT GRISETTES

Several other Amanitas in the subgenus *Vaginatae* are all edible, lack rings, and have striate cap margins and pronounced volvas. Fairly easy to tell apart is the very similar and common Constricted Amanita, *Amanita constricta*; its distinctive volva has a gray interior and is tightly wound around the lower stem base and flares up widely.

Our region features at least a half dozen other grisette species, whose taxonomy is still evolving thanks to genetic work. Alaska Grisette, *A. alaskensis*, is found from Washington northward. Farther south in the subgenus *Vaginatae* is the stocky, whitish Protected Grisette, *A. protecta*, which grows under willow and oak and features a pale gray cap with warty patches. Its striking similarity to deadly Amanitas should kill any thought foragers may have of consuming this *Amanita*! Many mushroom enthusiasts would like foragers to apply this same extremely cautious approach to all members of the *Amanita* genus.

Another southern grisette-like species that radiates into southern Oregon and grows in spring with oaks is the apricot to salmon to tan Springtime Amanita, *A. velosa*, a midsize mushroom without a ring that has a sticky-when-moist cap with a grooved edge and cottony white patch or large warts. The gills are white to dull pinkish when old and free to adnexed. The volva is saclike. It has thick, white flesh and tastes sweet and nutty—one of the most delicious mushrooms I have ever eaten. But make absolutely sure it is not the deadly look-alike found at the same time in the same habitat with oak, the Western Destroying Angel, *A. ocreata*, named for the last words of a mushroom fool: "Here I come a bit early, oh, Creator!"

Springtime Amanita, Amanita velosa

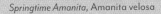

Caps (*A. phalloides*) are color-shifters, sometimes their caps have a yellow or brownish hue. However, *phalloides* has a ring (which can fall off!), no striations on the cap margin, and bulbous stem base. You really need to know what you are doing before you feed on grisettes, so that they are not your final meal!

NAME AND TAXONOMY: *Pachy-colea*, both Greek meaning "thick," as in pachyderm, and sheath, from *koleos* "sword sheath;" same meaning as Latin *vagina*, as in *A. vaginata*. *Amanita pachy-colea* D. E. Stuntz 1982. Amanitaceae, Agaricales.

⚠️ WESTERN YELLOW-VEIL

Amanita augusta

This majestic *Amanita* mirrors many of the characteristics of the Fly Agaric, but in a gorgeous yellow-brown color combination versus the loud red-white combo. This mushroom's beauty does not mirror its culinary quality. Unfortunately, its texture and flavor are dull. Still, it is a fairly easy-to-identify *Amanita* that is nontoxic to most. The edibility has only been recently figured out by David Arora, so we do not know how most people's GI tracts will react to the Western Yellow-Veil. The yellow ring on the stem, universal veil that forms yellow rings on the stem base, and yellow warts on the cap set this mushroom apart. With time, the veils lose their bright yellow shine, fading to gray, making it a bit trickier to correctly identify this mushroom.

IDENTIFICATION

- Smooth dark brown to brown to yellow-brown cap (Ø 3–15 cm), typically paler toward the margin, sometimes aging to yellow to grayish-yellow; occasionally ribbed near the cap margin.
- Powdery universal veil, yellow when fresh, aging to gray, breaking into concentric yellowish to grayish-white warts on cap.
- Narrowly attached to free, crowded gills, white to yellowish near the cap margin.

- Partial veil (ring) is yellow or, when white, at least with yellow patches on the tissue edge.
- The stem when young is yellow, fading to whitish, stuffed, widening to a sometimes bulbous base; stem above ring often marked by long vertical lines.
- Flesh is soft white or yellow tinged, mild tasting.
- Grows solitarily or scattered on soil in conifer and mixed forests.
- Fruits in summer and especially fall; common and widespread, especially in coastal regions from Alaska to Northern California.

LOOK-ALIKES: Darkish versions of the toxic Western Brown Panther, *Amanita*

Western Yellow-Veil, Amanita augusta

pantherinoides, with its brown warty cap, can be quite similar, but the Brown Panther usually lacks yellow tones; if there are any, they are only faint, pale yellow notes. Consuming a "Western Yellow-Veil" without clear yellow would be self-destructive. Yellowish-capped *Amanita gemmata* group have white partial and universal veils and no brown tones in their caps. Very similar can be the toxic, very rare, brown-capped Royal Fly Agaric, *Amanita regalis*, a close relative of *A. muscaria*, which can also have a pale yellow universal veil but is distributed in Alaska around tree line.

Coccora, Amanita calyptroderma

NAME AND TAXONOMY: *Augusta* Latin means "majestic, great, or venerable." *Amanita augusta* Bojantchev and R. M. Davis 2013; previously misapplied *A. franchetii* (Boud.) Fayod 1889.

COCCORA
Amanita calyptroderma

One of California's most appreciated edible Amanitas, the Coccora stoked the early interest of Italian Americans, since it is similar to the famous Caesar Mushroom of the Old World. However, many foragers keep their safe distance from these impressive Amanitas, a healthy practice for any careful person and anyone who has not spent a good deal of time studying

a genus that encompasses good edibles and devastatingly toxic members. The best practice is to study this mushroom carefully before eating it, pick it several times to familiarize yourself with its characteristics, and ideally have an expert help with identification. While the yellowish version is a bright yellow—quite distinct from the pale green-yellow of the Death Cap—do not attempt to eat very pale or brownish versions of Coccora unless you are extremely familiar with this mushroom. There is no room for error!

Mainly a fungal resident of California, *Amanita calyptroderma* is quite common in southwestern Oregon, and its distribution does extend into Western Washington, but only in rare sites in the Puget Sound area and in Victoria, BC. However, a warming climate might benefit this species, building up its presence in the Pacific Northwest.

IDENTIFICATION

- Caps (Ø 8–25 cm) are yellow, orange-brown to (dark) brown; smooth, but sticky when wet; often with yellow striate margin.
- Cap covered (at least in center) by large, thick, cottony white patch, a remnant of the universal veil that envelops the young mushroom completely.
- Adnexed to free gills are white to cream and close; white spores.
- Creamy to pale yellow (never white) nonbulbous stem, stains brownish, hollow, with center stuffed with cottony pith or clear gel; smooth surface but often etched with vertical lines above ring and powdery below.
- Fragile, felty, skirtlike ring colored like stem (7–25 cm long, 2–4 cm wide), fading to white in age from partial veil.
- Thick flesh, white to creamy-white, with narrow yellow band under cap surface.
- Mild flavor and odor, fishy in age.
- Grows scattered to gregarious in mixed forests with madrone, tanoak, and oak, but also with Douglas-fir.
- Fruits in fall in Pacific Northwest (mostly western Oregon), common in fall and winter in California.

LOOK-ALIKES: The Spring Coccora, *Amanita vernicoccora*, is paler than *A. calyptroderma* and fruits with oak in spring, just as the deadly and very similar *A. ocreata* does. Coccora is made distinct by the orange cap and orange flesh under the cap skin, but the cap color can vary, putting it close to the yellowish-green to brownish colors of the Death Cap, *A. phalloides*. The Coccora has pronounced striations, or furrows, on the cap edge; the Death Cap shows these only faintly. Also, the tissue of the universal veil of the Coccora is much thicker: it tends to form a single large central patch that peels off easily. The Death Cap's universal veil is thinner and holds to the cap surface. The Coccora's stem is hollow, the Death Cap's usually solid. Never use only one of these criteria to make the call—feast and funeral are simply too close.

NAME AND TAXONOMY: *Calyptra-derma* from Greek *kalyptra* "cover or veil" and *derma* "skin," a reference to the thick universal veil covering the entire mushroom. *Amanita calyptroderma* G. F. Atk. and Ballen 1909 = *A. lanei* (Murrill) Sacc. and Trotter 1925 = *A. calyptrata* Peck 1900.

‼ BLUSHING BRIDE
Amanita novinupta

The Blushing Bride is a shy beauty that usually will spring from fertile forest floors in April. *Amanita novinupta* is not much seen north of California, where she is usually engaged with Garry (and other oaks), but there are persistent rumors of an occasional liaison with Douglas-fir. Though first mistaken for the Blusher (*A. rubescens*), native to eastern North America and Europe, and recently found with European oaks and birches in Southwest BC, this mix-up gave many people the idea that the Blushing Bride is a good edible. However, eating the Blushing Bride raw will break your guts; the toxins must be cooked out.

A key feature in this group of Amanitas is the blushing or, less poetically, the red staining of its white flesh, be it gills or stem. I grew up picking the Blusher in the Alps, where it is very common, and the red staining enables a safe ID, separating it from the very common, and very similar, toxic Panthercap (*A. pantherina*). The blushing really helps to delineate it from poisonous and lethal Amanitas, especially Western Destroying Angel, *A. ocreata*, which also fruits in spring. Most foragers rightly play it safe and keep their hands off; you really need to know the *Amanita* clan very well if you want to get involved with the Blushing Bride!

IDENTIFICATION

- Cap (∅ 3–15 cm) at first hemispheric then opens to convex and flat; white with a rosy tint beneath the powdery surface and also covered in white, cottony warts that become finer toward the margin, which lacks striations. With age the chalk washes off, revealing a smoother (possibly cracking) cap that reddens and might turn brownish.

WAS I FED A DEATH CAP?

I was first offered *Amanita calyptroderma* at a Sonoma County Mycological Association foray at Salt Point State Park, and I was hesitant to cross the *Amanita* moat again, which I had done many times enjoying the Blusher (*A. rubescens*) in the old country. After a decade of mushroom studies as part of Seattle's Puget Sound Mycological Society, I'd been reconditioned to shun any *Amanita*. However, having run into Himalayan Caesars (*A. hemibapha*) during ethnomycological research in Tibet and thinking all these SOMA expert mushroom hunters must know what they were doing, I joined in trying Coccora. It was sliced and then stewed in fresh lemon juice, without involving any heat—just raw slices, with the citric acid doing the cooking, as raw fish is processed for ceviche. The dish did not trigger culinary bliss, but I thought, *Wow, it's interesting how the fishy Coccora flavor is perfect for this fungal ceviche.*

Well, the next morning I felt nauseous, had the runs and belly cramps, and got really worried. The timing was perfect for amatoxins, which usually kick in eight to twelve hours after ingestion. Was I fed a Death Cap instead? I did not really know the people who fed me the *Amanita*. Should I go to a hospital? I was in Berkeley, staying with dear but fungophobic friends who helpfully chimed in, "That's why we don't eat any mushrooms but the ones bought in the store." My rational mind told me, *This can't be the* Amanita *I ate*, but my fear was palpable, as was the cold sweat I produced. I called my wife, Heidi, who suggested I take homeopathic Arsenicum C200. "If that does not help," she said, "you better head to an emergency room."

The thought of the cost of medical treatment in the United States ratcheted my fear up another notch—not a worry in most other wealthy countries. I asked my friend to take me to the drugstore to get the remedy. When I was getting out of the car, he said, "You probably just have the same stomach flu we all had three days ago." And right he was!

- White, close to crowded gills are narrowly adnate to free, with short decurrent line on stem.
- Powdery, often with scales and cracking white stipe (2–18 cm tall x 1–5 cm wide).
- Membranous partial veil forming skirt-like ring, usually with reddish tints.
- Volva at stem base is often not well defined, but rings or patches atop the enlarged base are common.
- White flesh stains red in the whole mushroom.
- Normal fungal odor and mild flavor.
- Grows singly or in small groups in soil, mostly with oaks.
- Fruits mostly in spring; not common from southern British Columbia into Northern California.

LOOK-ALIKES: The very similar—also red-blushing—Bloody Ghost Amanita with the still unpublished name *A. cruentilemurum* is apparently hiding out under the bridal veil. All kinds of very dangerous *Amanita* species could be mistaken for the Blushing Bride; however, they do not all blush red in our region. Most dangerous is the also spring-fruiting Western Destroying Angel, *Amanita ocreata*, and white forms of the Death Cap, *Amanita phalloides*. Both of these lethal toxic species have well-developed, saclike volvas. Similar non-Amanitas that also have red-staining flesh are shaggy parasols (*Chlorophyllum* spp.) and the Prince (*Agaricus augustus*). The latter is very scaly and has chocolate-brown spores in addition to a strong almond fragrance.

NAME AND TAXONOMY: *Novi-nupta* means "newly-wed," and *rub-escens* "turning red," both Latin. *Amanita novinupta* Tulloss and Lindgr. 1994. *Amanita rubescens* Pers.: Fr. 1797.

Blushing Bride, Amanita novinupta

FLY AGARIC
Amanita muscaria group

Who doesn't know this iconic red mushroom with the cute white polka dots? Fly Agaric is the most storied of all mushrooms, with whole books written just about this emblematic toadstool with a long history of human use—universally known to be "poisonous," mistakenly declared deadly in reasonable amounts, used by Siberian and Finnish Sami shamans as entheogens to visit spirit realms, and sacred to the one-eyed Odin, a.k.a. Wotan, the Germanic god associated with wisdom, healing, sorcery, and frenzy.

KITCHEN AMANITA ALCHEMY

Years ago, an elderly Dutch friend told me that as a child she watched her grandma slicing Fly Agaric and frying it with other mixed mushrooms. I was in complete disbelief. Why would anyone risk poisoning? Researching *Amanita muscaria* edibility, I learned that by the end of World War II, to avoid starvation, Germans were eating *Fliegenpilz* by the ton, first peeling off the red cap skin, which allegedly contains most of the toxins. In 2008, I came across *A. muscaria* edibility research[49] by Wilhelm Rubel and David Arora when we all had articles in *Economic Botany*. Their research contained detailed instructions based on ten years of experimenting with how to render this toxic toadstool edible. The mushrooms are parboiled, then drained of all moisture and cooked like any other mushroom. Rubel and Arora's approach surely takes a bit longer than just peeling the cap before hard frying, as it might have been done in the old country—though I am not sure if the parboiling step is in fact missing from my friend's eyewitness account.

I always parboil my Fly Agaric, and I have been fine with one round. Some people parboil and drain twice. Parboiling seems crucial, since the troublesome compounds in Fly Agaric, muscimol and ibotenic acid, are not thermolabile but water soluble and therefore should be leached out. Not exactly advanced culinary alchemy, but ingesting *A. muscaria* that has not been parboiled will make you really sick.

TOP: *Thinly sliced Fly Agaric and the poisonous, orange-red water of the decoction after the first round of parboiling;* BOTTOM: *Fried, detoxified Fly Agaric*

POISONING BY *AMANITA MUSCARIA*

Poisoning by *Amanita muscaria* can be a very unpleasant experience. The toxic compounds ibotenic acid and the alkaloid muscimol profoundly upset the digestive system and induce hasty emptying of said system by both ends, sometimes simultaneously! I know this from experience. Young and curious, my twin brother, a friend, and I went for the ride.

We had grown up adoring the beauty of *Fliegenpilz*, German for the "fly mushroom," which our fungophile father called poisonous. He had no idea that *Fliegenpilz* was also known as *Narrenpilz*, the "fool's mushroom," for the intoxication it produces (and that is reportedly appreciated by lone livestock herders out of booze in the remote Bavarian Forest[50]). We encountered stories of its mind-altering powers in our late teens. Because we were curious but disinclined to hurt ourselves, we were inspired to visit a university library for the first time, where we found information on dose (one big and two small dried caps) and confirmation that the intoxication causes no serious organ damage and does not kill.

I experienced how it suppressed pain (as claimed in historical reports that the infamous Berserkers used it in battle) when I ran into a low wooden beam in an old farmhouse, hitting my forehead hard before falling and hitting the back of my head on the wooden floor. The next moment, I was standing on the balcony feeling euphoric. Later I recall sitting in a cow pasture and feeling terribly nauseous. Confronted with this incredible discomfort, I lay down and my mind decided to leave my body, thinking, *I am out of here*. I saw my body lying on the pasture from ten feet above, which was quite mind-blowing! Unfortunately, my Lutheran upbringing had not equipped me to make best use of an out-of-body experience. Instead of venturing into astral travel, I lived out the attachment to my human body, watching myself from above. Back in my body, that miserable stomach-wrenching feeling returned, but I was clearly past its peak.

Meanwhile, our friend hallucinated that he was eating his family, himself, and then the whole world in an endless loop; he said he felt terrible about it before falling into a deep sleep, typical, we learned, for a high dose. Unsurprisingly, none of us ever repeated this experience. At least I learned that mind and body are two units that are joined for this lifetime, parting ways at death, or as a Korean Zen master once said in broken but crystal-clear English, "Your body, rent-a-car, you drop off at airport!"

Fly Agaric, Amanita muscaria var. flavivolvata

His lost eye is visible in the freshly opening young mushroom.

Amanita muscaria supposedly inspired the modern-day Santa Claus dressed in red and white and relying on flying reindeer for transportation, the latter reputedly indulging in Fly Agaric themselves. Named for its power to kill flies, it was traditionally placed in milk so that flies would drink from the potion, pass out, and drown in the puddle. Last but not least, the Fly Agaric can be rendered into a good edible in times of crisis or curiosity by kitchen alchemists. But make sure to check for the subtle striation lines on the cap edge, the skirtlike white ring, and the arrangement of several concentric rings (the volva) above the bulbous stem base.

In the Pacific Northwest, there are several versions of Fly Agaric with caps ranging from scarlet-red in color to orange, yellow, and even white and brown forms. Even if the cap color is off, they can be recognized by several concentric rings or patches of creamy to yellowish tissue around the stem above the basal bulb. For kitchen alchemy, it is advised to work with the classic red- or reddish-orange-capped versions. So far the Fly Agaric most commonly encountered in western North America is referred to as *Amanita muscaria* var. *flavivolvata*. The variation name *flavivolvata* describes its yellow universal veil (or at least yellow traces on the upper stem base), yellow outer fringe of the ringlike partial veil, and yellowish polka dots. However, recent DNA work suggests that our Fly Agaric may need to be reclassified as a distinct species: its DNA matches that of *A. chrysoblema*, described by G. F. Atkinson in 1918, that was a North American species regarded, until now, as a white Fly Agaric; but according to the DNA work, it includes color variants ranging from white, yellow, and orange to red. The Eurasian *A. muscaria* has all-white

Fly Agaric, Amanita muscaria var. flavivolvata

universal and partial veils and tends to have a more solid red cap. It was introduced with its associated plant hosts to more southern regions in the Pacific Northwest but is native to Alaska.

- Caps (Ø 5–30 cm) bright blood-red or orange-red, often fading to orange; smooth, sticky when moist, with a slightly striated margin; covered in whitish to subtly yellow pyramidal warts that flatten with age and easily wash off.
- Gills free to narrowly adnate, somewhat close, leaving very fine vertical lines on top of stem, producing white spores.
- Tall, white stem with felty ring and several scaly concentric rings where the universal veil attaches above the bulbous base.
- Grows solitarily or in groups in soil with conifers, especially pine and spruce, and with hardwoods such as birch, oak, and madrone.
- Fruits in fall; widely distributed in the Pacific Northwest and common when its host trees abound.

LOOK-ALIKES: Alone in its class! There is no other orange-red mushroom with whitish warts in fall. However, yellow versions can be similar to the *Amanita gemmata* group and to *A. pantherinoides*, and orange versions to the spring-fruiting, stocky Sunshine Amanita, *A. aprica*, whose frosty universal veil is firmly attached to the cap. After losing their cap warts, white forms are similar to Death Cap (*A. phalloides*) and Western Destroying Angel (*A. ocreata*); both of these highly poisonous species have well-developed, saclike volvas, which *A. muscaria* lacks. The white *A. smithiana* lacks a proper ring. Juvenile Fly Agarics like most other Amanitas still enveloped in their universal veil I call "bluffballs," since they can be easily mistaken for puffballs when not cut in half.

NAME AND TAXONOMY: *Muscaria* from Latin *musca* "the fly." *Amanita muscaria* (L.) Hook 1797, *Amanita muscaria* var. *flavivolvata* (Singer) Dav. T. Jenkins 1977.

⚠ WESTERN BROWN AND WESTERN YELLOW PANTHERS
Amanita pantherinoides and *A. gemmata* group

Quite similar to the Fly Agaric in most ways, including toxicity, are the Western Brown and Yellow Panthers. Panthers have a rimmed stem base, a skirtlike ring, and more ragged tissue. Their cap color noticeably differs from Fly Agarics, ranging from warm dark and light brown to beige (typical for *Amanita pantherinoides*) and yellow to pale yellow for what we call "*A. gemmata*." DNA research reveals that our Brown Panther is not *A. pantherina*, a misapplied European name, but is most likely *A. pantherinoides*. The yellowish Amanitas have collectively been known as *Amanita gemmata*—another misapplied European name.

Panthers have a range of host trees and are common in both lowland and mountain conifer forests. *A. pantherinoides* (cap Ø 5–10 cm) fruits in both spring and fall and is especially fond of Douglas-fir, which anchors it to backyards and parks, where curious dogs happen to eat the Brown Panther and get badly poisoned, puppies or old dogs sometimes dying from it. Luckily, there are no known reports of deaths in people, and I have encountered anecdotal evidence that *A. gemmata* group and *A. pantherinoides* can be detoxified by parboiling: a small trial of *A. pantherinoides* did not make me sick.

NAME AND TAXONOMY: *Pantherin-oides* Greek "panther, spotted like a panther," *oides*-like—*A. pantherina. Gemmata* Latin "jeweled," for spots on cap. *Amanita pantherinoides* Murrill 1917, *A. gemmata* (Fr.) Bertill 1866. We don't have definitive names yet, so we still use *A. gemmata* group. There are several species involved. Some of them have small (cap Ø 3-8 cm); others have mid-size.

The Deadly Amanitas, Boogeymen of the Fungal Kingdom

There are other deadly mushrooms, but several of the *Amanita* reign supreme as the deadliest known mushrooms, so they take some special focus, especially if you intend to eat any gilled mushrooms that are even remotely similar.

☠ DEATH CAP
Amanita phalloides

The Death Cap is the head boogeyman of the fungal kingdom—and rightly so—given its lethal amatoxins! Originally from Eurasia, *Amanita phalloides* has now been

Western Brown Panther, Amanita pantherinoides

spread by humans to all continents in regions outside of the tropics. While it is much maligned, it is also the driving force behind mushroom education—a very good thing! The Death Cap was inadvertently introduced with European nursery trees, such as oaks, hornbeams, edible chestnuts, beech, and hazels. In the Pacific Northwest, it is still mostly confined to locations around human settlements. It is now notoriously common on the roots of street trees in Vancouver and Victoria, British Columbia, thus endangering kids, pets, and careless foragers.

The Death Cap has also been observed growing with native Garry oak (*Quercus garryana*),[51] but no "urban flight" has been observed yet. In California, it jumped to native oaks a while ago and is successfully invading the hinterland. Death Cap is now omnipresent in many busy California state parks, and unfortunately, now foragers get poisoned and even killed! Especially affected are immigrants from Southeast Asia, where several edible white Amanitas are widely beloved, one of them *Amanita princeps*[52]—another reminder that edible mushroom knowledge should often only be applied locally!

The cap of *A. phalloides* is usually yellow-green but also can be white or brown. The mushroom is better recognized by its deep, white, saclike volva, which must be carefully extracted to be seen. Always collect the whole stem when identifying unknown mushrooms. The Death Cap also usually has a bulbous stem base, a cap edge that lacks grooves or striations, and a partial veil that leaves a white ring on the stem. Unfortunately, the ring can fall off, and the volva can be lost during extraction or becomes less obvious when it dries out. These quirks are reasons why nonsuicidal mushroom hunters are scrupulously vigilant when collecting anything that looks like the Death Cap and other deadly Amanitas for the table.

Death Caps, Amanita phalloides

AMATOXIN POISONING—SYMPTOMS AND TREATMENT

Cooking does not destroy these pesky peptides. The first symptoms show up usually eight to twenty-four hours after consumption: vomiting, watery diarrhea, and abdominal cramps for a day. Then comes the strangely named "honeymoon phase" for another day, when people start feeling better and mistakenly believe they are recovering. Meanwhile, their liver is liquefying; amatoxins messing with messenger RNA cause the breakdown of cell membranes, and the overchallenged kidney soon fails. Untreated poisonings most often culminate in death. Contact the Poison Control hotline right away (1-800-222-1222)! The earlier treatment begins, the higher the chance of survival and recovery; toughing this one out is a certain-death strategy.

The most promising therapy as a potential antidote to amatoxin poisoning is injection of silibinin, a compound found in milk thistle, but this treatment is still being developed. Amatoxins are not limited to *Amanita*; they are also found in several skullcaps (such as Deadly Skullcap, *Galerina marginata*), in Ringed Conecap (*Pholiotina rugosa*) and in some little parasol mushrooms (including *Lepiota subincarnata*). Interestingly, recent genetic studies have shown that amatoxins in these two non-*Amanita* genera have not evolved in their evolutionary branches but were acquired by horizontal gene-exchange.[53]

IDENTIFICATION

- Classic, medium to big "cap and stem" mushroom.
- Smooth yellow-green cap (Ø 4–16 cm) that varies in color: yellow, yellowish-green, green to olive (but also white or brownish), usually darker in center, sometimes paler or browning with age.
- Cap has thin white patch (when not washed off by the rains) and neither warts on top nor clear striations on margin.
- Gills free, close, moderately broad, white becoming cream; producing white spores.
- Stem white or cap colored, normally solid, not hollow, with pendulous skirtlike ring on upper stem.

- White universal veil enveloping young mushroom and when grown leaving saclike volva around stem base and thin patch on cap.
- Odor mild when young, pungent when old (like raw potatoes or chlorine); mild flavor, but tasting is not recommended!
- Grows alone or in small groups from soil associated with hardwoods introduced from Europe (oaks, hornbeam, edible chestnuts, beech, and hazels) and possibly the Northwest's endemic Garry oak.
- Fruits mostly in fall, with northern distribution up into towns around the Salish Sea and Vancouver, BC.

LOOK-ALIKES: Foragers put their lives on the line confusing the following mushrooms with Death Caps: green Russulas, several

Agaricus spp., *Volvariella* spp., and especially edible Amanitas, such as grisettes and Coccora, as well as Fly Agaric.

NAME AND TAXONOMY: *Phalloides* Greek "like a penis," in reference to young, unopened fruiting bodies. *Amanita phalloides* (Fr.) Link 1833.

☠ SMITH'S AMANITA (FLOCCOSE AMANITA)
Amanita smithiana

Behind the harmless-sounding name Smith's Amanita and the cute white flakes hides an extremely toxic *Amanita*, the bane of novice and undereducated matsutake hunters. Unfortunately, again and again people pick this stately mushroom that can grow right in the middle of a patch of lovely matsutake. When harvesting and preparing it, the flavor will not give *A. smithiana* away, so you may overlook it mixed in with your other pine mushrooms, but its toxins (allenic norleucine and chlorocrotylglycine) will cause nausea, vomiting, diarrhea, and abdominal pain four to ten hours after ingestion. If you suspect you are experiencing these symptoms due to ingesting this species, visit a hospital with dialysis capacity immediately to save your kidneys; several weeks of hemodialysis could save your life. It is safely differentiated from Western Matsutake by several characteristics. It does not have the spicy cinnamon aroma of the matsutake, but the deep-rooting, often spindle-shaped stem base will release a toxic stench of bleach when crunched or sliced open. Some uninitiated would-be matsutake hunter might take that chlorine odor for the fabled matsutake aroma; don't be fooled!

Also, the flesh of this *Amanita* is lightweight and feeble in comparison to the heavy and super-tough matsutake. You can squish Smith's between your fingers; unless infested by insects, a matsutake is far too firm to break. Michael Beug points out that when slicing a matsutake,

Smith's Amanita (Floccose Amanita), **Amanita smithiana**

Western Woodland Amanita, Amanita silvicola

the flesh will squeak, while no *Amanita* will make a sound. The ring on Smith's Amanita is indistinct, mere flaky remnants of its partial veil, whereas matsutake has a very pronounced clasping ring and also distinct rust-colored scales on the stem and cap. In addition, matsutake tends to be squat and robust, while Smith's Amanita is taller and often thinner. All Amanitas are gravitropic; within several hours, their stalk bends into a new upright position, a shape-shifting wonder that matsutake cannot perform.

IDENTIFICATION

- Tall, white, robust mushroom with cap (Ø 5–17 cm) covered in cottony white tissue and warts that dry to tan.
- Veil warts are flaky and wash away easily, revealing smooth cap with no striations on edge.
- The close, white, free to narrowly adnate gills end before reaching outer cap margin; white spores.
- Ragged or scaly stem with nonfelty, flaky ring remnants; spindle-shaped bulb with concentric rings above; often has long rooting stem extension, frequently left behind.
- Thick, white, nonstaining flesh.

- Odor mild to intense—of bleach or ammonia—especially in stem base.
- Grows solitarily or scattered in soil of conifer forests in or near rotten wood.
- Fruits in fall west of Cascades and from southern British Columbia into Northern California.

LOOK-ALIKES: The very similar Woodland Amanita (*Amanita silvicola*) is another matsutake look-alike, as is the oak-associated Western Destroying Angel, *A. ocreata*, which has a pronounced volva, lacks the flaky tissue, fruits in spring, and is rare in the Pacific Northwest. White forms of the Fly Agaric, *A. muscaria*, have a pronounced ring.

NAME AND TAXONOMY: *Smithiana* is named for mycologist Alexander Smith. *Amanita smithiana* Bas 1969.

 **WESTERN WOODLAND AMANITA**
Amanita silvicola

Another common white toxic *Amanita* in the Pacific Northwest is the smallish but stout Western Woodland Amanita, *A. silvicola*, which sports a clearly bulbous base and, when young, a cap (∅ 4–10 cm) covered with flattened cottony or fluffy-powdery patches that easily wash off. The stem is powdery or covered with cottony scales; the fragile cottony partial veil often clings to the cap margin and leaves a slight ring zone. Its white flesh can have a mild to soapy to fishy odor but not bleachy like *A. smithiana*. It grows alone or in small groups in soil in conifer forests, especially with Douglas-fir and western hemlock.

Shaggy Parasols, Dapperlings, and Allies

Shaggy parasols (*Chlorophyllum*), dapperlings (*Lepiota*), and other mushrooms presented here all have a clear ring (some disappearing quickly) mostly around a slender stem. They are all saprobic, terrestrial, and mostly white-spored mushrooms. The Green-spored Shaggy Parasol (*Chlorophyllum molybdites*) breaks the spore color mold as does the pale-yellow-brown-spored Alaskan Gold (*Phaeolepiota aurea*). While there are some very good edibles, there are also plenty of questionable and toxic members.

Shaggy parasols are very common, impressive mushrooms that often fruit in groups around human habitat and appear from spring into late fall. They have an outstanding rich, savory flavor. Shaggy parasols would be offered in restaurants and stores were it not for a few people who have an allergic intestinal reaction to them—such a shame!

There are two other challenges. Shaggy parasols share some features with toxic Amanitas: they have white gills; a tall, slender stem; a pronounced ring; and a bulbous stem base. But shaggy parasols can be safely told apart by their cap scales and especially by their orange-staining flesh. And then there is the Green-spored Shaggy Parasol (also known endearingly as "the Vomiter"), which causes many poisonings. While this green-spored species

Gray Shaggy parasols, Chlorophyllum olivieri

PROCESSING, COOKING, AND CULTIVATING SHAGGY PARASOLS

As I pick shaggy parasols, I separate out the stems and stack the fragile caps like plates. When abundance strikes, I freeze the fried caps or dry them. Powdered or finely sliced, dried stems add a good dose of umami to sauces or soups. Do not eat shaggy parasols raw or undercooked: a still-unknown toxin upsets digestive systems and could be a contributing factor to reports of "allergic reactions." Shaggy parasols are also heavy metal accumulators, so pick from clean sites and do not overindulge (see Heavy Metal Concentrators in Part One).

Don't discard the dirty stem base! Take it home, and insert the cutoff bulbous base as soon as possible in a shady spot with a lot of dead biomass, such as compost or yard waste. If the conditions are right, the mycelium will grow from the stem base into the biomass and hopefully fruit next spring or fall. This is as easy as cultivation gets and an effective way to introduce other saprobic mushrooms like blewits, Princes, and others. You could also try to introduce Fairy Ring Mushrooms by transplanting a piece of turf with mycelium into your lawn. Bringing home infected wood chip clusters in which Wine Caps or Wavy Caps are growing and spreading them in fresh hardwood chips may introduce them into your own yard.

is still extremely rare in the Pacific Northwest, and was absent here until recently, it is apparently slowly spreading, encouraged by hotter summers.

Originally *Chlorophyllum* (Greek for "green gills") was applied only to the Green-spored Shaggy Parasol. However, genetic work by Berkeley-based Dutch mycologist Else Vellinga revealed that there were also white-spored *Chlorophyllum* that had been part of *Macrolepiota*. Before the DNA revolution, mushrooms with differing spore colors were not lumped into the same genus.

The rich umami aroma of shaggy parasols shines best when frying them. It is one of my all-time favorites! As a kid, I loved mushroom hunting but did not really enjoy them much cooked, though I loved the sauces my mother made with them. Even so, I was in love with parasols. As a six-year-old, when finding them in the woods with a schoolmate's family, I insisted on collecting them. Back home in their kitchen, I showed them how to fry them up, and we all enjoyed the meal. Half a century later, I still love to fry the caps gill-side down in olive oil or butter until they are nicely caramelized. This way a lot of tasty juiciness is retained in the cap—a true mushroom steak. They are perfect served on toast laced with herb butter and topped with Parmesan or as the centerpiece of your burger. Sometimes I fry the top too, but only briefly, since it will burn quickly. Shaggy parasols can handle a generous dose of garlic, since they hold their ground well even when used in tomato sauce or lasagna. Some books suggest throwing out the stems, but I like their crunch.

TABLE 3. COMPARING SHAGGY PARASOLS

Chlorophyllum SPECIES	Gray Shaggy Parasol Chlorophyllum olivieri	Brown Shaggy Parasol Chlorophyllum brunneum	White Shaggy Parasol Chlorophyllum rhacodes
CAP	Ø 6–18 cm Reddish-brown center on white base, less contrasting, often just grayish to light brown	Ø 5–20 cm Big brown center on white to cream base	Ø 10–20 cm Brown center on white base, high contrast
SCALES	Small, red-brown scales around center, transitioning to big gray shaggies on a cream background	Few bigger, smooth brown central scales and white shaggies, sometimes no scales when young	Few bigger, smooth brown central scales and white shaggies
RING	Double edged	Single edged, brownish below	Double edged
STEM BASE	Abrupt onion or club shape, no shoulder, solitary growth	Abrupt-shouldered bulb with flattened top (especially when young, less with age), sometimes fused stems	Onion- or club-shaped, no shoulder, solitary
STEM DIAMETER	0.8–1.5 cm	2–3 cm	1–2.5 cm
STATURE	Slender appearance	Stem 1.5 times as long as cap diameter Thicker stem	Stout stem height close to cap diameter
PRESENCE	Introduced, common around coastal settlements	Native, widest distribution from southern British Columbia southward	Introduced, widespread around settlements
HABITAT	Often with conifers, in litter in gardens and parks	Forest paths, all types of litter, compost piles	Nutrient-rich meadows, gardens, compost piles

IDENTIFICATION

- Big mushrooms with scaly caps (Ø 8–20 cm); smooth dark brown skin of young buttons expands and breaks into shaggy scales around a brown center on white to cream background.
- Close white gills are two- or three-tiered, free from stem; bruising red or brown; white spores (or very pale yellow for Chlorophyllum olivieri!), never green spored.
- Slender, smooth white stem (10–20 cm tall, 1–2.5 cm across), hollow with stuffed center, widening to a swollen base; no volva!
- Thick fleshy ring, movable up and down on stem when older.

Gray Shaggy Parasol, Chlorophyllum olivieri

- Flesh—especially stem—stains orange to red when scratched or handled, darkening slowly to red and brown.
- Fruits spring into late fall, fall being the most productive season.
- Grows often in groups in nutrient-rich habitats in parks and gardens, edge of woods, around compost piles, along paths; widespread distribution.

LOOK-ALIKES AND POISONOUS PEERS: The edible true parasol, *Macrolepiota procera,* does not occur naturally in the Pacific Northwest, though we think it might have escaped from Paul Stamets's farm in Olympia and other cultivators. It is easily recognized by a snakeskin pattern on the stem, immense height, and cap diameters of 25 to 40 cm. Besides the Green-spored Shaggy Parasol, there are a range of somewhat similar-looking, parasol-like mushrooms, including the questionably edible *Leucoagricus leucothites* (Smooth Dapperling) and inedible, rare *L. americanus* (American Dapperling), which stains first yellow then red, has a spindle-shaped stem, and grows on wood chips.

Some of the Lepiotas are similar, but they all tend to be smaller. Since some of the small Lepiotas contain deadly amatoxins, safest for the beginning collector is to avoid any shaggy parasols below a cap diameter of 4 inches (10 cm). None of the look-alikes share the combination of features that make shaggy parasols unique: a scaly cap, orange-changing flesh, movable ring, and white spores. Some people confuse Shaggy Manes (see Gilled Mushrooms with Black Spores) when young with shaggy parasols—however, the cylindrical cap of the Shaggy Mane is distinctive from the round drumstick of not-yet-open shaggy parasol.

 GRAY SHAGGY PARASOL
Chlorophyllum olivieri

The Gray Shaggy Parasol is a precious European import distinguishable when mature by its big gray scales on a cream background and the brown, smooth center of the cap. Overall it has a tall, slender appearance, though it is the smallest of the Northwest's three edible shaggy parasols. Of the three, it is the most common one

Brown Shaggy Parasol, Chlorophyllum brunneum; *note the abrupt stem bulb on right.*

White Shaggy Parasol, Chlorophyllum rhacodes

ring, which appears brown from below, and has the most abrupt stem base. Stouter than the other shaggy parasols, it is suspected to cause more digestive trouble than they do.

NAME AND TAXONOMY: *Brunneum* Latin "brown." *Chlorophyllum brunneum* (Farl. and Burt) Vellinga 2002, *Macrolepiota rhacodes* var. *hortensis* (Pilat) Wasser 1980.

found around human settlements west of the Cascades.

NAME AND TAXONOMY: *Olivieri* after M. J. Olivier, collaborator of mycologist J. P. Barla. *Chlorophyllum olivieri* (Barla) Vellinga 2002, *Macrolepiota rhacodes* var. *rhacodes* (Vittad.) Singer. Agaricaceae, Agaricales.

BROWN SHAGGY PARASOL
Chlorophyllum brunneum

The Brown Shaggy Parasol is our native species. Note the big brown center on a white to cream base. It has a single-edged

WHITE SHAGGY PARASOL
Chlorophyllum rhacodes

The White is the biggest of the shaggy parasols! It also arrived with European settlers. It is distinguishable when mature by its high-contrast cap, featuring a dark brown center surrounded by white shaggies with brownish tips over a whitish background. Of the Northwest's three edible shaggy parasols, it is the most common one found around human settlements east of the Cascades.

NAME AND TAXONOMY: *Rhacodes* Greek "rags," a reference to the scales; the original

Green-spored Shaggy Parasol, Chlorophyllum molybdites

rachodes is from an Italian name of a skin disease, but spelling was recently changed to *rhacodes*. *Chlorophyllum rhacodes* Vellinga 2002, *Macrolepiota rhacodes* var. *bohemica* (Wichanský) Bellú and Lanzoni 1987.

 ## GREEN-SPORED SHAGGY PARASOL
Chlorophyllum molybdites

Also known as the "the Vomiter," the Green-spored Shaggy Parasol is very similar to the edible shaggy parasols. It used to be absent in the Pacific Northwest, since it thrives in hot summers. Only very recently, ignoring climate change deniers, it arrived up here. There have been sightings in Walla Walla, southwest Washington State, the southern Willamette Valley in Oregon, and strangely a single report from Seattle. It fruits mostly in lawns and can make monster fairy rings. It tends to have fewer scales than the edible parasols, and its cap is very white. However, every parasol pursuer should know about

this green sheep of the family and be on the lookout for it. In many years, this mushroom causes more poisonings in North America than any other species.

Within a couple of hours, *Chlorophyllum molybdites* produces severe GI poisoning, including bloody stools and other symptoms that can last for several days. It is especially wicked when eaten raw, but far too many people are poisoned even after it has been cooked thoroughly. And then there are others who dare to eat it after two parboils. If you forage in California, Hawaii, the Midwest, or Southeast, or anywhere else comparable on Earth with hot summer climes, always check your parasol for spore color!

The safest approach is to make a spore print before attempting to cook a shaggy parasol. If the spore print turns out greenish-gray, find a different protein source! If you encounter a suspect shaggy parasol still in its drumstick phase before it produces spores, do not eat it; though if you put a drumstick in a glass with some water, it might open overnight and allow a spore print.

Smooth Dapperling, Leucoagaricus leucothites

Fatal Dapperling, Lepiota subincarnata

NAME AND TAXONOMY: *Molybdites* Greek "pertaining to lead." *Chlorophyllum molybdites* (G. Mey.) Massee 1898.

SMOOTH DAPPERLING
Leucoagaricus leucothites

The dapper Smooth Dapperling is as close as you can get to eating a Death Cap and surviving just fine. It has no volva, so make sure to get the bottom of the stem to ID it correctly. This elegant saprobic mushroom grows in lawns, and in some years it can be quite common and even form fairy rings. In

general, it is rightly shunned as an edible due to its uncanny resemblance to Western Destroying Angels and other white deadly Amanitas. Also, there are reports that *Leucoagaricus leucothites* can cause allergic reactions with flu-like symptoms. It is reputed to take up pesticides and fertilizer, so be very careful where you pick it. In short, don't let this dapperling smooth-talk you into a romantic dinner!

NAME AND TAXONOMY: *Leucoagaricus leucothites* (Vittad.) Wasser 1977; formerly *L. naucinus* Singer. Agaricaceae, Agaricales.

FATAL DAPPERLING
Lepiota subincarnata

Though fragile and small (cap ∅ 1–3 cm), this attractive mushroom is deadly, due to its amatoxin load, a pesky peptide shared with the notorious Death Cap. The reddish to vinaceous skin of the cap breaks up into scales as the cap grows, and its flesh turns slightly red. It was formerly known as *Lepiota josserandii.* Somewhat similar is the edible Fairy Ring mushroom (*Marasmius oreades*) that also grows in grass, but has neither cap scales nor fleeting ring in contrast to the Fatal Dapperling. Luckily, there are very few similarly sized, edible look-alikes, but a few other small *Lepiota* species are also suspected to contain lethal amatoxins, though scientific research recently cleared our *Lepiota castanea* of such a designation. Several other dapperlings or parasols, including species of *Lepiota, Leucocoprinus,* and *Leucoagaricus* (all saprobic), may produce different, nonlethal toxins, which explains why people shun this

Alaskan Gold, Phaeolepiota aurea

class of attractive mushrooms. Somewhat similar but thankfully much bigger are the shaggy parasols, *Chlorophyllum* species (see above).

Small Fatal Dapperlings usually have scaly caps with scales on a whitish background and caps that are much darker in the center. Center colors include tones of brown, red, gray, black, and yellow. They are all fragile, white spored, and free gilled, with more or less slender stems. Most have a ring. Their lifestyle is saprobic, meaning they usually grow from soil, dead leaves, or wood chips. They love to be admired for their beauty and are best left alone.

 ## ALASKAN GOLD
Phaeolepiota aurea

Not all that shines is gold, but Alaskan Gold is very impressive in coloration and size, often coming up in big groups or even giant fairy rings! Seeing it growing in disturbed areas, such as along logging roads with alder and cottonwoods, you will slam on your brakes for this golden mushroom and rush out of the car. Some people love it as an edible, some see it as mediocre, and others have painful memories and would not dare to eat it again. It turns out that it contains hydrocyanic acid, but it is unclear if that substance is what causes extended gastrointestinal problems, since the acid is thermolabile and should mostly cook away. The Northwest Mushroomers Association's Buck McAdoo and Jack Waytz, intrigued by Alaskan Gold and the contradictory reports, held culinary trials[54] and found that 15% of participants experienced upset stomachs and/or diarrhea and vomiting. With one in six diners feeling really bad after a mushroom meal, can it still be regarded as an edible mushroom?

A minimum precaution would be to limit an Alaskan (fool's) Gold dinner party to five people! However, any fool might have a hard time finding company after disclosing the whole story. But for the hardy folk who have carefully tested their reaction to the mushroom, finding this beauty might feel like striking gold. In Buck's report, it's interesting also that for a couple of the people who had eaten *Phaeolepiota aurea* in Alaska without trouble, eating it years later collected from the Cascades really upset their systems.

- Orange to golden-brown, completely dry, smooth cap (Ø 6–30 cm) covered in mealy granules when young.
- Attached to free, close gills, first pale yellowish then browning with age; yellow-brown spores.
- Slender, downward enlarging stem (5–25 cm tall x 2–6 cm wide), stuffed center turning hollow when old, colored as cap, stem base with white mycelial strings.
- Thick partial veil flaring out from below, ring and stem below covered in mealy granules.
- Flesh pale whitish or yellowish, firm, unchanging.
- Mild flavor, but also reports of slightly astringent or sweet flavor.
- Odor mild or, for subtle noses, of bitter almonds, due to presence of hydrocyanic acid.
- Grows in groups or clusters in rich humus and soil, in disturbed areas, and along forest roads with alder, poplars, and conifers.
- Widespread generally, but uncommon in the lower Pacific Northwest, fruits in fall; common in Alaska, where it fruits in summer and fall.

LOOK-ALIKES: Some inedible big rustgills (e.g., Big Laughing Mushroom, *Gymnopilus ventricosus*) are similar, but they always grow on wood, often have small scales, and taste bitter. Western Yellow Panthers (*Amanita gemmata* group), another poisonous group, have a volva, white ring, and white spores. The edible Ringed Webcap, *Cortinarius caperatus* (see Gilled Mushrooms with Warm Brown Spores), is usually smaller and has rusty-brown spores and a nonflaring ring.

NAME AND TAXONOMY: *Cysto-derma aurea*: Greek *kystis* meaning "bladder" and *derma* meaning "skin," named for the grainy, mealy skin, and *aurea* Latin "golden." The synonym *Phaeo-lepiota* is from Greek *phaios* "dusky," as in dusky-spored and looking like a *Lepiota*. *Phaeolepiota aurea* (Matt. ex Fr.) Maire 1928 = *Cystoderma aureum* (Matt.) Kuehner and Romagn 1953. According to recent DNA studies, the latter could be used as its valid name. Agaricaceae, Agaricales.

WHITE-WOOLLED SHAGGY STEM
Floccularia albolanaripes

When fresh and sticky, that shiny golden cap catches the eye right away! A closer look reveals these attractive mushrooms. I am tempted to call them cute because that white valance-like fringe around the cap margin loads them with granny energy, and the fuzzy stem completes the heartwarming look. And their culinary quality invokes a savory meal your grandma might cook for you. The White-woolled Shaggy Stem has long been underrated in mushroom guides, but the tide is turning. Buck McAdoo, author of *Profiles of Northwest*

Fungi, raves, "Besides having just the right ratio of crunchy exterior and juicy interior, the flavor had a gourmet aftertaste that brought up visions of manna!"[55]

Also known as Scaly Bracelet, they are a special treat should you be lucky enough to chance upon them. More widely distributed, but not found in the Pacific Northwest, is its sister, the Yellow Bracelet (*Floccularia luteovirens*). I first fell in love with Sersha (Tibetan for "yellow mushroom") when I found it in the vast grasslands of Tibet. There, it has been a sought-after specialty for centuries! It is also beloved in Europe and the Colorado Rockies (where it was formerly known as *Armillaria straminea* var. *americana*). *Floccularia albolanaripes* is recognized by its yellow-brownish cap with drooping white veil fragments, cream-white gills, and stipe covered below the veil in yellowish-tipped white scales.

TOP: *White-woolled Shaggy Stem,* Floccularia albolanaripes; **BOTTOM:** *Sersha, yellow mushroom at a restaurant in Tibet*

IDENTIFICATION

- Yellow to golden cap (∅ 5–15 cm) with brownish center, moist to slightly sticky, covered with flattened fine fibers or scales that darken to brown with age.

- Flaky partial veil remnants (often white) hang from cap margin, but no warts on cap.
- White to pale yellow, close gills; attached, adnexed, or notched; in age become serrated like a saw; white spores.

- Ragged white ring represents upper end of flaky stem sheath.
- Fairly stout white to yellow stem; above ring silky smooth, below ring covered in ragged, cottony white veil remnants, often with yellowish to brown scale tips in chevron shape.
- Fairly firm, fibrous flesh, cream to yellow.
- Mild flavor (sometimes sour to acrid), slightly mealy odor.
- Grows solitarily, in fused pairs, or in small groups in duff under alder and conifers, often in compressed soil along forest roads.
- Fruits from southern British Columbia into California; mostly in fall, but sometimes in spring and during mild winters.

LOOK-ALIKES: The closely related and also edible Yellow Bracelet, *Floccularia luteovirens*, tends to be paler yellow and has crowded yellow scales on its cap and lower stem. The Western Yellow-Veil, *Amanita augusta*, and the poisonous Brown Panthers like *Amanita pantherinoides* have warts on their caps and free white gills. When young, the lower stems of the Amanitas can be scaly with multiple "rings" marking the volva rim, resembling the cottony lower stem sheath of *Floccularia*. Yellow brittlegills (*Russula* spp.) have no scales or flakes on their caps or stems.

NAME AND TAXONOMY: *Floccu-laria* from Latin *floccus* "wool tuft," *albo-la-nari-pes* Latin "white woolly stem," all references to the white flakes. *Floccularia albolanaripes* (G. F. Atk.) Redhead 1987; formerly *Armillaria albolanaripes* G. F. Atk. Squamanitaceae, Agaricales.

Knights or Trichs and a Mixed Basket

This section contains white spored, terrestrial, ectomycorrhizal mushrooms with attached gills. Most belong to the Tricholoma family (*Tricholoma & Catathelasma*) or are closely related like the Fried Chicken Mushroom (*Lyophyllum*). The Deceivers (*Laccaria*) and Woodwaxes (*Hygrophorus*), however, share these characteristics without being closely related.

Tricholoma and Allies

Tricholoma is one of the more challenging genera of gilled mushrooms to get a handle on, but white spores, firm flesh, attached and often notched gills, a terrestrial, tree-associated lifestyle, and a range of interesting, often strong odors point the right way. DNA evidence is helping a great deal to straighten things out, as uncovered in a new study by Steve Trudell, Drew Parker, and others, but it also shows that there is much we do not know yet. It is not always helpful to know that the commonly used European name is misapplied, especially when we do not yet have a replacement name, or that we may be dealing with several species. Luckily, *Tricholoma murrillianum*—the flagship *Tricholoma* species commonly known as the Western Matsutake— soars above such trouble. Yes, it went through a few name changes itself, but it is an easily recognizable mushroom for those familiar with its unique fragrance and exquisite firmness. Several Tricholomas release interesting aromas, the worst being the Sulfur Knight's aroma of tar gas.

MATSUTAKE AROUND THE WORLD

Matsutake is beloved in Japan, and Japanese demand for this highly prized mushroom has spawned commercial collection around the world, from mountain forests in Baja, Mexico, to Washington's Pacific coast to the northern interior of British Columbia. Its fame has spread far and wide. It is also collected in Bhutan, Tibet, southwest and northeast China, Korea, Northern Europe, and Morocco, largely for the Japanese market! Our species is different from that of East Asia, but I cannot tell a difference in flavor among matsutake I have eaten in Tibet, Bhutan, the Alps, or the Pacific Northwest.

In the heyday of commercial collection and peak popularity in the late 1980s and the 1990s, matsutake was referred to as "white gold," since the first pounds of a season would fetch several hundred dollars each. Nowadays, in retail, a fresh pound usually costs around $30. However, at peak production in a rare banner year, a collector may receive a pathetic buck for a pound from a field buyer, and in an average year $5 to $10. While its peculiar spicy aroma is adored in Japan, when *Tricholoma matsutake* was first described as a species in Sweden, it was stigmatized with the name *Armillaria nauseosa* and didn't make it into an edible mushroom field guide until the 1990s. That's when Swedish mycologists realized that their Nauseous Knight was identical to the iconic matsutake and graciously yielded their taxonomic naming rights. Today, Japanese matsutake enthusiasts are spared from having to pay $50 to $80 for *nauseosa* but still pay that for a prime specimen (one with a still-closed partial veil) of *T. matsutake*.

Matsutake is not the only interesting edible in this big ectomycorrhizal genus, but unfortunately for the forager, although no Trich is known to be deadly, things get murky quickly thanks to some fairly toxic Tricholomas like the Leopard Knight, *Tricholoma* aff. *pardinum*. The Man on Horseback, *T. equestre* group, a fairly easily recognized yellow Trich, once considered edible, has now received the mark of dubious edibility. And then there are the Streaked Knight, *T. portentosum*, a very enjoyable edible, and several species close to the Gray or Mouse Knights (*T. terrerum* group), which are hard to differentiate in the first place and have toxic look-alikes.

Even experienced mushroom hunters just let these species go about their spore-release business without abducting them to their kitchen.

The scientific name *Tricholoma* is derived from Greek *tricho* for "hair" and *loma* for "edge or border," for their hairy cap edges, although a particular specimen may not have many of them. In the United States, we lack a common name for the genus, and some people just call them Trichs. In the UK, they are known as the "Knights." They received that name from their former fellow, *Tricholoma gambosum*, Saint George's Mushroom, a choice edible spring mushroom often found in pastures and grass verges. Saint George

Western Matsutake, Tricholoma murrillianum. *Note the tapered, always sandy stem base and how the partial veil protecting the gills is still intact in the youngest specimen and turns into a big felty ring. The misapplied species name,* Tricholoma magnivelare, *now reserved for the Eastern Matsutake, referred to this feature, since* magnivelare *means "big veil" in Latin.*

is not known as a mushroom lover but rather as dragon slayer, patron saint of soldiers and England, always depicted mounted high on a horse fighting the dragon—real knight business! He conquered the mushroom's naming rights by mastering synchronicity, showing up in the church calendar on April 23, the famous Saint George's Day, around the same time when *Calocybe gambosa* pops up. Unfortunately, we have no equivalent to Saint George's Mushroom in the North American funga.

 **WESTERN MATSUTAKE**
Tricholoma murrillianum

Matsutake, literally "pine mushroom," is one of the most storied mushrooms in the Pacific Northwest. Knowledge of matsutake as a choice edible has been a key contribution of the Japanese community to mushroom hunting. Finding them in a high-end

Asian food store or at a mycological event lets you learn its exclusive scent; the closest match is Red Hots cinnamon candy. Its spicy cinnamon-like scent makes this a safe edible mushroom, since the otherwise very similar toxic Amanitas, especially *A. smithiana,* lack the spicy fragrance (Smith's Amanita can have an intense bleach stench in its spindle-shaped stem base). While you can crush the stem of an Amanita when you squeeze it between your thumb and index finger, the very solid matsutake stem will stand up to your pressure. Though the combination of fragrance and firmness makes matsutake a safe-to-identify mushroom, unfortunately, each year a few people cook up Amanitas and fry their kidneys! Use your discerning nose to stay safe. And remember: "Missing the matsi aroma might induce kidney coma!"

Matsutake are mycorrhizal and occur in a wide variety of habitats from coastal dune pine forests to mountain conifer forests. They love well-draining, often gravelly or sandy soils. Such sites are regularly dominated by a variety of pines, one of their mycorrhizal partners. Farther inland, matsutake grows under mixed conifers and second-growth Douglas-fir. Along the coast in southern Oregon and California, they are common under tanoak, madrone, and manzanita. The fruiting bodies of matsutake are often hidden in a thick duff layer, arising slowly to form "mushrumps." An interesting indicator plant is candystick (*Allotropa virgata*), which parasitizes matsutake mycelium. Matsutake start fruiting in summer or fall after enough rain; the season comes to a halt after the first serious frost. Cold weather and light frosts often induce peak season in the Cascades and luckily reduce worm infestation.

A Bhutanese mushroom hunter with true matsutake, T. matsutake

MEDICINAL: It seems that no medical study has used the Western Matsutake or its compounds, but there are a plethora of studies of Eurasian Matsutake (*Tricholoma matsutake*) that show antioxidant, anti-inflammatory, and immune-stimulating capacity, as well as antitumoral and anti-arteriosclerose propensities. Traditionally in East Asia matsutake is used to ease difficult childbirth, acute gastritis, fevers, and convulsions.

IDENTIFICATION

- Big white mushroom with sparse red-brown scales, spots, streaks, or fibers on cap ($\varnothing$ 5–30 cm); cap convex, at first with inrolled cottony margin, becoming flat with maturity.
- Crowded gills, adnate to adnexed or notched, developing spots or discoloring to rusty-brown in age; white spores.
- Tough and solid stem, upper part all white, below ring scaly or fibrous, similarly colored as cap. Stem base tapered and extended, always "rooting" in sand or volcanic ash that clings to base.
- Immature mushrooms have a membrane between cap edge and stem that envelops the gills; with maturity this partial veil turns into a thick, fleshy, membranous, upward-flaring white ring that becomes appressed with age.
- Very firm, solid white flesh with strong aromatic fungal scent of cinnamon and a hint of dirty socks. Not recognizing the scent can be deadly!
- Grows solitarily or in groups in ground, with ponderosa, lodgepole, and sugar pines, Douglas-fir, noble fir, and Shasta red fir, as well as broadleaf trees such as tanoak and Pacific madrone.
- Widespread and common; with summer rains it can start fruiting in July but most commonly fruits into late fall surviving light frosts well, all around the Pacific Northwest.

LOOK-ALIKES: The Northwest has a lot of gilled white mushrooms a beginner could confuse with matsutake. Besides the very similar and potentially deadly *Amanita smithiana* (see The Amanitas) and the probably less devastating *A. silvicola*, many people become mistakenly excited about finding Short-stemmed Brittlegill, *Russula brevipes*. Two other species of *Tricholoma* are very similar. *Tricholoma dulciolens* (for-merly mistakenly called *T. caligatum*) is usually darker, smaller overall, rarer, and less robust appearing, and ranges from bitter tasting to edible. While *T. dulciolens* may have the typical matsutake scent, *T. focale* lacks it and smells mealy. It has an orangey-red to green streaked cap, and its pale forms can be confusing. It is not rec-ommended as an edible because it could cause severe GI symptoms.

PROCESSING AND COOKING MATSUTAKE

Uncut, matsutake keep longer than many other gilled mushrooms, up to a few weeks. Carefully pull the whole mushroom from the ground without break-ing off the extended stem base. Store it in a paper bag in your fridge. If you dry matsutake, most of their special fragrance dissipates. Some people freeze whole fresh matsutake in water for later use. You could also brine them (see Preserving Mushrooms in Part One), as is done in East Asia in years with over-abundant harvests.

However you want to prepare your matsutake, it is always important to find a good match for its strong aroma and flavor. One of the best allies is fermented soy, such as soy sauce, tamari, or miso. My favorite preparation, which I learned from Puget Sound Mycological Society's Milton Tam, is to slice the matsutake into ⅛-inch slices and then steam it for fifteen minutes in a mixture of soy sauce, sake, and a bit of butter. To keep the aroma, fold the ingredients into aluminum foil that you then put in a water bath. Simplifying the process a bit, I often use white wine instead of sake and a pot with a good lid. Matsutake have plenty of aroma to share.

One very traditional way of using matsutake is to add it to miso soup or add some slices on top of or mixed into white rice when boiling it, which infuses the rice with its lovely flavor to different degrees. My favorite approach when frying them is to space them well so that all the pieces brown nicely, and then deglaze them with a mixture of white wine or sake and soy sauce—so yummy!

When camping or at home, matsutake can also be just grilled as they are, or you could add some butter or oil to the gills when you are roasting them cap down. The stems are better removed and grilled separately. Matsutake can also be eaten raw, for example, perhaps adding its crunchy texture and intense flavor to a salad. I have not heard of anyone having gastrointestinal problems (I never have), but matsutake is still not widely enjoyed raw, which would give us a bigger sample.

Man on Horseback, Tricholoma equestre *group*

Also often mistaken for matsutake are the big cats (*Catathelasma* spp.), impressive mushrooms that are even bigger than matsutake, have decurrent gills and a double ring, and also lack the matsutake aroma. The inedible, often also decurrent but ringless Large White Leucopax (*Leucopaxillus albissimus*) is less hefty and is usually bitter.

NAME AND TAXONOMY: *Murrillianum* is named for American mycologist William Alfonso Murrill (1869–1957). *Tricholoma murrillianum* Singer 1942; previously misapplied names include *T. magnivelare*, *T. ponderosum*, and *Armillaria ponderosa*. *T. dulciolens* Kytöv. 1989. Tricholomataceae, Agaricales.

 MAN ON HORSEBACK
Tricholoma equestre **group**

Appreciated at least since medieval times as a good edible, the Man on Horseback has recently been stabbed in the back. It all started when some French fungal gluttons completely overindulged and sadly died. Two more people were suspected to have died from it in Poland and Lithuania,[56] and other people who ate it became seriously sick, experiencing fatigue, muscle pain, nausea, profuse sweating, respiratory insufficiency, and an increase in serum creatine kinase, possible indicators of rhabdomyolysis.

All known cases involved people who ate big portions or several meals a day for at least three consecutive days. When mice were fed powdered *Tricholoma equestre*, similar symptoms manifested, confirming the connection, so case closed—or not? Poisoning of innocent mice was evident at daily doses corresponding to 8.8 pounds of fresh Yellow Knights, another common name, for a 133-pound adult human. At 6.6 pounds per day, however, researchers recorded no significant effects.[57] It has been argued that such doses are extremely high and that anyone who eats a serving or two is in no danger at all![58] To paraphrase the revolutionary Renaissance doctor Paracelsus: The dose makes the poison. After all, we can die from drinking too much water.

Finding Man on Horseback and identifying it successfully is aided by its attractive yellow cap, yellow gills, and yellow stem (the stem can also be whitish), hence another common name: the "Yellow Threesome." Add fine, darker scales in the cap center, a cap that becomes sticky when moist (which results in debris becoming glued to the cap when it dries afterward), and the absence of ring and scales on the stem, and you should be looking at a Yellow Knight. It has a great firm texture and good fungal flavor that make it versatile in the kitchen. To reflect current taxonomy we should be talking about "Men on Horseback." We have at least three different Yellow Threesome species, and *T. equestre*, which has been described in Europe but needs taxonomic work, has not been found here.

- Yellow cap (Ø 5–20 cm) with lighter edge and brownish center (cap browning with age); sticky when moist, mostly smooth, possibly with small scales at center; margin inrolled when young, flattening in maturity.
- Close gills, narrowly attached or clearly notched evenly pale yellow to yellow; white spores.
- White to pale yellow stem with whitish to lemon-yellow flesh, no ring, sometimes with slightly bulbous base with darker stains.
- Thick, firm, white flesh (to lemon-yellow in stem), but pale yellow near cap surface.
- Mild farinaceous (mealy) odor, sometimes sweet flavor.
- Grows solitarily or in groups in ground, often with pines in sand, but also in Douglas-fir–western hemlock forests and with madrones and oak.
- Widespread, fruits in fall all around the Pacific Northwest and beyond.

LOOK-ALIKES: Very rare and also edible is the similar Middle-class Knight, *Tricholoma intermedium*, which differs in having all-white gills. The inedible Sulfur Knight, *T. sulphureum* group, has a dry cap and an intense offensive coal tar odor. Also inedible, the Deceiving Knight or Green-streaked Trich (*T. sejunctum* group) has blackish radial fibers at the cap center and gills are often only yellowish near cap margin, the latter trait also shared by the rare and inedible *T. arvernense*. There are other yellow-capped mushrooms, but most do not have the Yellow Threesome; the closest are probably yellow Russulas, which can be told apart by their stems that break like chalk. The most dangerous look-alikes to Man on Horseback are the yellow-trending Amanitas, like Death Cap and Western Yellow Panther and allies (see The Amanitas), which both have ringed stems and a volva, be it a cuplike volva or rings on the stem base. The edible Ambiguous Roundhead, *Stropharia ambigua*, is much more slender and has a white-fringed cap edge and purplish-black spores.

NAME AND TAXONOMY: *Equestre* Latin "horse," *flavo-virens* Latin "yellow-green," hence it is also commonly known as the Canary Trich, *Tricholoma equestre* (L.) P. Kumm. 1871 = *Tricholoma flavovirens* (Pers.) S. Lundell 1942.

STREAKED KNIGHT
Tricholoma portentosum

This intriguing mushroom is hardly known in the Pacific Northwest, although it is a very good edible and is apparently a Northern Hemisphere denizen of many forest types (until DNA research might split the one species into several). I have rarely encountered it and was unaware of its quality, but Terra Fleurs' James "Animal" Nowak kept insisting I check out the Coalman, or le Charbonnier as it is known in France.

Having searched in vain for some years, I finally found it in coastal Douglas-fir and hemlock forest in Oregon and was pleased by its firm flesh. Once cooked, its farinaceous odor is gone and a nice, mild fungal flavor is revealed. I just wished the yellowish tones would be better developed in Pacific Northwest Streaked Knights; sometimes the yellow is missing, making identification arduous. Unless you have eaten this mushroom many times without trouble, make sure there is the revealing yellow stem coloration plus a sticky cap. They handle frost quite well and fruit into late fall.

IDENTIFICATION

- Midsize gray mushroom with sticky, streaked cap (Ø 4–12 cm) ranging from pale gray to dark gray with brownish or purplish tinges, or with olive tints.

- Cap center often pointed or raised, always darker than edges, which can become wavy with maturity.
- Closely spaced white gills, attached or notched, become grayish or pale yellow with age; white spores.
- Dry, stout, white stem, sometimes tinged yellow; scale-free.
- Firm flesh with nicely mild to farinaceous or cucumber odor and flavor.
- Grows in groups in soil with pine, Douglas-fir, hemlock, and hardwood forests.
- Fruits in fall in coastal forests and also in the Rockies.

LOOK-ALIKES: There is no shortage of inedible and toxic *Tricholoma* look-alikes, though only the Streaked Knight has a sticky cap (but dry stem) and gills that have yellow areas or an overall yellow hue. Lacking all yellow tones are the similar *T. mutablile*, *T. nigrum*, and the oak-associated *T. griseoviolaceum*. Probably toxic, *T. subacutum* (often referred to by its European sister's name, *T. virgatum*, or Silverstreaks) lacks the sticky cap, has

Streaked Knight, Tricholoma portentosum

Cottonwood Mushroom, Tricholoma ammophilum

more clearly defined streaks, and also has an acrid or bitter taste that might be slow to develop.

Overall, the similarity to the toxic Leopard Knight (*T. aff. pardinum*) rightly scares many a forager; the Leopard Knight can cause persistent gastroenteritis that often requires hospitalization. It is often bigger, and the cap has fine, regular-spaced pale gray to dark gray scales on a whitish background. Leopard Knights have neither a sticky cap nor yellow stains or hues.

NAME AND TAXONOMY: *Portentosum,* Latin, meaning "monstrous, ominously prophetic, or marvelous." It is speculated that Elias Magnus Fries referred to its culinary quality as "marvelous." Our West Coast Coalman may turn out to be genetically quite similar to the European species. *Tricholoma portentosum* (Fries) Quel. 1873. Tricholomataceae, Agaricales.

COTTONWOOD MUSHROOM
Tricholoma ammophilum

The Cottonwood Mushroom is also known as the Sandy, since it is often found growing in sandy soils along creeks and rivers. In Europe, it is known as Poplar Knight, but the Northwest has a genetically distinct sister species, as revealed by Lisa Grubisha, Drew Parker, and Steve Trudell. It could also be called the "Popular Knight," since east of the Cascades people have valued this highly productive mushroom since time immemorial. It is one of the few edible mushrooms we know to have been a staple for Interior Salish people, who know it in their native languages as the Cottonwood or Sticky Head Mushroom. In the past, Salish people dried them on strings and rehydrated them overnight for venison or salmon stews, but nowadays cooking and freezing are preferred storage methods.[59]

EDIBLE MUSHROOMS AND NATIVE AMERICANS

Given the vastness of Indigenous peoples' plant knowledge and the abundance of choice edible mushrooms in the Pacific Northwest, there are surprisingly few references to traditional edible mushroom use by Native peoples. Perhaps early explorers raised in a fungophobic culture simply ignored Native mushroom use. However, it seems that fungophobia was already well established among some Native people long before European arrival. For example, the Alaskan Iñupiat people disdained all mushrooms. According to Anore Jones in a guide to edible plants, "Traditionally, the local mushrooms were never eaten. The local Inupiat word for mushrooms means 'that which causes your hands to come off.'"[60] Another factor could have been that the main mushroom season is synchronous with the return of salmon, and hence mushrooms were relegated as an inferior food source. Or perhaps Indigenous people had no inclination to share their mushroom knowledge, some of which may also have been lost in the recurring epidemics that wiped out more than two-thirds of the Indigenous population, a loss further undermined by forced acculturation.

In any case, not all Native cultures were fungophobic, and some tribes used a variety of mushrooms, including puffballs, as medicine, and several conks as pigments, tanning agents, insect repellents, tinder, and so on. Some conks were even used to carve spirit figures for rituals. Maybe most famous is the Tree Fungus Man in the Haida creation story, who steered Raven's canoe to the intertidal zone to obtain female genitalia and thus bring people to this world.[61] Documented traditional edible mushrooms in the Pacific Northwest include chanterelles, meadow mushrooms, matsutake, morels, and oyster mushrooms. Appreciated edibles, mostly east of the Cascades, are the Cottonwood Mushroom, Thunderstorm Head, and Lightning Mushroom;[62] however, the taxonomic status of the latter two is not yet clear.

Use of mushrooms by Nlaka'pamux people, formerly the Thompson River Salish people, living in the North Cascades, is well documented.[63] Interestingly, sometimes the Nlaka'pamux people "washed new babies in an infusion of mushrooms to make them strong and independent, since mushrooms can push through hard ground and move rocks when they grow," according to Nlaka'pamux plant expert Mabel Joe, as quoted by ethnobotanist Nancy Turner.[64]

The widespread neglect of culinary mushrooms in the PNW seems to contrast starkly with research indicating that Indigenous peoples in California commonly ate all the best culinary species and frequently burned forests to optimize collection and fungal habitat.[65] With the development of a lucrative global culinary mushroom trade, some Pacific Northwest tribes are also involved in commercial collection, especially of matsutake. In addition, some members of tribes seem to have been introduced to "new" edible mushrooms from European immigrants who explored and began to inhabit the region.

As David Arora has reported, in regions where poplars are prevalent[66] and other trees are scarce, Cottonwood Mushroom is one of the last mushrooms to fruit in fall and one of the few white-spored ones growing in sand. It develops underground, then surfaces in rings or densely packed masses. Since many *Tricholoma* species are quite similar, including poisonous species, habitat is key for safe identification; its mycorrhizal partners, cottonwoods or aspen, must be close by. The Cottonwood Mushroom, like most knights (*Tricholoma* spp.), has a firm, enjoyable texture. In addition, it has a strong mealy flavor that mostly cooks away.

IDENTIFICATION

- Sticky cap (Ø 6–15 cm), pale pinkish-brown to reddish-brown, darker at center, sometimes with netting of fine hairs; inrolled margin flattening in age with uplifted edges.
- Close, whitish gills, adnexed or notched, developing reddish-brown spots and stains, as is typical for many *Tricholoma*; white spores.
- Fibrous fat stem, white to dull whitish when young, developing reddish-brown stains and yellowing toward wide base.
- Firm, thick, white flesh, rosy near cap surface.
- Strongly farinaceous (mealy) odor.
- Grows scattered or in dense flushes or clusters in sandy soil; always associated with cottonwoods and aspen.
- Fruits in summer and fall from Alaska south to California.

LOOK-ALIKES: Don't be tricked! Many other Trichs, like pale versions of the Red-brown Knight (*Tricholoma pessundatum* group)—and too many others to list—can be quite similar and share characteristics like a sticky cap, mealy odor, spotted gills, etc. Bottom line, unless you are a *Tricholoma* expert, they are only safe when found in their cottonwood habitat with no other tree species (like Douglas-firs) around that could indicate the presence of other species.

NAME AND TAXONOMY: *Ammo-philum*, Greek "sand-loving." *Tricholoma ammophilum* A. D. Parker, Grubisha, and S.A. Trudell 2022; misapplied European name *Tricholoma populinum* J. E. Lange 1933. Tricholomataceae, Agaricales.

LEOPARD KNIGHT
Tricholoma aff. *pardinum*

Here we have another dangerous mushroom invoking a big cat, the animal name solely inspired by the spots on the cap. The devastating blow is always triggered by a careless forager. When the Leopard Knight, a.k.a. Tiger Trich, is erroneously ingested, the digestion system will revolt—and not just for a few hours as with many of the other poisonous Tricholomas. The *pardinum* poisoning resulting in vomiting and diarrhea can last for many long days!

IDENTIFICATION

The Leopard Knight is recognized by its dry whitish cap marked with small and regularly spaced, pale to dark gray fibrous scales that invoke small spots. It differs from other dry, grayish Tricholomas by its larger, fleshier stature and often paler color. It shares the mealy or farinaceous odor and fibrous stem tissue with other

Leopard Knight, Tricholoma aff. pardinum

Tricholomas. Fall- and early-winter-fruiting Leopard Knights are ectomycorrhizal and team up with a range of conifers and some hardwoods, at least madrone and tanoak. Recent DNA research reveals that what we call *T. pardinum* is instead a complex of species, with one species very closely related to the European species, hence "aff. *pardinum*," and another recently described as *T. venenatoides* by mycologists Steve Trudell, Drew Parker, and Matt Gordon. However, it should be just as wickedly poisonous.

VANISHING CAT
Catathelasma evanescens

The biggest and heaviest of our gilled mushrooms is the Vanishing Cat. It is not caught too often, and there are only a few of these giants lurking in our mountain conifer forests. But when caught, the Vanishing Cat will feed a whole family. It takes some creativity to add value, or rather great flavor, to this mushroom. But this cat brings impressive firmness to the table, a quality frequently lacking in fungal fare. So, think of it as myco bamboo shoots that you flavor by marinating or pickling. There are cat lovers out there, and the closely related Imperial Cat (*C. imperiale*, a name formerly misapplied in North America) is found at mushroom markets in the Himalaya and Tibet.

A second, slightly smaller, Pacific Northwest cat species, *C. ventricosum*, is not helping to raise the culinary standing of the Cats. This mushroom, also known as the Swollen-stalked Cat (its stilted but verbatim translation from Latin), though edible, can have a disagreeable odor and flavor, an annihilating assessment I found to be on target when I tasted it. The cats are distinct: big and hard fleshed, with decurrent gills, on a tapering stem diving deep underground. They have two stem rings: the remnant of a universal veil, the outer ring winds around the base of the much bigger upper (partial veil) ring. Interestingly, their family (Catathelasmataceae or Biannulariaceae, depending on whom you ask) also includes the warmth-loving, giant

Vanishing Cat, Catathelasma evanescens

Macrocybe titans—the biggest "toadstool" in the New World.

IDENTIFICATION

- Orange-brown to hazel-brown smooth cap (Ø 10–40 cm), slightly sticky but soon dry, lighter in color as it expands to reveal whitish areas between darker patches.
- Whitish, decurrent gills, crowded (sometimes forked); white spores.
- Pale brown stem (12–18 cm tall x 3–8 cm wide), whitish above double rings; tapering below, often partly buried.
- Thick, hard, white flesh.
- Mealy odor and flavor that can turn bitter with age.
- Grows solitarily or in groups in soil with conifers, especially fir and spruce.
- Fruits in late summer and fall in coastal and mountainous regions.

LOOK-ALIKES: Edible *Catathelasma ventricosum* is a bit "smaller" (cap Ø 10–35 cm), its cap grayer than brown and dry (not sticky) when young; its stem tends to be widest in the middle and the flavor is unpleasant. Edible Western Matsutake, *Tricholoma murrillianum,* and the kidney-killer *Amanita smithiana,* two big, deep-"rooting," whitish mushrooms, both have their own unique odor with neither the double ring nor, more importantly, decurrent gills. Also big to huge, white, and with decurrent gills are the edible Short-stemmed Brittlegills, *Russula brevipes* group, but their caps are centrally depressed and their flesh brittle, breaking chalklike. The inedible, often also decurrent but ringless Giant White Leucopax (*Leucopaxillus albissimus*) can have a cap as wide, but it is much less hefty and usually bitter.

NAME AND TAXONOMY: *Evanescens* Latin "fleeting, vanishing," possibly a reference to its ring, which dries quickly, *Cata-thelasma* Greek "downward-suckling," probably a reference to the teat-like shape of the freshly exposed young stem. *Catathelasma ventricosum* (Peck) Singer 1940, *C. evanescens* Lovejoy 1910. Catathelasmataceae, Agaricales.

This section is a catchall for white spored, terrestrial, root-associated mushrooms, that do not have rings and did not fit with Knights (*Tricholoma*), Funnelcaps (*Clitocybe*), Amanitas, Brittlegills (*Russula*), and Milkcaps (*Lactarius*). Some, such as Fried Chicken Mushroom (*Lyophyllum decastes*), used to be part of the Tricholoma family until their DNA results made it clear that they belong elsewhere.

FRIED CHICKEN MUSHROOM

Lyophyllum decastes group

Every pot hunter aspires to find this monster cluster with such a tempting name. However, they are not as ubiquitous as you may hope, and it is difficult to feel sure about identifying them the first time or two you find them. The main characteristic that will help you ID them is that they occur in big, dense clusters in conifer forests (they are mycorrhizal with conifers) along gravel roads or other disturbed places, never on wood. Very few other mushrooms have that lifestyle.

The cap color varies from whitish to pale tan to yellowish-brown, grayish-brown, or dark gray. This wide range has inspired several scientific names, but most people will stick with *Lyophyllum decastes* group until molecular studies clear up the situation. DNA has actually showed that our Fried Chicken Mushroom is closely related to Hon-shimeji, a.k.a. Clustered Domecap (*Lyophyllum shimeji*), a sought-after Japanese mycorrhizal edible that can also

be cultivated. Surprisingly, DNA analysis just revealed that we have *L. shimeji* growing with three-needle pines in the Northwest!

Besides its clustered growth and habitat, the distinguishing traits of Fried Chicken Mushroom are subtle: smooth, continuous, greasy-feeling caps, often with a darker center; bundled stout, sometimes off-center stems; and close white gills. To be on the safe side, make sure the cluster has at least ten fused stems. Fried Chicken Mushroom offers firm-textured fungal flesh in abundance and can be prepared in many ways, making it a desirable edible, but it takes culinary creativity to make it into an exceptional mushroom dish.

IDENTIFICATION

- Gray-brown to yellowish-brown, rounded caps (⌀ 3–12 cm) with silvery streaks; surface smooth, soapy feeling; inrolled margins mature to convex or flat with sometimes wavy edges.
- Close white to grayish gills that are soft and waxy, attached to slightly decurrent, producing tons of white spores.
- White to dingy white clustered stems (often off-center), stout and fibrous; no ring.
- Rubbery white flesh; mild to indistinct flavor and odor.
- Grows in (big) clusters on disturbed ground in conifer forests.
- Fruits summer and fall and into mild winters.

LOOK-ALIKES: The most dangerous look-alike is the truly white, ground-growing, rightly infamous Fool's Funnel (*Clitocybe rivulosa*), which usually grows in grass but is smaller and never grows in monster clusters. Also growing in the woods

Fried Chicken Mushroom, Lyophyllum decastes *group*

along old gravel roads and paths is the bright white, but light grayish–dusted, smaller Clustered Whitecap, *Leucocybe connata*. Falsely mistaken for the Crowded White Funnelcap (*Clitocybe dilatata*), it has recently lost its seriously toxic label. The dull pinkish–spored Clustered Funnelcap, *Lepista* (*Clitocybe*) *subconnexa*, is otherwise edible but hardly recommended for eating because of its similarity to *C. rivulosa*. There are several dome caps (*Lyophyllum* spp.) without multiple stems that are hard to identify. Big clusters of honey mushrooms (*Armillaria* spp.) might be superficially similar, but they grow on exposed or buried wood, and most have scaly caps and ringed stems (see Dark Honey Mushroom, *Armillaria ostoyae*, p. 275).

NAME AND TAXONOMY: *Lyo-phyllum* Greek "free-leaves," supposedly a reference to gill attachment, but it does not fit well; *decastes* Latin means "in tens," as in "decades," a reference to the multitude of mushrooms. *Lyophyllum decastes* (Fr.) Singer 1951; previously also *Clitocybe decastes* (Fr.) P. Kumm. 1871. Lyophyllaceae, Agaricales.

COMMON DECEIVER
Laccaria laccata

These small to rarely midsize slender mushrooms are probably known as the "deceivers" because their rather dull cap hides colorful gills in a range of hues: intensely purple for the Western Amethyst Deceiver (*Laccaria amethysteo-occidentalis*); light purple for the Two-colored Deceiver (*L. bicolor*); and pale pinkish to brown-orange in the case of the Common Deceiver (*L. laccata*). Turned over, the deceivers reveal their colorful, wide-spaced, waxy gills (they are also known as tallowgills)—you will not be deceived about their identity. They all have firmly textured flesh, which makes them interesting edibles. With maturation, their nicely colored gills lose some intensity, and their hygrophanous caps dry and fade. Deceivers are mycorrhizal with many conifers—including pines and

Douglas-firs—as well as birches and members of the oak family.

The deceivers are the classic junior partner in a mixed mushroom dish, adding color and texture if not much flavor. Some people shun their stems for being a bit tough. What deceivers lack in aroma can easily be added, be it plenty of garlic or a rich finish with port wine, or they can help stretch a slim picking of porcini. What this *Laccaria* lacks in flavor beyond fungal, it offers in structure.

However, a new narrative is evolving as well. Are we being deceived by these harmless-looking mushrooms that turn out to be vectors for delivering high arsenic loads? Research from Eurasia implicates some Laccarias as scary arsenic accumulators, especially the purple species, while other species like *L. bicolor* do not accumulate or only do so in small amounts, like *L. laccata*—comparable to several other choice edibles.[67] For North America, we know only that our purple deceiver is laced with arsenic and should be avoided; more research is needed to sort out the toxicity of the other deceivers.

The Common Deceiver is widespread in conifer and hardwood forests, but often overlooked, as expressed by its alternative common name, Lackluster Laccaria. The gills, however, are not lackluster, displaying a nice pinkish color. The latest DNA research indicates that the Common Deceiver contains several species—deceived again! So technically, we should be talking about the "Common Deceivers."

IDENTIFICATION

- Dull pinkish to orangish-brown cap (Ø 1–5 cm) fades to pinkish-ochre or buff, dry with very fine hairs, often with depressed center, margin smooth and even or lined to grooved.
- Well-spaced, broad, waxy gills are pale pinkish to the color of Caucasian flesh, often darkening with age to reddish-tan; adnate to slightly decurrent; white spored.
- Slender, tough, fibrous stem colored as cap with white mycelium on base.
- Flesh is thin, tough, fibrous.
- Mild flavor and indistinct odor.
- Usually grows in small groups or tufts in soil with conifers, oaks, and birches.
- Fruits mainly in fall and early winter, but pops up in spring and summer.

LOOK-ALIKES: Also edible and fairly common in the Pacific Northwest are Noble Deceiver, *Laccaria nobilis*, a look-alike found in mountain forests, and *Laccaria*

Common Deceiver, Laccaria laccata

bicolor, which has violet mycelium on the stem base of young specimens, hence the name Two-colored Deceiver. Older Western Amethyst Deceivers can look similar from above, with faded caps turned buff, but their all-purple gills, stems, and mycelium show the difference (see below).

NAME AND TAXONOMY: The most common variety in the Pacific Northwest is probably *Laccaria laccata* var. *pallidifolia*. *Lacca* Latin for "lacquer," referring to the cap surface, but apparently the author saw fit to apply two coats! *Pallidi-folia* Latin for "pale-leafed," a reference to the gill color. *Laccaria laccata* (Scop.) Fr. Grevillea 1884. Hydnangiaceae, Agaricales.

ⓧ WESTERN AMETHYST DECEIVER
Laccaria amethysteo-occidentalis

When it is young, the purple color of the Western Amethyst Deceiver is eye-catching, sometimes outright drop-dead gorgeous! Unfortunately, the just-as-gorgeous Eurasian *Laccaria amethystina* has been shown to contain extreme loads of arsenic, ranging from 33 to 320 mg per kg dry biomass,[68] while other *Laccaria* and *Agaricus* in Europe contain less than 10 mg per kg and many other mushrooms contain less than 1 mg per kg of this heavy metal. For comparison, requirements for soil levels requiring arsenic cleanup are 0.38 to 0.41 mg per kg for Oregon and 7 to 40 mg per kg for Washington.[69] The absolute ace at accumulating arsenic is the Violet Crown Cup (*Sarcosphaera coronaria*), which hyperaccumulates arsenic to levels between 150 and 7,100 mg per kg. This Asco is not described as an edible in this field guide. If you were

Western Amethyst Deceiver, Laccaria amethysteo-occidentalis

to eat arsenic-laced fungi, you would not drop dead on the spot, but continuous intake is a serious health hazard that can cause cancer, cardiovascular disease, and other health problems. So it is probably better to experience the purple deceivers visually, not viscerally! Hopefully there will soon be some studies that look into arsenic concentrations in all our common western deceivers.

Woodwaxes and Waxgills

Woodwaxes and waxgills are white-spored mushrooms with thick, waxy gills and caps that are frequently waxy and slimy. The two main groups of what used to all be called waxcaps can easily be distinguished: Woodwaxes (*Hygrophorus* spp.) have medium-sized to large caps that are convex, slimy, and whitish or more often dull colored, with tones of gray, pink, orange, red, or brown; they are all ectomycorrhizal. Waxgills (*Hygrocybe* and others) have smaller, thin-fleshed caps that are convex to conical, slimy or dry, and are often impressively brightly colored; their lifestyle, or rather nutritional base, is often still a mystery.

While *Hygrophorus* has held up pretty well under DNA scrutiny, *Hygrocybe* has been divided up into several new genera. For example, the pale yellow, dry-capped Meadow Waxcap (*Hygrocybe pratensis*) is now *Cuphophyllus pratensis*, and the stunning green, slimy-capped Parrot Waxcap (*H. psittacinus*) is now *Gliophorus psittacinus*. Both these mushrooms are edible, but ignored. Both have turned out to be species groups that have not been worked out yet. There are also inedible or toxic *Hygrocybe*.

In contrast, all *Hygrophorus* are regarded as edible according to a 2021 review encompassing edible mushrooms around the world.[70] In the Pacific Northwest, the most interesting *Hygrophorus* is the Subalpine Woodwax, since it fruits when few other gilled mushrooms do. Many other fall-fruiting woodwaxes, though edible, are only rarely collected, such as the common Blushing Woodwax (*H. pudorinus*), Purple-red Woodwax (*H. purpurascens*), and the almond-scented Mount Baker Woodwax (*H. bakerensis*), which loses its aroma when cooked. One of the best woodwaxes is the Pink-mottled Woodwax (*H. russula*), which is widely eaten in the Himalaya and Tibet, where it shares pine, oak, and spruce habitat with matsutake. The latter is collected for cash income, while the woodwax is a local culinary mainstay, often soaked overnight in cold water to reduce its bitterness and improve its flavor.

SUBALPINE WOODWAX
Hygrophorus subalpinus

The Subalpine Woodwax is one of the famous "snowbank mushrooms"—fruiting bodies that start popping up at snowmelt, right on the edge of a melting patch of snow. Western North America's climate has triggered this globally unique fruiting pattern. A wide range of mushrooms will fruit when the soil in the mountains is reliably soaked each spring. Summer rains are unreliable, and in the fall, precipitation soon turns into snow. Most morel hunters are aware of this fruiting pattern, and the Subalpine Woodwax will appear even before morels. This snow-white mushroom is fairly easy to identify due to its early fruiting and robust, stout stature, with a thick, dry stem ringed by a partial veil. The thick, white flesh is soft with a mild fungal flavor. It is best to dry sauté the mushroom to firm it up.

IDENTIFICATION

- White, smooth cap (Ø 5–20 cm), sticky when moist, sometimes with veil remnants on margin.
- Soft and waxy feeling, close narrow gills; adnate to decurrent; aging from white to cream, often with dingy yellowish areas.
- Thick, stout, dry, silky stem (3–10 cm tall x 1–6 cm wide at top), bulbous base that straightens out with age.
- Sheathlike partial veil leaves fleeting ring (thin, nearly fibrous, white) on top of the basal bulb.

Subalpine Woodwax, **Hygrophorus subalpinus**

- Thick, white, mild-tasting flesh; at first firm, then softening with age.
- Grows singly or in groups in soil under mountain conifers.
- Fruits in spring right after snowmelt all over the Pacific Northwest and in the Rockies.

LOOK-ALIKES: There are several whitish funnelcaps with decurrent gills, like the White-stranded Funnelcap, *Clitocybe*

March Mushroom, **Hygrophorus marzuolus**

albirhiza, and Snowbank Funnelcap, *Clitocybe glacialis.* Both lack the bright white splendor of the Subalpine Woodwax, have much thinner flesh, lack waxy softish gills, and have stem bases with mycelial strands that hold tightly to the forest litter. Sordid Woodwax, *Hygrophorus sordidus,* is similar to the Subalpine Woodwax but seems to lack culinary value (and a veil), and most importantly, it is an oak associate and does not grow in spring in subalpine conifer forests.

Another snowbank woodwax, *Hygrophorus marzuolus,* is known in Europe as the March Mushroom, where it is appreciated by connoisseurs as a choice spring edible. It is often quite rare and is protected in Switzerland and Germany. In western North America, this species grows with Engelmann spruce and true fir, fruiting mostly around snowmelt, but also sometimes late into fall. It has a robust fruiting body with a streaked, moist to sticky, dark gray cap (Ø 4–10 cm) that can have a brownish hue but is paler in the center; its waxy gray gills are adnex to adnate. The stem can be moist or dry and tinted gray, with a white zone at the top or bottom. The flesh is thick, water-soaked, and whitish to gray, and it lacks flavor. Apparently our March Mushroom needs a new name, which might explain its lack of culinary fanfare out West.

NAME AND TAXONOMY: *Hygro-phorus* Greek from *hygrophoros*, "bearing water," a reference to the sticky, moist caps; *subalpinus* Latin refers to growing in the subalpine zone. *Hygrophorus subalpinus* A. H. Sm. 1941. Hygrophoraceae, Agaricales.

LARCH WOODWAX
Hygrophorus "speciosus"

This tasty larch-associated woodwax is a bright orange *Hygrophorus* species with a sticky smallish umbonate cap (Ø 2–5 cm). The adnexed to decurrent waxy gills mature from creamy white to yellow. DNA suggests that *H. "speciosus"* might be identical to pine-loving *H. siccipes* (meaning "dry stem," though its stem is often sticky). Its dark (red) brown cap center contrasts with a paler grayish tan to dingy orange cap margin. Luckily all Woodwaxes are reported as edible.

Funnelcaps and Allies

The funnelcaps and other ringless white-spored mushrooms are a big and diverse group with several genera, including the original genus *Clitocybe* and also *Lepista*, with a few known edibles, many difficult-to-discern mushrooms, and a few poisonous mushrooms. They are called funnelcaps for their fruiting bodies, often shaped like a funnel around a depressed center. Fatalities caused by *Clitocybe* are very rare; the Fool's Funnel (*C. rivulosa*) is the most notorious of them, containing a high dose of nasty muscarine, which could kill small children, the elderly, and pets.

The tightly spaced gills are usually decurrent or adnate and produce white to yellow, cream, or light pinkish spores. Many *Clitocybe* grow from a dense mycelial network that welds debris to the stem base. Also, many funnelcaps have diverse aromas, and the common spicy *Clitocybe* fragrance unique to funnelcaps, is best learned at mushroom forays.

Larch woodwax, Hygrophorus "speciosus"

WOOD BLEWIT
Lepista nuda

The fact that this beautiful purple mushroom has its own traditional common English name tells us that the Wood Blewit is special! The fruity aroma in connection with firm flesh and its cool color spruces up many a dish late in the mushroom season. The lack of striking features prevents the Wood Blewit from being an easy, beginner's mushroom; there are similar-looking purple toadstools out there. However, the fruity aroma that David Arora describes as reminiscent of "frozen orange juice concentrate," added to its pinkish spore color, is very helpful for confirming its identity. Unfortunately, not every nose seems to be able to detect this pleasant smell.

Blewits should not be eaten raw—they contain thermolabile hemolysin, which attacks red blood cells and, according to Michael Beug in his book *Mushrooms of Cascadia*, can result in bloody vomit and bloody diarrhea. Cooking neutralizes this toxin. Blewits seems to really like people, or rather places transformed by humans,

including parks and yards adorned with compost, brush, or other piles of organic debris. Blewits wait for chilly nights to fruit. A lack of cold nights seems to be the only limit for their otherwise global distribution; however, a closer look at their genetics may reveal several species worldwide.

Wood Blewits are saprobic and can be cultivated. European-grown Blewits are exported to the United States and sold for a pretty penny in high-end food stores. Their mild flavor and silky texture ask that they be sliced and seared before being added to stir-frys or soups.

IDENTIFICATION

- Midsize purple mushroom with smooth brownish-purple or gray-purple cap (Ø 5–15 cm), fading to brownish.
- Irregularly shaped convex cap with inrolled margin when young, raised and wavy margin when mature.
- Close, purple to bluish-purple gills that fade to gray in age, notched, adnate to adnexed (or decurrent), producing dull pinkish-white (not rust-brown!) spores.

Wood Blewit, Lepista nuda

- Purplish fibrous stem, with base of same diameter or fatter; base covered by purple mycelium fuzz.
- Odor subtly fruity, flavor mild to slightly bitter.
- Grows in woods, brush, gardens, compost piles—wherever there is organic debris.
- Fruits solitarily or fused in small to big groups, often in rings or arcs, from fall into early winter.

Aniseed Funnelcap, Clitocybe odora

LOOK-ALIKES: Edible after cooking and of similar appearance are two other blewits: The much rarer and inconspicuous, similarly sized Pale Blewit (*Lepista glaucocana*) is whitish when young with pale gray-blue to tan tones and sometimes a subtle hue of purple. The Grass Blewit (*Lepista tarda*) is smaller (Ø 1–5 cm) with a slimmer fibrous stem and lighter purple hue (most likely to show purple colors in its gills) and lacks the fruity aroma. Purple webcaps (*Cortinarius* spp.), told apart easiest by their rust-brown spores and cortina (a weblike "curtain" functioning as partial veil), are visible in young specimens. The Gassy Webcap (*C. traganus*) usually has a fruity smell!

Only one of the purple webcaps is known to be edible: the Violet Webcap, *C. violaceus* (see Gilled Mushrooms with Warm Brown Spores). For other purple look-alikes see the Western Amethyst Deceiver, *Laccaria amethysteo-occidentalis* (see species listing, earlier in this section).

NAME AND TAXONOMY: *Lepista* Latin for "bowl" or "goblet," a reference to the shape of the cap, and *nuda* Latin "naked," for the smooth cap surface. *Lepista nuda* (Bull.: Fr.) Cooke 1871; synonymous with *Clitocybe nuda* (Bull.: Fr.) H. E. Bigelow and A. H. Sm.

ANISEED FUNNELCAP
Clitocybe odora

What an awesome aroma *Clitocybe odora* effuses! Sniffing the rich fragrance of these mushrooms in the woods can be enough of an enjoyable treat. There are actually two Anise Funnelcaps marked by their anise odor and the blue-greenish color that fade with age and dryness. The Aniseed Funnelcap, *Clitocybe odora*, has more whitish gills and stem; *Clitocybe odora* var. *pacifica* flashes greenish-blue gills and stem.

Eating them raw reportedly causes poisoning. Cook them first: even reduced, the intense aroma will overpower most other ingredients. Pair them skillfully and dose them carefully unless you love a serious punch of anise. Frying them too long will reduce if not annihilate the anise flavor. Infusing vodka produces a drink reminiscent of Greek ouzo or French pastis; remove the caps as soon as the liquor reaches the intensity you desire. If you leave them too long, your vodka will taste weirdly pungent and more suited for use as antifreeze!

IDENTIFICATION

- Blue-green to dull greenish-whitish streaked cap (Ø 3–10 cm), smooth, dry or greasy; often with dingy center, graying (losing color) with age and dryness, often turning dingy beige.
- Close gills colored as cap (but more intensely colored in var. *pacifica*), broadly attached to decurrent; pinkish-cream or buff spores.
- Flesh whitish or with shades like cap; thin, except at cap center.
- Strong, sweet scent of anise, licorice, or fennel; slightly sweet and mild flavor.
- Stem stuffed to hollow, colored like cap (or white to buff) with white mycelium at base.
- Grows scattered or in groups in duff in conifer or mixed forests.
- Widespread, mostly fruits in fall in the PNW or earlier in the Rockies.

LOOK-ALIKES: There are no other blue-greenish mushrooms with an anise odor. However, there are at least two smallish (Ø 1–6 cm) white to off-white anise- or licorice-scented funnelcaps, usually referred to as the Anise Mushroom, *Clitocybe deceptiva*, and Fragrant Funnel, *C. fragrans*—neither are blue-greenish. Both are regarded as edible with care out West. In Europe, *C. fragrans* is reported to contain toxic muscarine, common in many other Clitocybes.

NAME AND TAXONOMY: *Clito-cybe* Greek "sloped" (as in "incline") and *cybe*

"cap" or "head"; *odora* Latin "fragrant." *Clitocybe odora* (Bull. ex Fr.) P. Kumm. 1871, *Clitocybe odora* var. *pacifica* Kauffman 1928. Tricholomataceae, Agariclaes.

CLOUDY FUNNEL
Clitocybe nebularis

The stately Cloudy Funnel has a storied history of heated discussion in France about whether *le nuage* (the Cloud) is an excellent edible or a disgusting toadstool. This big funnelcap used to have that special *je ne sais quoi* fragrance—I simply can't describe it! With a little help from American friends, the mystery can finally be solved, and the beholder's association makes or breaks this *champignon*! Do you detect the stench of a malodorous skunk, or are you bathing in a pungent but delicate cloud of sinsemilla? The skunky marijuana fragrance, possibly derived from prenylthiol,[71] persists when cooking this fleshy mushroom and will surely inspire interesting table conversations.

This mushroom, as with all *Clitocybe* and allies, cannot be eaten raw and needs to be heated to be detoxified. *Le nuage* fruits under a big Douglas-fir in our yard, and I admit to enjoying its *goût unique* fried up in butter once in a while. However, there are reports of the Cloudy Funnel upsetting not only sensitive noses but also sometimes digestive systems. While the causes are not clearly understood, it has been suggested these problems could be caused by nebularine, a much-researched purine ribonucleoside (and adenosine analog) named for the mushroom. It has shown antibiotic, anti-amebal, antiviral, and antiparasitic but also cytotoxic activity. However, nebularine concentrations in the Cloudy Funnel

Cloudy Funnel, Clitocybe nebularis

seem far too small to be the sole cause of digestive upset. Still, caution is advised; don't be the person to prove that reckless overindulgence causes serious medical problems! By the way, nebularine is highly water soluble, so parboiling will eliminate this toxin, but there are apparently other toxins as well.

IDENTIFICATION

- Large, grayish-brown, dry cap (Ø 6–25 cm), darker at center; often with grayish-white bloom and/or watery spots, cap color matures to whitish-gray.
- Attached to decurrent, close gills colored as cap producing cream-colored spores.
- Thick, tough but pliant, whitish flesh.
- Strong musty odor of skunk and marijuana.
- Whitish stem with dingy brown fibrils, and often-widened base with white mycelium.
- Grows scattered or in groups or rings in duff under conifers.
- Widespread, mostly fruiting in fall.

LOOK-ALIKES: The skunky odor is unique, although there are other mushrooms that smell unpleasant and disagreeable, like the darker-gilled (but very rare) *Clitocybe* (*Harmajaea*) *harperi*. Bigger and bitter, but sometimes also slightly skunky, is the Giant Leucopax (*Leucopaxillus giganteus*), and less offensive smelling but still very bitter and whiter is the Giant White Leucopax (*Leucopaxillus albissimus*). Other decurrent gilled look-alikes include Short-stemmed Brittlegill (*Russula brevipes*) and White Chanterelle (*Cantharellus subalbidus*).

NAME AND TAXONOMY: *Nebularis* Latin "cloudy." *Clitocybe nebularis* (Batsch ex Fr.) P. Kumm 1871.

❓ WHITE-STRANDED FUNNELCAP
Clitocybe albirhiza

If you happen to find yourself stranded and hungry on the dry side of the Cascades in spring, this mushroom, also called Snowmelt Funnelcap, could offer relief. It is not a mushroom with culinary history but has been eaten by some members of Bellingham's Northwest Mushroomers Association and myself without ill effects. Buck McAdoo reports enthusiastically that *Clitocybe albirhiza* is a darn good edible, with an interesting bit of bitterness but not offensively so!

Did I mention that this mushroom has basically no history as an edible? Try it at your own risk—only a few brave (or stupid?) souls went before you. It should be consumed only in small amounts by adventurous, iron-stomach-type foragers. Make sure you found it in spring in the eastern Cascades and Sierra; there are too many look-alikes elsewhere in other seasons. It must have clear rhizomorphs on the stem base, and concentric cap rings will help you confirm the ID as well.

IDENTIFICATION

- Dry, smooth (or with whitish fuzzy down) cap (Ø 2–10 cm), watery brown to buff, often with cinnamon or pinkish tones; turning pale buff when dryish; when fresh often with concentric rings.
- Attached to decurrent, close gills colored as cap; cream-colored spores.
- Thin outer cap (fleshier in center), tough but pliant whitish flesh.
- Stem colored like cap (or lighter), often hollow, with dense mycelium and rhizomorphs at base.
- Odor mild to disagreeable; mild to bitter flavor.
- Grows scattered or in groups under conifers.

- Fruits in spring after snowmelt in the eastern Cascades and Sierra.

LOOK-ALIKES: Most common in spring in the same habitat is the (probably) harmless Snowbank Funnelcap, *Clitocybe glacialis*, which lacks rhizomorphs and has a grayish-white cap surface when young. In general, there are a number of other whitish to buff-colored *Clitocybe* species, some with rhizomorphs, that are labeled "toxic" (like the Fool's Funnel, *C. rivulosa*; see below) or "edibility unknown."

Also similar is the cavalier *Melanoleuca cognata*, which fruits in early spring. Though it has white mycelium on the stem base, it lacks rhizomorphs and tastes rather sweet not bitter. It is reputed to be edible, as are several other cavaliers, *Melanoleuca* spp.

NAME AND TAXONOMY: *Albi-rhiza* from Latin "white" and Greek *rhiza* "root." *Clitocybe albirhiza* H. E. Bigelow and A. H. Sm. 1962. Might soon turn into *Rhizocybe albirhiza*.

White-stranded Funnelcap, *Clitocybe albirhiza*

Fool's Funnel, Clitocybe rivulosa

☠ FOOL'S FUNNEL
Clitocybe rivulosa

Let's call this species a "Class 2 deadly mushroom," since most healthy grown-ups normally survive should they feed on this innocent-looking and mild-tasting funnelcap. Deadly poisonings have happened: children have been killed from the Fool's Funnel due to its high concentration of the heat-resistant toxin muscarine, bane of funnelcaps and fiberheads (*Inocybe* spp.).

Fool's Funnel (a.k.a. the Sweater or Sweat-producing Funnelcap) is recognized by its white to grayish to pink-tinged (when wet) fruiting body, small size, adnate to decurrent, close, gray-white to pink-buff gills, and white to (rarely) creamy spores. Its thin flesh is brittle when dry but pliant when wet, and it has a mild odor and flavor. They love growing scattered or in groups, often in arcs and rings in grass or forest openings in summer and fall.

IDENTIFICATION

- Small cap (Ø 2–5 cm).
- White, grayish, or pinkish fruiting bodies.
- White to creamy spores.
- Adnate to decurrent, close, gray-white to pink-buff gills.
- Thin flesh and mild odor.
- Grows scattered or in groups.

LOOK-ALIKES: Extremely similar and also found in a grassy habitat is the edible Sweetbread Mushroom (*Clitopilus prunulus* group), which can be differentiated by its pink spores and strong, mealy odor. Otherwise it is a total dead ringer— pun intended! The grassy habitat is also shared by the tasty Fairy Ring Mushroom (*Marasmius oreades*). However, caps of Fairies are light brownish and not as white or light gray, they do not have adnate to decurrent gills, and their stems are slender and appear much rounder in cross-section. Very similar is the edible-with-caution Clustered Whitecap, *Leucocybe connatum*, which grows in clusters.

NAME AND TAXONOMY: *Rivu-losa* Latin "little river or channel," a reference to the cracks that develop in the cap with age. *Clitocybe rivulosa* (Pers.) P. Kumm. 1871; formerly also falsely called *C. dealbata* (Sowerby) Gillet 1874.

! FAIRY RING MUSHROOM
Marasmius oreades

Though most of us have not seen any fairies, some of us have been lucky enough to find the fungal orb they reportedly use as a portal to commute between dimensions. Similarly, the mushroom seems to traverse dimensions. After completely drying out, it can revive in the next rain to live again and continue to produce spores—an extremely rare accomplishment among mushrooms! It's also nice for the soup pot. It rehydrates in minutes and so can be tossed in at the last minute to add its fungal umami magic.

While there is no magic trick to identifying this little brown mushroom, there are key signs. It grows in lawns, feeding on grass roots. Its stem is very tough: shake it and it will not break, unlike most other lawn LBMs. It is frequently clustered, and caps often overlap each other, dropping white spores for a natural spore print; both features help differentiate it from the quite different and deadly rusty-brown-spored Ringed Conecap, *Pholiotina rugosa*. Other helpful characteristics for identifying *Marasmius oreades* are the light brown, hygrophanous cap color and the widely spaced, broad (up to 0.5 cm deep), whitish gills. If you are new to identifying Fairy Ring Mushrooms, it is best to get expert confirmation. Lucky

Fairy Ring Mushroom, Marasmius oreades

you, they will come up in the same spot over the years; these little nuggets are extremely tasty, with a rich umami taste reminiscent of Shaggy Parasols, and it is strange that they are not cultivated widely. For a beginner, learning how to identify them takes much longer than tediously harvesting them on your knees with scissors.

A few warnings: Do not eat raw *M. oreades*, as they contain toxic cyanohydrin,[72] employed by the Fairy Ring to successfully fight off slugs. Also, take care that the hunting grounds have not been violated by fertilizers, pesticides, or pollution. If uncertain, allow them to grow, and hope the fairies are happy to keep their portal. (Increased fungal predation by hungry "big people" might explain the reduced sightings of fairies these days.)

IDENTIFICATION

- Smooth, dry, light brownish hygrophanous cap (Ø 2–5 cm), often with an umbo; inrolled margins when young, uplifted in maturity.
- Cap color varies depending on age and water saturation, from reddish-tan to light brown or even whitish when older and dry.
- Free or adnexed whitish or cream and well-spaced gills (at first shallow but maturing to a depth of up to 0.5 cm) producing white spores.
- Tough but thin, smooth, equal stem, whitish to light brown, darker at base.
- Flesh white or buff with a tough, pliant texture; pleasant (possibly faintly almond) odor.
- Grows in groups, arcs, and fairy rings in lawns or grass, including dune grasses.
- Fruits in spring, summer, and fall on watered lawns; a cosmopolite.

LOOK-ALIKES: Of the LBMs most similar to fairy rings, few grow in grass. The most dangerous look-alikes are the highly toxic Fool's Funnel, *Clitocybe rivulosa*, which is bigger and bright white with a bit of a gray cast and decurrent gills; the Fatal Dapperling, *Lepiota subincarnata*, which has a scaly cap and scaly stem; and the deadly Ringed Conecap, *Pholiotina rugosa*, which is orange-brown with rusty-brown gills, the latter more a danger to a grazing toddler than to a Fairy Ring chaser. Common lawn mushrooms like the Haymaker (*Panaeolus foenisecii*, possibly poisonous to toddlers and usually harmless) and brittlestems like *Psathyrella* or *Candolleomyces* (see Gilled Mushrooms with dark spores) have fragile stems and dark spores.

NAME AND TAXONOMY: *Marasmius* from Greek *marasmos* "to waste or shrink," alluding to its capacity to dry out and rehydrate; *oreades* is a reference to Oreads, the Greek nymphs of mountains and hills. *Marasmius oreades* (Bolton: Fr.) Fr. 1838 = *Scorteus oreades* (Bolton) Earle ex Redhead 2015. Omphalotaceae (synonymous with Marasmiaceae), Agaricales.

 **RED VELVET PINWHEEL**
Marasmius plicatulus

You will not miss spotting a young Red Velvet Pinwheel when you come across it! The red velvety cap is an eye-catcher, and the stem is on fire contrasting against the white gills. With age it is a little less striking, but this slender mushroom with broad, distant gills and a dark, thin stem is easily recognized. Though a single specimen does not offer much biomass, they

Red Velvet Pinwheel, Marasmius plicatulus

often grow in groups and sometimes in big troops. Portland's mushroom huntress and educator Leah Bendlin introduced me to their edibility. The cap offers a great crunch full of umami, reminiscent of a shaggy parasol; the stems are too tough to enjoy.

IDENTIFICATION

- Velvety dry cap (Ø 1–5 cm) wine-red to reddish-brown (sometimes pinkish or yellow-brown); at first bell-shaped, expanding to convex, sometimes uplifted when old.
- Adnate to nearly free, distant, broad gills, colored whitish, buff to yellowish, or tinged pink.
- Long, shining, smooth (and brittle) reddish-black stem (5–12 cm tall x 0.2–0.4 cm wide), no ring, stem attached to mycelial mat.
- Thin, pliant flesh, indistinct taste and odor.
- Grows solitarily or in groups on the ground or in grass under conifers and hardwoods.
- Fruits in coastal areas from Alaska down to California, mostly in fall and into early winter.

LOOK-ALIKES: Cucumber Cap, *Macrocystidia cucumis*, named for its fragrance (edibility unknown), also grows in rich soils, but its cap has a clear white edge and its crowded gills produce pinkish spores.

NAME AND TAXONOMY: *Plicat-ulus* Latin "pleated or folded" and *-ulus* "small." *Marasmius plicatulus* Peck 1897.

SALAL VAMPIRE'S BANE

Mycetinis salalis

You might well smell Vampire's Bane, a.k.a. Garlic Pinwheel, in the forest before you find them, especially after having accidentally stepped on these slender mushrooms that show up in fragrant flocks. They are not much of a source of protein but provide a cool fungal spice that will cause the dinner table conversation to circle around these pungent pinwheels. Be careful not to over-cook them because it might kill the garlic fragrance. The dipeptide glutamyl maras-mine is responsible for the garlic odor in all *Mycetinis* and allied species as well as in the aftertaste of Shiitake (*Lentinula edodes*). Although they deliver a serious garlic punch to any dish—a single cap can go a long way—once ingested, they do not linger in breath or sweat.

In France, they are used in haute cuisine and their price per pound is higher than the price of truffles! Admittedly, we are comparing dried apples to fresh oranges. Big amounts of these tiny mushrooms grow only in spe-cial years. However, eating them in bigger amounts has reportedly caused digestive upset, so it's best to use them just as a potent spice. Vampire's Bane that is too old will impart a rotten garlic flavor.

Detecting the garlic odor first is key to identification. We have several quite similar species of *Mycetinis* growing in different habitats. *M. salalis* favors dead leaves and twigs of Oregon grape and salal; *M. copelandii* grows on leaves of oak, tanoak, and chinquapin (all in the beech family); and both have a light brown cap, darker at the center, white at the margin, pale white gills, and a stem covered with tiny hairs. *M. scorodonius* grows on conifer debris and has a smooth, hairless stem. Biggest of all is the Garlic Parachute, *Mycetinis alliaceus*, with a cap up to 4 cm in diameter and a dark, tough, tall but thin stem (15 cm tall x 0.3 cm wide). It also grows on dead wood and leaves of beech family trees.

IDENTIFICATION

- Buff, dry cap (Ø 1–2 cm) with a lighter striate margin.
- Gills adnate, thin, moderately closely spaced, first cream-buff, darkening to tan-buff, producing white spores.

Salal Vampire's Bane, Mycetinis salalis

- Slender fibrous stem (2–6 cm long), finely haired, white on top, darkening via reddish-brown to nearly black at base.
- Strong odor and flavor of garlic.
- Grows solitarily or in groups on dead twigs and leaves of salal and Oregon grape.
- Fruits in coastal areas from BC down to California, mostly in fall.

LOOK-ALIKES: There are too many similar little brown mushrooms to mention them all; but if we limit LBMs to the ones that effuse a clear garlic fragrance, there are two notable non-*Mycetinis* look-alikes. Stinking Parachute, (*Para-*)*Gymnopus perforans*, is often shorter—reaching only 3 cm in height—with a dainty cap ($\varnothing$ 0.8–1.2 cm), and it brings a much less attractive odor and flavor to the table (figuratively speaking here). It is "strong of putrid water, with a hint of garlic," according to the late, great Gary Lincoff. Also edgy smelling is the quite similar Stinking Gymnopus, *Gymnopus brassicolens*, with an odor of rotting cabbage, sometimes reported as garlicky or like sewer gas. Bon appétit (but edibility is unknown)!

NAME AND TAXONOMY: *Myce-tinis* Greek "pertaining to bonnets," (*mycena*); *salalis* as in growing on salal. *Sarcodonius* and *alliaceus* Greek and Latin for "garlicky." *Mycetinis salalis* (Desjardin and Redhead) Redhead 2012; *Mycetinis copelandii* A. W. Wilson and Desjardin 2005, formerly *Marasmius copelandii* Peck. Omphalotaceae, Agaricales.

Oysters and Allies Growing on Wood

Gilled mushrooms with laterally attached or eccentric stems growing out of wood are known as oyster-like or pleurotoid. There are dozens of inedible tiny to small oyster-like mushrooms, including the brown-spored oysterlings (*Crepidotus*, not edible) and white-spored oysterlings (*Panellus*). Look for at least hand-size oyster-like mushrooms growing in shelflike clusters from dead wood, and the field narrows substantially; there is an excellent chance that your collection is one of the choice edible oysters (*Pleurotus* spp.): Aspen, Pale, or Classic Oyster. Although differentiating to the species level can be challenging (see Name and Taxonomy, below), luckily, all the big oysters are edible and choice, and additional look-alike species are easily delineated, including Winter, Elm, Veiled, and Shoehorn Oysters, as well as Angel Wings. If you are new to oyster harvesting, it is safest to only pick oysters when some of their fruiting bodies are at least 10 cm across. In the smaller (2–7 cm) class, there are simply too many look-alikes.

OYSTER MUSHROOMS
Pleurotus ostreatus group

Oysters are some of the most popular edible mushrooms, due to their subtle but pleasant anise-like fungal aroma and how easy they are to prepare and digest. They also offer bonus medicinal benefits. Oysters usually fruit in abundance and are easily recognized by midsize to large, shelflike clustered fruiting, mostly on hardwoods. They have decurrent, smooth-edged gills and laterally attached, short stems covered in a fine fluff. Oysters offer great fungal protein, especially in abundant spring fruitings on dead alder and cottonwoods, at times

OYSTER ODDITIES AND FACTS

Oyster species are grown worldwide and make up a quarter of the global market of cultivated mushrooms. They are also used in bioremediation for their capability of breaking down many pollutants, including toxic hydrocarbons. Oyster mycelium is also used as the main structural agent in innovative packaging and construction products.

Oyster mushrooms must feel like they don't get enough nutrients out of the dead wood they feed on, so they have come up with a side hustle. The fungus sets mycelial traps for nematodes tunneling in the dead wood. A noose made of three hyphal cells awaits the tiny nematode. When the worm crawls through the loop, the cells instantly inflate (think airbags in a car), locking the nematode in place. Next, the oyster grows hyphae into the arrested nematode, releases digestive enzymes, and absorbs the nutritious insect proteins (otherwise a rarity in dead wood) into the mycelium.[73]

when other edible mushrooms are scarce. However, oysters should never be eaten raw. Cooking will neutralize their toxin pleurotolysin, which attacks membranes of red blood cells and can lead to cell death.[74]

Wherever dead red alders are present, such as in more coastal areas, the most common oyster is the Pale Oyster, *Pleurotus pulmonarius*. Often white or with a pink or tan hue especially when young, it tends to have a more pronounced stem and fruits under warmer conditions, hence its alternate name, the Summer Oyster. The Aspen Oyster, *P. populinus*, is best differentiated by the fact that it grows on aspen and cottonwoods (both trees belong to the genus *Populus*); it is more common east of the Cascades. A stem is nearly absent or rudimentary, but otherwise the Aspen Oyster looks very similar to the Pale Oyster. Both have caps with a diameter of 5 cm to 15 cm (and sometimes bigger), but only the Pale Oyster grows rarely on conifers.

Aspen Oyster, Pleurotus populinus

TABLE 4. COMPARING PACIFIC NORTHWEST OYSTER SPECIES

Pleurotus SPECIES	Pale Oyster P. pulmonarius	Aspen Oyster P. populinus	Oyster Mushroom P. ostreatus group	Veiled Oyster P. dryinus
SUBSTRATE	Especially red alder (rarely conifers)	Cottonwood and aspen	Hardwoods, less commonly pine and fir	Maple, oak, birch, cotton-wood, and fruit trees
MAIN AREA & SEASON	More coastal; year-round, but especially spring	Widespread; year-round, but especially spring	Riparian habitat, cool weather	Widespread; fall
CAP WIDTH	5–15 cm	5–15 cm	5–25 cm	5–20 cm
CAP	White or with a pink or tan hue, especially when young	Ivory white to pinkish-gray or pinkish-buff	White, yellow-ish, gray, tan to brown, often dark with age	White, cream to grayish with tan notes
GILLS	Decurrent, close, white to slightly cream color when old	Decurrent over the short point of attachment, close, two tiers of subgills	Adnate to mostly decurrent, close, sometimes form-ing a network near base	Decurrent, close, often veined or forking on the stem; white, yel-lowish when old
SPORE COLOR	Whitish, grayish, or lilac	Whitish to buff (never lilac)	Whitish to lilac tinged	White
STEM & RING	More pronounced, often bristly to fluffy, no ring	Nearly absent or rudimentary, often hairy, no ring	Nearly absent or rudimentary, often hairy, no ring	Long stem with partial veil
CULINARY NOTE	Delicious, most tender oyster, turns softish when overmature	Delicious, very tasty and tender when young	Delicious, stem and cap can be tough when overmature	Toughest and hence reputedly least desirable

Due to lack of taxonomic clarity, *P. ostreatus* group was used as a catch-all taxon, though in 1993 (Vilgalys et al.) researchers recognized that *P. ostreatus* (described first in Austria) was restricted to Europe and mostly eastern North America. However, I recently found it growing in Kirkland, Washington, on introduced Lombardy poplars, a discovery confirmed by DNA analyses. *P. ostreatus* is often bigger and tend to have darker, especially brown toned caps (Ø 6 cm–25 cm) than our other oysters. Also, their fruiting is triggered by cold weather.

- Midsize to big, smooth cap (Ø 5–25 cm) growing shelflike, oystershell to fan-shaped; white, cream, pinkish to gray, bluish-gray, or brown; smooth, with margin inrolled at first.
- White to cream or pinkish gills, broad, close (to crowded), running down the stem; spores whitish to lilac tinged.
- Absent or short white stem covered in fine fluff. Stem is eccentric (off-center) but can move to center if fruiting body grows straight up from top of a fallen trunk.
- Soft, thick flesh with mild, pleasant aroma, possibly anise-like.
- Grows in shelflike clusters, groups, and dispersed singles on dead hardwoods (especially alder, cottonwood, aspen), sometimes conifers.
- Fruits potentially year-round in mild weather; on alder, especially in spring.

Pale Oyster, Pleurotus pulmonarius

Pleurotus ostreatus

a beta-glucan shown to be effective in several clinical trials in reducing upper respiratory tract infections. A topical cream containing pleuran significantly improved symptoms in patients suffering from atopic dermatitis.[76] However, oysters produce prodigious numbers of spores that can cause fairly severe adverse effects for sensitive people when inhaled in quantity.

LOOK-ALIKES: The also edible Shoehorn Oyster (*Hohenbuehelia* "*petaloides*") has similarly white to gray or dark brown caps but with a powdery layer when young and a moist surface. The stem is also laterally attached, and the cap shape ranges from shoehorn to funnel-shaped when upright or fan-shaped when more shelflike. The cap tapers into the stem base. The underside has crowded pale gills that produce white spores. Shoehorn Oysters grow in spring, summer, and fall on dead wood, often woody debris, including wood chips and sometimes on planter wood.

MEDICINAL: *Pleurotus* species have received a lot of interest from researchers and are known to have antibiotic, antiviral, anti-inflammatory, and anticancer potential. Oyster mushrooms (*Pleurotus* spp. and *Hypsizygus tessulatus*) are so far the only known food sources that contain statins,[75] bad-LDL-cholesterol–reducing compounds. An animal study found oyster mushroom–derived chrysin to be almost as effective as the drug lovastatin. As with many other edible and medicinal mushrooms, oysters are rich in ergothioneine, a powerful antioxidant and cytoprotective. Plants and animals are not able to produce this amino acid, but some fungi are.

There is ongoing research into oyster-derived anthroquinone, which seems to have potential for treating both Alzheimer's and Parkinson's. Clinical trials have shown that oysters reduce blood pressure. They also contain the polysaccharide pleuran,

Another good-size, laterally attached wood decayer with white to yellowish spores is the edible but tough Hairy Oyster Mushroom. This gilled polypore has had many names, but it is currently known as *Panus lecomtei*, a.k.a. *Panus rudis* and *Lentinus strigosus*. It loves birch and often has purple tones when young, and its cap is really hairy.

Also oyster-like shaped, but clearly off in color, and with a rotten smell and flavor as well as tough flesh, is the inedible Orange Mock Oyster, *Phyllotopsis nidulans*, which

VEILED OYSTER
Pleurotus dryinus

Unless you are foraging in oak, maple, or birch forests, the least common of our oysters is the Veiled Oyster. It is easy to tell apart from other oysters by its long stem, partial veil (when it still has one), and some fluff, at least in spots, on the cap. It is just as edible as—though maybe a bit less enjoyable than—the other oysters, but it usually fruits alone or with scant fruiting bodies that can each grow quite big.

IDENTIFICATION

- White, cream, to grayish dry cap (∅ 5–20 cm) with tan notes and soft fibrils or scales; inrolled edge when young, becoming wavy with age.
- Close gills, decurrent, often forking on stem; discoloring yellowish with age, white spored.
- Long, off-center, whitish, dry stem.
- Features a partial veil (or at least a ring zone) when young; veil remnants might be attached to cap edge or disappear completely.
- Firm, rubbery flesh, aroma slightly anise-like.
- Grows solitarily or in small groups on (usually) living hardwoods, such as maple, birch, oaks, and fruit trees.
- Fruits summer into winter from Alaska down into the Andes and all over North America and Eurasia.

TOP: *The Orange Mock Oyster,* Phyllotopsis nidulans, *an oyster look-alike that is not edible;* **BOTTOM:** *Edible Shoehorn Oyster,* Hohenbuehelia *"petaloides"*

sports pale orange to yellow-buff, fuzzy, fan-shaped caps and grows in shelflike groups on wood. Close, narrow gills match the cap in color and produce pinkish-orange to creamy-peach spores.

NAME AND TAXONOMY: *Pleurotus* means "sideways," referring to the lateral attachment, *ostreatus* "oysterlike," *pulmonarius* "related to the lungs" probably based on shape, and *populinus* "belonging to poplar"—all Latin. *Pleurotus ostreatus* (Fr.) P. Kumm. 1871, *Pleurotus pulmonarius* (Fr.) Quel. 1872, and *Pleurotus populinus* O. Hilber and O. K. Miller 1993.

LOOK-ALIKES: Veiled Oysters are frequently mistaken for the similarly long-stemmed Beech Oyster, *Hypsizygus tessulatus,* a choice edible also known in the trade by its Japanese name, Buna Shimeji, meaning "beech mushroom." It can be told

Veiled Oyster, Pleurotus dryinus

apart easily from the Veiled Oyster by its gill attachment: the gills are not decurrent (typical for *Pleurotus*), but are evenly notched, offering a visual break between stem and gills. They grow in relatively long-stemmed clusters, mostly from poplars such as cottonwood and aspen in the Pacific Northwest, but occasionally from other hardwoods. In general, they are much more

Beech Oyster, Hypsizygus tessulatus

an eastern American species. The cap is uniquely marbled.

Also similar is the nontoxic Poplar Scalycap, *Hemipholiota populnea* (formerly *Pholiota destruens*), a big, fleshy, brown-spored, whitish cottonwood decayer with a scaly cap and stem.

NAME AND TAXONOMY: *Dryinus* from Greek *drys* "oak." *Pleurotus dryinus* (Pers.) P. Kumm. 1871. Pleurotaceae, Agaricales.

❗ ANGEL WINGS
Pleurocybella porrigens

These enjoyable, bright white, fan-to-shoehorn-shaped mushrooms, often found in dark, mossy corners of the forest feeding on conifer cadavers, especially hemlock, have been a staple edible for many Pacific Northwest foragers for decades, thanks to their firm structure and a solid fungal flavor. Often confused with oyster mushrooms, these fan-shaped

THE PERFECT FUNGAL STORM

In 2004, a perfect storm hit Japan in the form of several early typhoons, delivering an extended rich mushroom season that brought forth Angel Wings in great abundance and stretched them to impressive sizes. Many foragers went on to feast on Sugihiratake (meaning "autumn mushroom" in Japanese). However, some people got sick from this mushroom they'd eaten regularly before without a second thought. All victims were elderly and had a history of serious kidney problems; some were on dialysis treatments. Symptoms usually occurred two weeks after a meal. For some people, death followed two to three weeks later, caused by acute encephalopathy, degenerative lesions in the brain.

The suspected toxin is the unstable amino acid pleurocybellaziridine.[77] Strangely, no other cases have been reported previously in Japan or anywhere else. Angel Wings are widespread in temperate Eurasia and North America. No reports of poisonings are known from the Pacific Northwest, and many people (this author included) continue to enjoy Angel Wings feeling secure that *Pleurocybella porrigens* will not serve them an untimely set of wings to go to heaven. The message is to not overindulge, and if you are elderly or have kidney issues, to abstain.

mushrooms are fairly easy to tell apart by their thinner and more pliant nature. When held against light, they appear semitranslucent. Debris, such as fir needles lying on the Angel Wings for some time, cannot be brushed off; it will become embedded as the cap tissue grows around it and will have to be cut out.

However, the reputation of Angels has taken a serious hit in recent years due to tragic poisonings in Japan that killed one person in 2009 and seventeen people five

Angel Wings, Pleurocybella porrigens

years earlier. Therefore, today, this species is considered a fringe edible. Please, if you are so inclined, enjoy it responsibly.

IDENTIFICATION

- Smooth white cap (Ø 4–8 cm), fan- or lobe-shaped, partially translucent.
- Crowded, shallow, white, or yellowish gills, running all the way to point of attachment; white spored.
- Stem usually missing, sometimes stub present.
- Firm, thin, pliant flesh with mild flavor and odor; possibly tinged yellow with age.
- Grows in shelflike groups or overlapping clusters on old rotting conifers, especially hemlock.
- Fruits from late summer through fall in the whole region.

LOOK-ALIKES: The most similar mushrooms are light-colored oysters, especially *Pleurotus pulmonarius*. However, oyster flesh is less pliant and more fragile, also thicker, and the gills are deeper. Oysters usually have a clear short stem, often with fine hair. White and also growing on hardwoods is the quite rare Mealy Oyster (*Ossicaulis lignatilis*), with a more prominent central stem and a less tongue-shaped fruiting body, a more traditional cap and stem appearance, tough flesh, and an often mealy odor. Edibility of the Mealy Oyster is unknown. And then there is a sea of inedible whitish oysterlings (*Crepidotus* spp.), all brown spored.

NAME AND TAXONOMY: *Pleuro-cyb-ella*: *pleuron* for "rib" or "side," as in lateral, and *cybe* "head," both Greek, plus Latin -ella diminutive, making it small; *porrigens* Latin for "stretched," referring to the gills stretched all across the fruiting body. *Pleurocybella porrigens* (Pers.) Singer 1947. Marasmiaceae, Agaricales.

LATE OYSTER
Sarcomyxa serotina

The late fruiting of the Late Oyster is what this chameleon brings to the table, fruiting when most mushrooms are already in hibernation or mushed by frost. Few mushrooms exhibit such a range in cap coloration as this conch-shaped wood decayer that often grows in layers. Most commonly it is greenish, but there are also gray, blue, and purple tones and, often with advancing age, loud orange coloration, which can include the gills and the laterally attached stem. But more often than not, the fuzzy stem is cream

Late Oyster, Sarcomyxa serotina

to dull brownish and displays a striking transition where the dense gills meet the hairy, stubby stem.

Also known as the Green or Winter Oyster, *Sarcomyxa serotina* handles frost quite well and seems to go back to spreading spores once released from the grasp of father frost. That does not necessarily improve its culinary consistency, but fresh mushrooms in winter can be a treat! When young and fresh, the Late Oyster has a pleasing texture and flavor. In Japan, where they are known as Mukitake, the "peeling mushroom," they are cultivated and researched for their liver-protecting qualities! With age, Late Oysters can become tough and bitter: extended cooking can help. Removing the cap skin (hence the "peeling mushroom") also removes bitterness. Some people first soak them overnight or parboil them for a short time, but often the skin comes off nicely when you start peeling from above the stem, without special treatment. Very soggy or overly soft fruiting bodies benefit from dry frying—or if they're too far gone, disposal.

IDENTIFICATION

- Shiny, shell-shaped cap (Ø 4–15 cm), sticky when moist, greenish, with tones of gray and blue, inrolled margin when young, aging to wavy edges and lobes (turning brownish-orange in age).
- Crowded, cream to pale orange gills, sometimes forking; cream to yellowish spores.
- Stubby lateral stem covered in hairs, sometimes scalelike toward gill edges.
- Flesh whitish, thick, firm, thinning to margin; mild tasting, sometimes bitter in age.

- Grows in groups or clusters, mostly on dead alder and maple (rarely on conifers).
- Fruits late fall and early winter.

LOOK-ALIKES: The combination of greenish color, shell shape, and fruiting on dead wood makes this mushroom unique. The much rarer, edible, but tough Lilac Oysterling, *Panus conchatus*, has a lilac-brownish (never greenish) cap and fruits earlier in the year. Its toughness is not surprising given that it is a gilled polypore. Otherwise, check out look-alikes for the other oysters, above.

NAME AND TAXONOMY: *Sarco-myxa* Greek (and not flattering) "flesh-slime," *serotina* Latin for" late," as in the fruiting season. *Sarcomyxa serotina* (Pers.) P. Karst. (1891); synonymous with *Panellus serotinus* (Pers.) Kühner (1889).

❓ STICKY OYSTERLING
Scytinotus longinquus

This small, sticky, and locally abundant little oysterling is not a traditional edible but is included in this guide partially because it is an Oyster look-alike, though it is much smaller. However, Buck McAdoo reports of trials by David Arora and Jack Waytz, in which they appreciated the flavor of this little winter mushroom. Due to its sticky cap and small size, cleaning it is a bit tedious, but I would describe the flavor as nicely fungal, after an initial dry sauté and finished with a bit of olive oil and salt and pepper. However, there is no guarantee that it is truly a recommended, edible mushroom; the only data we have is from a few daredevils.

Sticky Oysterling, Scytinotus longinquus

The Sticky Oysterling has an unusual distribution, documented in the coastal Pacific Northwest, coastal Chile, Madagascar, New Zealand, and southeastern Australia.

IDENTIFICATION

- Fan- or kidney-shaped, lobed, sticky cap (Ø 0.5–4 cm) mostly light pinkish but ranging from whitish to brown.
- Short decurrent gills, whitish to pale peach, with yellow-cream spores.
- Short, tough stem, blackens in age (sometimes absent).
- Firm, fibrous flesh with indistinct odor.
- Grows in big groups, mostly on alder, in late fall and winter.
- Fruits from Alaska into California in coastal climates.

LOOK-ALIKES: Being a small, whitish, laterally attached wood decayer, it has a wide range of look-alikes. If it wasn't so small and sticky, it would be easy to confuse with the Pale Oyster (*Pleurotus pulmonarius*), which also sometimes has a pink tinge. The Elastic Oysterling (*Panellus mitis*) grows on conifer wood, is a bit smaller, and has gelatinous gill edges. The Bitter or Luminescent Oysterling (*Panellus stipticus*) has a bad flavor, ranging from bitter to spicy to astringent, and is possibly poisonous and not perceptibly luminescent in the Pacific Northwest. Similar oysterlings from the species-rich genus *Crepidotus* (Inocybaceae) are all brown spored and inedible.

NAME AND TAXONOMY: *Scytinotus* Greek "leathery," a reference to the tough stipe, and *longinquus* Latin "longish." *Scytinotus longinquus* (Berk.) Thorn 2012 = *Panellus longinquus* (Berk.) Singer 1951. Porotheleaceae, Agaricales.

Other White-spored Mushrooms That Grow on Wood

Oddly, besides oysters, there are few edible mushrooms that grow on wood and have white spores. Most outstanding are the often growing in clusters

Honey mushrooms (*Armillaria* spp.) and Velvet Feet (*Flammulina velutipes*). Also included below is the Giant Sawtooth (*Neolentinus ponderosus*) and Plums and Custard (*Tricholomopsis rutilans*).

ⓘ DARK HONEY MUSHROOMS
Armillaria ostoyae

With a name seemingly straight out of the fungal kingdom PR department for a mushroom with real personality, honey mushrooms often pop up in impressive abundance in clustered fruitings on living trees or dead wood. They are both loved and hated! It's easy to imagine the excitement and gratitude a forager experiences in hungry times at finding honeys to harvest for a starving family at home. While honeys are edible, some are also destructive. Foresters and tree lovers know that some honey mushrooms can parasitize roots and kill a previously healthy tree in an irreversible death spell. It was forest pathologists

who mapped out the mind-blowingly big "humongous fungus" (see sidebar) that knocks off trees, initiates forest rejuvenation, and can be controlled neither by affordable mechanical means nor toxic fungicides. Yet honey mushrooms are part of the natural cycle.

Honey mushrooms are shape-shifters, making them tricky to identify sometimes. They are midsize to large, whitish-gilled, white-spored mushrooms with brown scaly caps, usually well-developed rings, and white pith in the stem. The unique, shiny, and tough black rootlike rhizomorphs (bundled mycelium) in the wood they parasitize and beyond are a great clue to their identity. If you are unfamiliar with *Armillaria*, it is best to stick with specimens growing in clusters on wood and not pick single fruiting bodies.

In the Pacific Northwest, we have at least half a dozen species. The most common is the Dark Honey Mushroom, *Armillaria ostoyae* (a.k.a. *A. solidipes*), described below. The very similar Honey Mustard Mushroom (*Armillaria sinapina*) is named for the mustard-yellow fibrous tis-

Dark Honey Mushrooms, Armillaria ostoyae

HONEYS AS EDIBLES

Besides their abundance and firm texture, as edibles, honeys offer a rich fungal flavor with maybe a bit of spiciness. You can find them brined in jars in European specialty stores, imported from Poland and labeled as "*opieńka*." To render honey mushrooms safe for consumption, parboil them for ten minutes; they might turn mucilaginous or jellylike, which is easily remedied with dry frying. Many lucky pot hunters can enjoy Honeys without parboiling them, but to figure out if you are one of those consumers, ingest only small portions at first. Without detoxing them, some people might experience very unpleasant digestive upset with flu-like symptoms, including fever. (A few people experience an allergic reaction even after parboiling.) Raw honey mushrooms can be troublesome. The German name *Hallimasch* derives from *Höll im Arsch*, or "hell in the arse," but you might vomit as well. Yet there are no known cases of really serious poisonings with long-term health issues.

Norway's famous forager Pål Karlsen showed me how to use them as a pizza crust, which turned out delicious (see Honey Mushroom Pizzette in Part Three). I have never had any problems, but strangely, my daughter Sophia felt awfully sick for several hours, and had to vomit; she will not eat them again. She was eating honeys with her friends, specimens I had picked from a cluster growing on apple tree roots in our yard. I had eaten them the evening before without any problems (we did not parboil them either time), and her three friends sharing the meal were all fine. People's reactions clearly vary and the causation is still a mystery.

An impressive and mysterious experience is witnessing foxfire, the green glow of bioluminescent *Armillaria* mycelium that lights up roots and trunks penetrated by this fungus, on a very dark night. When I worked to help local people generate income in rural areas of Tibet, we supported a project that cultivated honey mushrooms: the objective was not the fruiting bodies, but the rhizome production from the precious medicinal plant *tianma*, the parasitic orchid *Gastrodia elata* that feeds on *Armillaria* mycelium.

sue on the stem and yellow ring. However, in our region the "mustard" is often lacking, and this species can be more reliably distinguished by its tendency to grow in smaller clusters or alone, a ring that is more cobwebby than membranous, smaller brownish caps (Ø 2–6 cm) with a reddish tinge, and round, not flattened, rhizomorphs.

The Olympic Honey Mushroom (*Armillaria nabsnona*), first described from the Olympic Peninsula, has orange-brown caps (Ø up to 7 cm) with fine black hair, but lacking scales, and whitish to pinkish-tan gills, and it grows from hardwoods such as maple and alder. The Honey Mushroom, *A. mellea*, is rare up in the Pacific Northwest, but it grows in California; *mellea* is Latin for "honey-like," which describes the cap color well. The cap is also smoother and without small scales, but has fine black hair in the

Dark Honey Mushrooms, Armillaria ostoyae

center, and it also has a felty ring that it shares with *A. ostoyue*.

- Light to dark brown, dry or slightly moist cap (Ø 5–12 cm) with radiating silky fibers or small scales (especially in the center) that can disappear with age; paler cap margin often fringed or splitting.
- Close whitish to cream gills, attached to somewhat decurrent, producing white spores often visible on other caps in the cluster.
- Whitish to cream stem (5–20 cm long x 1–1.5 cm wide at top), often with vertically lined surface with whitish to grayish-brown patches and scales; white felty partial veil leaves a ragged ring with brownish edge, staining brown from handling, darkening to black with age.

THE HUMONGOUS FUNGUS IN OREGON

The Dark Honey Mushroom growing in the Malheur National Forest in eastern Oregon's Strawberry Mountains is regarded as the world's largest known organism, covering (or rather burrowing under) an incredible 2,385 acres (9.6 square km).[78] This organism is estimated to be 2,000 to 8,000 years old and possibly weigh up 35,000 tons—not that anyone could ever get the underground mycelium, its rootlike rhizomorphs, and tissue in infected trees on a scale!

Forest pathologists insist that DNA analysis of this Shoestring Root Rot (as some foresters call *A. ostoyae*) consistently demonstrates that the whole area is infested—or maybe more myco-centrically stated, "managed"—by one single parasitic fungal individual. Oregon Public Radio's Vince Patton made an interesting short film on the "Humongous Fungus,"[79] as it is called. The world record is claimed for the size of the spread. When it comes to weight, it is probably outdone by clonal aspen groves, such as Pando in Utah's Fishlake National Forest, which is surely home to innumerable Aspen Scaberstalks.

- Stem wider at base, with yellow mycelium at bottom and black belt-like rhizomorphs; stem with rind-like surface and fibrous white stuffing.
- Firm, white flesh, thick at cap center; mild fungal flavor and odor.
- Grows on wood, especially conifers, but sometimes maple, birch, oak, and fruit trees, and kills its hosts.
- Common and widespread; fruits mainly in fall.

LOOK-ALIKES: The honey paradox informs us that honey mushrooms are difficult to identify to the level of species, but are fairly easy to tell apart as a group from other wood-digesting mushrooms: they are white spored, and most other somewhat similar wood decayers are dark spored (e.g., *Gymnopilus*, *Pholiota*, *Galerina*, *Hypholoma*). In this way, honeys really stand out. A reminder: with any wood-growing mushroom, be aware of the Deadly Skullcap (*Galerina marginata*), which is much smaller and thin stemmed with rusty-brown spores.

NAME AND TAXONOMY: *Armillaria* Latin for "bracelet" or "ring" and *solidi-pes* Latin for "solid foot," as in the stem; *ostoyae* may be a reference to Ostoja, a region in Ukraine. *Armillaria ostoyae* (Romagn.) Herink 1973, synonymous *Armillaria solidipes* sensu Burdsall and Volk, formerly included in *Armillaria mellea* (Vahl) P. Kumm. 1871. Physalacriaceae, Agaricales.

!! VELVET FOOT
Flammulina velutipes

A good edible that does not get cold feet when temperatures dip is the Velvet Foot. It contains antifreeze-like xylomannan, a beta-glucan that acts as a natural cryoprotectant.[80] It can handle being frozen, and when the temperature warms back up, it will get back to producing its white spores. Though more easily found in the winter when fewer mushrooms compete for attention, this wood decayer can fruit from October through May. It is most easily recognized by its nearly black velvety stem (hence another of its names, Velvet Shank). However, that trait is not visible in young specimens, and unless you are very familiar with Velvet Feet, it is safer not to pick

Velvet Foot, Flammulina velutipes

them unless the black-ish velvet shows at the base of the stem. When young, the sticky, bright, mostly yellow-orange but highly variable color of the smallish caps, usually clustered growing from dead hardwoods, attracts the eye.

Aspen Velvet Foot, Flammulina populicola

White spores and the absence of a ring help distinguish it from the Deadly Skullcap. Talking about toxins, raw Velvet Foot contains the hemolytic protein flammutoxin; do not eat them raw. Fortunately, the toxin is neutralized when cooked.

Very similar, also edible, and even sweet tasting is the Aspen Velvet Foot, *Flammulina populicola*, which grows from the base of aspens and cottonwoods. Also very similar and great tasting, but smaller and with a lighter-colored cap is Lupine Velvet Foot, *F. lupinicola*, which fruits in small clusters on dead bush lupines, especially on beaches.

Velvet Foot is very closely related to all-white commercial enoki mushrooms, which look quite different, since they are cultivated in the dark. Astonishingly, enoki have been cultivated since the eighth century! Recent DNA studies have revealed that Enokitake (*-take* is Japanese for "mushroom") is a different species, now called, *Flammulina filiformis*. *Flammulina* are great edibles, and people in Central Europe praise Velvet Foot as one of the best edible winter mushrooms. Its stems can be too tough, but it makes a great mushroom powder.

Not surprisingly, Enokitake is also used in traditional Chinese medicine and has been researched a lot in relation to cancer. Most notable are reports from Japan's Nagano Prefecture, where remarkably low cancer rates perplexed researchers in the 1970s. Enoki farmers even had a 60% lower death rate from cancer than the general Japanese population. These much lower deadly cancer rates were explained by high enoki consumption in an area with an astounding density of enoki farms. Further lab research, including at Bastyr University in Kenmore, Washington, near Seattle, as well as human trials, have confirmed enoki's anticancer capacity.[81]

IDENTIFICATION

- Yellow-orange to reddish-brown or dark brown cap (Ø 1–6 cm), bald and rounded (not conical), striations might show with aging, sometimes spotting or staining dark brown.
- Cap surface smooth, sticky to slimy (when moist), and can be peeled.
- Broad, moderately close, adnexed gills; white to cream (slowly bruising brown); white spores.
- Slender, tough, often curved hollow stem; initially colored like cap but becomes covered from base upward

with a dark rusty-brown to blackish velvet as it matures; top of stem remains yellowish; no ring.

- Very firm, white to yellowish flesh tastes pleasantly fungal, even sweet.
- Grows often on stumps alone or in tufts or clusters on hardwoods like alder.
- Relatively common and widespread in Northern Hemisphere, fruits from fall into spring.

LOOK-ALIKES: Do **not** confuse Velvet Foot with Deadly Skullcap, *Galerina marginata*, which is somewhat similar but has brown spores and a ring (or at least a ring zone) on the stem. Growing in very similar conditions on dead wood are two *Kuehneromyces* species: the common and inedible Spring Scalecap (*K. lignicola*) and the edible Sheathed Woodtuft (*K. mutabilis*), which has a sheathed, scaly lower stem that flares into a conspicuous ring.

There are other small brown clustered mushrooms that grow on wood like the very common Pinewood Gingertail (*Xeromphalina campanella*), which has a similar but much thinner stem, decurrent gills, and usually occurs in huge groups, mostly on conifers. Conifer Tuft (*Hypholoma capnoides*) and Sulfur Tuft (*H. fasciculare*) have dark purple-brown spores and a ring or partial veil.

NAME AND TAXONOMY: *Flammu-lina veluti-pes*, "little flame" for its bright cap, and *velutipes* meaning velvet foot, both Latin. *Flammulina velutipes* (Curtis) Singer (1951). Physalacriaceae, Agaricales.

GIANT SAWTOOTH
Neolentinus ponderosus

Talk about a great common name! "Giant Sawtooth" gives it all away. Some caps can be a foot or more wide. Also very cool and helpful for identification are the close, sawtooth-like gill edges. The scientific species epithet *ponderosus* literally means "heavy," but it is also an indicator of the Sawtooth's love of ponderosa pines, or at least their wood. Though it often seems to fruit from the ground, it is usually connected to dead wood or very near to it,

Giant Sawtooth, Neolentinus ponderosus

and it also loves to fruit on pine stumps. It has an excellent flavor, but the Sawtooth is a tough guy; it often dries out over many weeks before rotting. The taxonomy reveals it came by its toughness honorably; it is a member of the Gloeophyllales, an order that contains many tough wood decayers.

Instead of feeding on young or old fruiting bodies, which both tend to be very tough, stick with middle-aged mushrooms. Thorough cooking will soften up these firm mushrooms! A good trick is to cover the sliced cap pieces with water in a frying pan and boil the water away. Once the water is gone and the tasty residue not yet burned, add oil or butter to sauté the slices. David Arora suggests barbecuing preboiled 1-inch pieces. *Neolentinus ponderosus* shows up first in late spring, and with surprisingly little moisture, they keep fruiting through fall. Their tough consistency (and ability to withstand drought) well equips them to fruit in light, sunny locations like thinned forests.

IDENTIFICATION

- Light-colored, tough cap (Ø 10–45 cm); inrolled margin when young, flattening with maturity; breaking up into large, light brown, flat scales over cream base.
- Serrated, narrow, close gills; adnate to decurrent; cream colored turning buff to tan-brown.
- Stubby, hard, whitish ringless stem (5–20 cm tall x 2–8 cm wide), aging yellowish-brown or rusty-orange and often with brownish patches or scales and narrowed, rooting base.
- Thick, tough, white flesh, often aging or bruising yellow.

- Grows singly, or in small groups or clusters on or near dead conifers, especially pines.
- Fruits late spring and, with enough moisture, into fall.

LOOK-ALIKES: The also edible, tough, whitish-capped Scaly Sawtooth (*Neolentinus lepideus*) is smaller (Ø 5–15 cm), with a thinner stem (3–5 cm tall x 1–3 cm wide) and a clear ring. It has thickened, fine scales on the cap and stem. More poetically, it is called the Train Wrecker for digesting railroad ties, which are usually made of conifers. It outsmarts toxin-soaked timber; do not consume any Scaly Sawtooths found growing on treated wood.

Other huge whitish mushrooms are the edible, double-ringed Vanishing Cat (*Catathelasma evanescens*) and the hard-to-digest, but not physically tough, Western Giant Funnel, *Aspropaxillus giganteus* (formerly known as *Leucopaxillus giganteus* a.k.a. *Clitocybe gigantea*). However, both giants lack sawtooth gills and grow in soil. Somewhat similarly capped but much smaller is the brown-spored Poplar Scalycap, *Hemipholiota populnea*. It is harmless to eat, but often tastes bad; it also lacks serrated gills and grows on cottonwoods and aspen, never conifers. There are several inedible, much smaller sawtooth-gilled wood decayers in the genera *Neolentinus* and *Lentinellus*, most of them bitter or spicy, often with laterally attached stems, and growing in clusters.

NAME AND TAXONOMY: *Neo-lentinus*: *neo* Greek for "new," *lentus* Latin meaning "tough, pliable." *Neolentinus ponderosus* (O. K. Mill.) Redhead and Ginns 1985; formerly *Lentinus ponderosus* O. K. Mill. 1965. Gloeophyllaceae, Gloeophyllales.

Plums and Custard, Tricholomopsis rutilans. *What a beautiful contrast between the scarlet-red to purplish cap and stem and the golden gills!*

PLUMS AND CUSTARD
Tricholomopsis rutilans

Most people leave it unpicked, taking with them only the joy of observing this beauty. Plums and Custard is edible, although the common name does not allude to its culinary profile; it will not make anyone's top forty. Plums and Custard can be recognized by its red cap and stem and contrasting yellow gills growing on or near conifer wood or humus rich in lignin. It is a widespread cosmopolite, and future DNA work may separate it into several species.

IDENTIFICATION

- Purple-red to dark red, dry cap (Ø 4–12 cm) with dark scales or fibrils on yellow background, thinning toward margin in maturity.
- Cap edge at first inrolled, flattening with age.

- Egg-yolk-yellow to pale yellow gills, close, adnate or notched; white spores.
- Thick, firm pale yellow flesh with slight flavor of radish.
- Grows alone, in tufts, or in small groups on or near rotting conifers as well as wood chips.
- Fruits from summer into fall, widespread and cosmopolitan.

LOOK-ALIKES: There are at least two similar rustgills, which share somewhat similar cap and gill colors. Both of these *Gymnopilus*—the Yellow-gilled Rustgill (*G. luteofolius*) and Magic Blue Rustgill (*G. aeruginosus*)—have rusty-orange spores and duller stipes with rings. They have a very bitter flavor.

NAME AND TAXONOMY: *Tricholomopsis* Greek "like a *Tricholoma*" and *rutilans* Latin "reddening." *Tricholomopsis rutilans* (Schaeff.) Singer 1939. Tricholomataceae, Agaricales.

GILLED MUSHROOMS WITH PINKISH-SALMON SPORES

Most pink-spored mushrooms belong to the Entolomataceae family, infamous for being hard to differentiate, but famous for the obscenely blue *Entoloma hochstetteri*, the only mushroom to date featured on a banknote—a fifty-dollar note in New Zealand. Some pinkgills are laced with poisonous muscarine. Others are good edibles and fairly straightforward to identify: three of these are described in this field guide. Two other edible pink-spored mushrooms in the Plutaceae family, *Volvariella bombycina* and *Volvopluteus gloiocephalus*, are presented with images and short descriptions that reveal all you need to know for identification.

A few other genera have spores with a pinkish hue, including such choice edibles as oysters (*Pleurotus* spp.) and the Wood Blewit (*Lepista nuda*). Since they belong to families dominated by white-spored mushrooms, they are covered earlier in this section. Mother Nature is just not into these straight lines we modern humans like to project onto organisms.

‼️ SWEETBREAD MUSHROOM
Clitopilus prunulus group

The innocuous Sweetbread Mushroom (a.k.a. Miller) languishes in culinary obscurity. However, Bellingham-based megamycophile Buck McAdoo informs us that "fabulous flavor and fine texture should place this mushroom in everyone's top five." MykoWeb's Michael Wood loves to tell me how at a California foray, the Miller beat out porcini in a blind taste test! Luckily, the strong mealy odor cooks away and reveals a rich fungal flavor reminiscent of shaggy parasols. However, the lack of general appreciation by most mushroomers can be partially explained by the sheer abundance of look-alikes, some of them seriously poisonous. Crucial to a safe ID is the presence of pinkish spores (see print on cap in lower right of photo, next page), decurrent gills, and the mealy sweetbread fragrance (also described as similar to a watermelon rind) that can be tasted in the flesh. What looks like the same mushroom, but lacking the fragrance and having white spores, is the vicious and potentially deadly Fool's Funnel—be careful!

Molecular analyses have so far not turned up *Clitopilus prunulus* in the Pacific Northwest, but according to Danny Miller, our sequences match the very similar *Clitopilus cystidiatus*, recently described in Europe. However our taxonomy shakes out, many fabulous foragers have enjoyed our Miller! In Germany, expert mushroom col-

New Zealand's Entoloma hochstetteri

Sweetbread Mushroom, Clitopilus prunulus *group; note the pink spore print on the lower right.*

lectors—who reject the idea of being mushroom hunters, since this Anglicism invokes for them the ridiculous idea of firing a gun at their beloved mushrooms—know that the Miller often grows in the same spot as *Steinpilz* (King Bolete). I have actually found them together with Fiber Kings (*Boletus fibrillosus*) in grass under Sitka spruce and in Douglas-fir–hemlock forest. Buck has found them growing in a King Bolete spot as well.

IDENTIFICATION

- White to gray (sometimes with beige hue), dry, slightly felty cap (Ø 3–10 cm) with inrolled fragile margin, often turning wavy in age.
- Cap center often a bit darker and umbonate, with raised knob.
- Decurrent (sometimes attached) close gills, whitish to gray when young, maturing to pinkish; pink spores.
- Stout, naked, whitish stem with often wider, finely hairy or felty base.

- Thinnish, fragile, white flesh.
- Odor and flavor strongly farinaceous, pleasantly mealy or like sweetbread, sometimes an additional whiff of cucumber.
- Saprobic, in conifer or hardwood forest; grows on ground singly, rarely in bigger groups in grass.
- Widespread from Alaska down to California, fruits in late summer and fall.

LOOK-ALIKES: Make sure you do not fall for the Fool's Funnel (*Clitocybe rivulosa*) and its high dose of nasty muscarine. It can be told apart by odor and spore print; Fool's Funnel lacks the mealy fragrance and has white spores. There are also other similar-looking toxic Clitocybes, some of which, like the Frosty Funnel (*Clitocybe phyllophila*), can carry a sweet, tempting odor. The pink-spored *Entoloma* genus contains toxic white mushrooms that look like the Miller. They can be told apart by their notched (rarely decurrent) gills that do not have the mealy fragrance. There are many, many more whitish, decurrent, gilled mushrooms, but none that are pink spored with a mealy fragrance.

NAME AND TAXONOMY: *Clito-pilus* Greek "sloped and felty," since the base of the stem is often wider and is finely haired; *prunu-lus* Latin "little plum," *Clitopilus prunulus* (Scop.) P. Kumm. 1871. *Cystidiatus* for the presence of microscopic cystidia cells, missing in *C. prunulus. Clitopilus cystidiatus* Hauskn. and Noordel, 1999. Entolomataceae, Agaricales.

MIDNIGHT PINKGILL
Entoloma medianox

The Midnight Pinkgill is a beautiful, fleshy mushroom that can be fairly easy recognized by its midsize bluish-gray cap, a stem with a bluish-gray tinge and yellow in its base, and the pink spores it produces (visible on the top of the lower mushroom on the left in the photo). *Entoloma medianox* is not really rare in the woods of the coastal Pacific Northwest but is hardly seen in kitchens, even though it's an enjoyable fleshy mushroom for foragers who want to diversify their catch beyond the mainstay prey. Santa Cruz's citizen mycologist Christian Schwarz, who first described *E. medianox*, told me when asked about its edibility, "I have eaten *medianox* with no issue. Some people like it very well, more than me. It's okay, but cooks down quite a lot."

- Dark bluish-gray, sticky cap (Ø 5–15 cm), wrinkled to radially fibrous-streaked.
- Center somewhat umbonate; the often-lighter-colored margin is incurved, often remaining so.
- Gills adnexed to notched, close, white to pale blue, becoming pinkish as spores mature; pink spores.
- Solid stem (4–13 cm long x 1–3 cm wide), streaked; upper part similar color to cap, paling downward to whitish with distinct yellow-orange tinge in base.
- Flesh thick, white, and unchanging; odor and flavor farinaceous.
- Saprobic, grows on ground singly, scattered, or in groups in conifer and hardwood forests.
- Widespread from southwestern BC down to California, mostly in coastal areas; fruits in fall.

Midnight Pinkgill, Entoloma medianox

LOOK-ALIKES: Pine Pinkgill (*Entocybe nitida*) is similar but with a much smaller, dry cap (Ø 2–4 cm) and a mild to radish-like odor. Other bluish-black *Leptonia* spp. are also much smaller and less fleshy. Edibility is unknown for all these pink-spored mushrooms. The edible Wood Blewit, *Lepista nuda*, is dull pinkish-buff spored, but is more purple than blue and has a fruity odor. For other similar purple mushrooms, see Wood Blewit in Funnelcaps and Allies. The edible Streaked Knight, *Tricholoma portentosum*, has a similar cap, but it is white spored, as is the also similar, toxic Leopard Knight, *Tricholoma pardinum* group. Some brittlegills (*Russula* spp.) have a similar gestalt, but mature caps are often centrally depressed, and all have chalklike flesh as well as white to yellow spores.

NAME AND TAXONOMY: *Ento-loma* Greek "inner-margin," alluding to the inrolled cap margin, and *media-nox* Latin "mid-night," referring to the cap color. *Entoloma medianox* C. F. Schwarz 2015 (previously misapplied *Entoloma bloxamii* [Berk. and Broome] Sacc 1887) and *Entoloma madidum* Gillet 1876.

DEER MUSHROOMS
Pluteus exilis and *P. cervinus*

These good-sized mushrooms can be found growing from dead wood long before more attractive mushrooms will entice you, since one good rain shower seems to spur the mycelium of these wood decayers to crank out fruiting bodies. It is fairly easy to recognize from its habit of growing on wood, its largish size, free gills, and pink spores. Deer mushrooms start fruiting in early spring and can go into late fall. Recent DNA studies at first suggested that all our *Pluteus cervinus* should be *P. exilis*, but then revealed that both species are present in our region. A kissing-cousin species is the Black-edged Deer Mushroom (*P. atromarginatus*) with its finely drawn, dark-edged gills, a feature retained even after spore fall. There are also other similar (and similarly edible) *Pluteus* species; microscopy is required to distinguish among them.

Uneven culinary qualities between deer mushroom collections might reflect the taxonomic differences in this species complex. In general, we are extra careful with pink-spored mushrooms, especially the ones that belong to the Entolomataceae family, since they entail a range of toxic mushrooms, but the deer mushroom complex in the Plutaceae is safe if all its identification features are met, including growing on obvious wood and having free gills. The flesh is fragile and the cap thin, especially at the margins. Buck McAdoo suggests frying deer mushrooms until at least the thinner parts are crispy. The interior will still be moist and give off a fine flavor of roasted parsnips. Sometimes with this species, it seems that much of its biomass is in the swarm of microscopic springtails that will jump out of your frying pan!

IDENTIFICATION

- Midsize to sometimes big mushroom, dark brown to tan; white-edged, smooth cap with low umbo (always the darkest spot on the cap) and sometimes fine matted hair at the center.
- Close free gills, off-white when young, darkening in age to the color of the pinkish-brown spores.
- Whitish stem typically adorned with long, brown vertical fibers.

Deer Mushrooms, Pluteus cervinus

- Mild flavor, sometimes with a note of radish, commonly detected in its odor.
- Always grows on dead wood, on fallen trees, or in wood chips or sawdust.
- Fruits in spring and fall and other seasons when conditions are conducive; widespread and common.

LOOK-ALIKES: Similar, but growing in soil with a dark- to gray-brown cap is the edible, cream-spored Common Cavalier (*Melanoleuca melaleuca*). Its bulbous stem has dark gray-brown hairs.

NAME AND TAXONOMY: *Pluteus* Latin "umbrella roof," reflecting the shape of the cap, *exilis* Latin "thin" or "small," and *cervinus* Latin "deer." *Pluteus exilis* Singer 1989, *Pluteus cervinus* (Fries) P. Kumm. 1871. Pluteaceae, Agaricales.

🅴 SILKY ROSEGILL
Volvariella bombycina

The Silky Rosegill, *Volvariella bombycina*, is a cosmopolitan (pictured on following page growing in Colombia) *Amanita* look-alike with an impressively big, whitish to dirty gray-brown volva. It has a silky whitish cap (Ø 5–20 cm) with free, crowded, white gills producing spores that range from pinkish-salmon to brownish-rose. The often-curved white stem (1–3 cm wide) can reach from 6 to 20 cm in length.

Silky Rosegills grow in fall mostly on hardwoods and extremely rarely on conifers. They are very rare in the Pacific Northwest. The pinkish spore print and growth on wood set it clearly apart from dangerous Amanitas. It is reputedly a good edible but a rare treat. The closely related

Silky Rosegill, Volvariella bombycina

Paddy Straw Mushroom, *Volvariella volvacea,* is widely cultivated and commonly found in delicious dishes in your local Thai restaurants.

NAME AND TAXONOMY: *Volva-riella* "little sheath," for its huge volva, and *bombycina* "silky," for its cap surface (both Latin). *Volvariella bombycina* (Schaeff.) Singer 1951. Pluteaceae, Agaricales.

!! STUBBLE ROSEGILL
Volvopluteus gloiocephalus

The Stubble Rosegill grows from spring to fall in rich soils in fields (hence "stubble"), in landscaped or disturbed sites, often mulched or covered in wood chips, and also in greenhouses. It is not a famous edible, but rather closer to famine food. Decades ago, it was wrongly regarded as poisonous, "perhaps due to confusion with Amanita," speculates David Arora, continuing, "Should you try it, be sure to take a spore print—some Amanitas have pinkish gills in old age!"[82] The fact that it grows from soil with a clear volva—and often the whole mushroom is dull white (but also gray to grayish-brown)—rightly scares many foragers away. It's too darn close to the Death Angels! However, the pink spore print, as well as the absence of a ring left by a partial veil (but rings do sometimes disappear), exorcises the *Amanita* spirit.

The sticky cap (Ø 5–15 cm) protects free, crowded, and broad white gills from the

Stubble Rosegill, Volvopluteus gloiocephalus

Deadly Webcap, Cortinarius rubellus

weather. The whitish, smooth stem (5–20 cm tall x 1–2.5 cm wide) grows from a white to pale grayish, saclike volva.

NAME AND TAXONOMY: *Volvo-pluteus* Latin from "sheathed," *gloio-cephalus* Greek "sticky-head." *Volvopluteus gloiocephalus* (DC.) Vizzini, Contu, and Justo 2011; formerly *Volvariella speciosa* (Fr. ex Fr.) Singer 1951. Pluteaceae, Agaricales.

GILLED MUSHROOMS WITH WARM BROWN SPORES (RUST TO CINNAMON)

After white, the second most common spore color of gilled mushrooms is brown, all here with attached gills. The dark color is caused by melanin, a dark pigment that is highly protective against UV light. The biggest genus is the rusty-brown-spored webcaps—*Cortinarius*—with hundreds of species in the Pacific Northwest and only a handful of known edibles (e.g., Ringed Webcap, *C. caperatus*). More Corts are picked for pigments than for food. Mushroom dyers especially love red- and yellow-gilled web-

caps in the subgenus *Dermocybe* for the dyes they yield.

Most brown-spored mushrooms, however, are nonedible, and some are even lethal, like the Deadly Skullcap (*Galerina marginata*), the Ringed Conecap (*Pholiotina rugosa*), and the extremely rare Deadly Webcap (*Cortinarius rubellus*). There are also toxic fiberheads (*Inocybe* spp.) and nasty poison pies (*Hebeloma* spp.). Many brown sporers grow saprobic on dead wood, such as scalycaps (*Pholiota*) and rustgills (*Gymnopilus*); the latter are notorious for their bitterness, such as the gorgeous psychedelic Yellow-gilled Rustgill, *Gymnopilus luteofolius*. There are reports of several scalycaps being edible, i.e. Flaming and Bristly Scalycap (*Pholiota flammans* and *Ph. squarrosoides*); however, the latter is hard to differentiate from toxic Shaggy Scalycap (*Ph. squarrosa*).

RINGED WEBCAP
Cortinarius caperatus

The Ringed Webcap is one of the most noticeable of the Corts, and we are not talking about its yellowish-brown cap and rusty spores! Firstly, it is the only webcap widely known to be an enjoyable edible. Secondly, it is a webless webcap with a membranous ring instead. Ninety-nine percent of the all-ectomycorrhizal *Cortinarius* have a fibrous net, partial veil, called a cortina (its Latin root means "curtain"). *Cortinarius caperatus* belong to the one percent! For a while they were bequeathed their own genus, *Rozites*, but they have since been folded back into *Cortinarius*. The splitters in the global *Cortinarius* community are just starting their work.

Ringed Webcap, Cortinarius caperatus; *detail of felty ring*

By petting its wrinkled (in Latin *caperatus*) but smooth cap surface, you will get a sense for identifying your Ringed Webcap before picking it. The cap is distinctive; those wrinkles are real! Once you have picked it, check it for rust-brown gills and the felty ring. There is no other *Cortinarius* out West that has a felty ring. Growing up in Germany, we knew this mushroom by two names, *Zigeuner* or *Reifpilz*, the "gypsy" or "hoar mushroom," but it can be translated as "ring mushroom" too. Here in the United States, I used to call them "gypsies," associating them with Europe's only nomadic people and one of Europe's most outstanding jazz guitarists, the Belgian-born Romani Django Reinhardt. However, in searching for the source of the name "gypsy," I found several references making a connection to its "uncombed" head. Romani have been denounced and

badly abused for too long, so I am going with Ringed Webcap.

Ringed Webcap is possibly the best of more than two thousand webcaps worldwide. This honor is achieved with a good texture and an enjoyable fungal flavor. And yet, Ringed Webcaps may be ignored in favor of king boletes fruiting in the same forest—unless you are collecting medicinals. Some in vitro and rodent tests have shown them to be active against the otherwise uncompromising *Herpes simplex* virus,[83] although there are no clinical trials that show relief for humans yet.

Tibetan mushroom hunters pick only the rounded, unopen young mushrooms, since the caps break easily once open and the broken pieces are not marketable. In Tibet and the Himalaya, in addition to *Cortinarius caperatus* grows a second species, *C.*

emodensis, with an overall purple hue on the cap, stem, and gills,[84] traits sometimes seen in young Pacific Northwest *Cortinarius caperatus*.

IDENTIFICATION

- Wrinkled tan to golden-brown bald cap (Ø 5–15 cm), with lighter outer edge; thin whitish coating over cap when young (remnants from universal veil).
- Attached, closely spaced gills, spore color pale tan maturing to rusty-brown.
- Slender, whitish to light yellow-brown stem; covered in fibers, above ring looking sometimes like upward-pointing whitish scales, possibly turning yellow-brownish with age; base sometimes wider (showing lower part of universal veil).
- Thick, white, membranous, double-edged ring in middle of stem.
- White, generally firm flesh but brittle caps; mild flavor.
- Grows in small groups on ground in conifer forests, often under blueberries or huckleberries.
- Fruits summer into fall, widespread in conifer forests of the Northern Hemisphere.

LOOK-ALIKES: There are several similarly colored webcaps (for example, *Cortinarius seidliae* and *C. vanduzerensis*) that may fruit in the same habitat. They have cortinas and nonwrinkly, sticky caps; edibility unknown. Look-alikes include toxic poison pies (*Hebeloma* spp.) and poisonous fiberheads (*Inocybe* spp., which are much smaller than Ringed Webcaps). Also similar are fieldcaps (*Agrocybe*; see below) and many other brown mushrooms.

VIOLET WEBCAP
Cortinarius violaceus

This webcap flashes its deep purple unabashed. What a mushroom! It's easy to identify by color together with its dry and finely scaly cap, cortina, and rust-brown spores. And it is edible, or shall we rather say quite digestible. None of my multiple attempts to enjoy the Violet Webcap culinarily on its own terms have succeeded, and unfortunately, the purple becomes even darker and less vibrant when fried. However, as a common mushroom and look-alike for blewits and all other purple mushrooms, it deserves its place in this field guide. And what a beauty it is!

IDENTIFICATION

- Dry, dark purple cap (Ø 5–15 cm), tufted-haired to finely scaly, slightly umbonate with inrolled margin when young.
- Close, dark purple gills, notched to attached, rusty-brown spores.
- Fibrous or woolly dark purple stem, firm and wider at base, with purple cortina when young, leaving a few discernable strands catching the rusty spores.
- Grows on ground, solitarily or in small groups, with older trees in conifer and mixed forests.
- Fruits in late summer and fall, widespread in humid forests from Alaska to Northern California.

Violet Webcap, Cortinarius violaceus. Note the lilac hue, which disappears with maturation.

LOOK-ALIKES: Dozens of purplish webcaps are similar (though none are deep purple and have tufted, hairy caps); one of these is the Gassy Webcap (*Cortinarius traganus*), with its pear fragrance, edibility unknown. Other purple mushrooms include the arsenic-loaded Amethyst Deceiver (*Laccaria amethystina*), the edible Wood Blewit (*Lepista nuda*), and the dark-capped and stemmed pinkgills (*Entocybe* and other Entolomataceae). None have rusty spores or cortina.

NAME AND TAXONOMY: Although *Cortinarius violaceus* is a European name, our species is the same. *Cortinarius violaceus* (L.) Gray 1821. Cortinariaceae, Agaricales.

PONDEROUS WEBCAP
Cortinarius ponderosus

This heavy webcap can grow very large, making it one of the biggest Corts. It is frequently encountered in western Oregon but more common in California. The high diversity of webcaps makes it tricky to identify any one definitively down to the species. But the Ponderous

Webcap is recognized by its large size and massive stem; sticky, scaly, orangish-brown and yellow cap; grayish-lilac young gills; and mostly white, very firm flesh with a peculiarly pungent, sour odor. Whenever you see a list with so many specifics, it tells you right away that you need to be *very specific* in your identification.

I have eaten it only a few times, but I really enjoyed the firm consistency and thought of it as an enjoyable fried fungal potato. Now, that will not generate acclaim in classy culinary circles, but that's solid, especially for a webcap. Some people, however, sense an unpleasant flavor even after cooking it.

IDENTIFICATION

- Sticky, scaly, orange-brown cap (Ø 10–35 cm) with inrolled lighter margin opening in maturity; sticky surface layer can have an olive metallic tinge.
- Notched to attached close gills, lilac to light gray, turning rusty-brown in spots from mature spores.
- Fat, very firm stem (lower part often wider), colored like cap; browning flesh and sticky surface.
- Thick white cortina, often lasting and sticking to upper whitish third of stem, collecting rust-brown spores.
- Hard, whitish flesh with grayish to light purple spots when young.
- Flavor mildly to lightly sour, odor often pungent.
- Grows solitarily, clustered, or in small groups on ground with ponderosa

Ponderous Webcap, Cortinarius ponderosus

pine, Douglas-fir, spruce, oak, and others.
- Fruits in late fall in Washington and Oregon west of the Cascades.

LOOK-ALIKES: There are dozens of brownish webcaps, including sticky ones like Largent's Webcap (*Cortinarius largentii*), which has a much more bulbous stem base. Plus, its sticky yellowish cap lacks the Ponderous Webcap's scaly tufts.

NAME AND TAXONOMY: *Ponderosus* Latin for "heavy" and also an allusion to ponderosa pine. *Cortinarius ponderosus* A. H. Smith 1939.

 **SHEATHED WOODTUFT**
Kuehneromyces mutabilis

The Sheathed Woodtuft is known as a good if not excellent edible in Europe but

is only mentioned here with a loud, clear warning not to fall for the very similar and highly toxic Deadly Skullcap (*Galerina marginata*). The caps, gills, and spore color of both of these wood-growing mushrooms are extremely similar. Distinctive only are the stems: the Sheathed Woodtuft sports fine, upward-pointing scales on the stipe and a clear ring that tells you edible, not deadly! Most foragers prefer not to rely on such subtle differences to keep deadly amatoxins out of their frying pans. Also, many Western field guides appear to avoid introducing the Sheathed Woodtuft at all, or at least do not recommend it as a good edible to their readers.

Strangely, I did not find it for years, but I probably did not check the Sulfur Tufts I found carefully enough; most likely, a few were actually Sheathed Woodtufts. Or did the mushroom elves not want me to fall for the Sheathed Woodtuft? No matter, I finally fell for it! It has a nice, nutty fungal flavor and good structure, but the stems can be very tough. Remembering to check for their rings and scaly surfaces below is your life insurance.

IDENTIFICATION

- Cap yellowish orange-brown to reddish-brown or tawny colored. Hygrophanous: cap color often two-toned, fading from center outward to yellowish-brown or ochre as it dries.
- Smooth cap (Ø 2–6 cm) sticky when wet, more or less peelable; with low umbo or even flat; cap margin when moist can be translucent-striate.
- Close gills, broadly attached to slightly decurrent, with age more decurrent; when young, pale white, turning brown when mature from cinnamon-brown-colored spores.
- Stuffed or hollow stem (3–10 cm tall x 0.2–1 cm wide) differing strongly below and above ring; whitish, smooth to silky-striate above the ring, below ring whitish when young, becoming brownish; base sometimes blackish-brown with age.
- Stem below ring covered with numerous small, often upward-curving scales (very crucial detail distinguishing it from Deadly Skullcap!).
- Ring high up on stem, membranous, sometimes fibrillose scaly on underside; white when young, browning with age and from spore deposit.
- Flavor fungal, nutty, and slightly spicy.
- Thin, white flesh, sometimes tinged brown.
- Typically grows in (large) clusters on dead hardwoods, conifer stumps, and trunks. When growing from woody berry canes (like salmonberry, as pictured), forms small clusters or solitary fruitings.
- Commonly fruits in fall, but also in late spring and summer; widespread.

LOOK-ALIKES: If we only consider wood-growing mushrooms, the Deadly Skullcap, *Galerina marginata*, is a very close match; the distinctive difference is only visible in the stem, since the Deadly Skullcap lacks a sheathed, scaly, lower stipe. So does the very common, slender, and inedible Spring Scalecap, *Kuehneromyces lignicola*. It is so strikingly similar to the Deadly Skullcap that it should not be tried. Also, some smaller honey mushrooms like *Armillaria nabsnona* are similar, but they have white spores and lack the hygrophanous cap. Many scalycaps (*Pholiota* spp.) have scaly caps and are mostly

Sheathed Woodtuft, Kuehneromyces mutabilis, *with close-up of distinct ring*

Deadly Skullcap, Galerina marginata

inedible to poisonous. The edible Conifer Tuft (*Hypholoma capnoides*) and poisonous Sulfur Tuft (*Hypholoma fasciculare*) are similar but have much darker, purplish-black spores.

NAME AND TAXONOMY: *Kuehner-o-myces* named for French mycologist Robert Kühner (1903–1996), plus *myces* Greek "mushroom"; *mutabilis* Latin "changeable," a reference to the hygrophanous cap. *Kuehneromyces mutabilis* (Schaeff.) Singer and A. H. Sm. 1946. Strophariaceae, Agaricales.

Toxic Brown-Spored Gilled Mushrooms

You don't need to know every toxic mushroom by heart before venturing out, but it is prudent to pay close attention to look-alikes and look-alike warnings, especially when hunting gilled mushrooms. Below are several warm-rusty- to cinnamon-brown-spored toxic species, such as the lethal Deadly Skullcap (*Galerina marginata*) and Ringed Conecap (*Pholiotina rugosa*), as well as very troublesome genera

like fiberheads (*Inocybe*) and poison pies (*Hebeloma*).

☠ DEADLY SKULLCAP
Galerina marginata

The embodiment of LBM, or little brown mushroom, is this Lethal Brown Mushroom. Most fungal foragers collecting for dinner ignore little brown mushrooms, which helps them avoid extremely dangerous mishaps caused by the deadly amatoxins some of those species contain. However, when on the search for the healing power of magic mushrooms, be aware of these dangerous look-alikes; they never stain blue as nearly all *Psilocybe* would. The experience the skullcaps (a.k.a. Funeral Bells) convey is a miserable one-way trip with the awful experience of organ failure.

Galerina marginata always grow on dead wood, sometimes buried. Their brown caps (Ø 1.5–5 cm) have different degrees of stickiness and the cap margin is striate—showing radial lines. The cap color changes quite a bit depending on water content; reduced water content brings out a dull yellowish-brown or tan coloration.

The thin stem, grayish-brown to brown (darkest at the base), bears a thin whitish ring that deteriorates into a nearly fibrillose ring zone about four-fifths of the way up the stem; this zone collects the rusty-brown spores.

Neither the mild flavor nor the slightly mealy odor indicates its treacherous toxicity. Based on DNA analysis, several *Galerina* species, including *G. autumnalis* and *G. oregonensis*, were concluded to be basically identical. All are now included in *G. marginata*. Other *Galerina* species also contain deadly amatoxins.

LOOK-ALIKES: Given that it is an LBM, there is a seemingly infinite assembly of look-alikes to *Galerina marginata*, though most are less lethal! Foragers need to discern two genera in particular: *Psilocybe* and *Kuehneromyces*. Several *Psilocybe* species also grow on dead wood, especially wood chips. However, magic mushrooms all leave dark purplish-brown spore prints, and their fruiting bodies stain blue from handling and exposure.

Growing in very similar conditions on dead wood are two *Kuehneromyces* species: the common and inedible Spring Scalecap (*K. lignicola*) and the edible Sheathed Woodtuft (*K. mutabilis*), which has a sheathed, scaly lower stem that flares into a conspicuous ring. Edible honey mushrooms are also brown and grow on wood, but they are much bigger, are white spored, and usually have a scaly cap.

NAME AND TAXONOMY: *Galerina* Latin from *galerus* "hat" or "helmet," *marginata* Latin "with a border." *Galerina marginata* (Fr.) Kuehner 1935 synonymous with *Galerina autumnalis* (Peck) A. H. Sm. and Singer 1964. Hymenogastraceae, Agaricales.

☠ RINGED CONECAP
Pholiotina rugosa

This tiny orange-brown mushroom (cap Ø 0.3–1.2 cm) is potentially deadly, since it contains amatoxins. It is best recognized

Ringed Conecap, Pholiotina rugosa

LEFT: *Lilac Fiberhead,* Inocybe pallidicremea *(formerly I. lilacina);* RIGHT: *Straw-colored Fiberhead,* Pseudosperma (Inocybe) rimosum

by a pronounced, felty striated ring quite high up on the stem that often collects the rust-brown spores. The ring can slide down or fall off with age. As Danny Miller warns in his online PNW Mushroom Pictorial Key, "Remember, rings can fall off, so any LBM could potentially be deadly." The dull orange gills have a flaky, white, fine-toothed edge, especially striking under a hand lens.

Ringed Conecaps grow in fertile or clay-rich soil in woodlands and along paths, as well as on lawns and in parks in summer and fall. Only mushroom nerds would pick anything that small, but inquisitive toddlers could be in danger. This mushroom was until recently known as *Conocybe filaris,* (Fr.) Kuehner 1935, but is now recognized as *Pholiotina rugosa* (Peck) Singer 1946. Bolbitiaceae, Agaricales.

FIBERHEADS
Inocybe species

Fiberheads are often a bit tattered looking from fibers that radiate from a lovely pointed cap center. They are small (cap Ø 1–5 cm), dry-capped, drab brown, whitish, or gray

mushrooms (colorful on rare occasion, such as the purplish Lilac Fiberhead, *Inocybe pallidicremea*). Their gills are first white and turn dark brownish. Some are graced by a cortina, a fibrous partial veil. Many have a spermatic odor, a helpful identifying trait as expressed in the newly split-off genus *Pseudosperma,* which now includes the eye-catching (as Inocybes go) Straw-colored Fiberhead, *Pseudosperma rimosum.* The very similar Corn Silky Fiberhead (*P. sororium*) smells strongly of green corn.

Fiberheads (family Inocybaceae, order Agaricales) are all ectomycorrhizal, many of them grow with Douglas-fir, and large groups grow in late fall or early winter with conifers. Many of the fiberheads have a high content of the toxin muscarine, which could be deadly in very high doses. Their odors can attract dogs; sadly, some small dogs have died from eating them.

POISON PIES
Hebeloma species

What a great descriptive name—poison pie! "Pie" alludes to their clean, attractive

Pale poison pie, Hebeloma velutipes

caps, "poison" to their ability to cause severe gastrointestinal upset. Mostly medium-sized and nondescript, their caps (Ø 3–10 cm) are sticky and cream to beige to tan to pale brown. Some have a cortina, a cobweb-like partial veil, like the webcaps, *Cortinarius.* But the spores of poison pies are dull cigar-brown, while those of Corts are bright rust-brown. Fine white flakes on the upper stem plus a unique odor—best described as radish-like—distinguish many Hebelomas. Poison pies (family Hymenogastraceae, Agaricales) are mycorrhizal and common in the woods but ignored by many due to their toxicity and the challenge of identifying them to species.

GILLED MUSHROOMS WITH COLD, DARK BROWN, PURPLE–BROWN, OR BLACK SPORES

This group entails quite a mixture of mushrooms; all its members, except the ectomycorrhizal gilled boletes, are saprobic. The overall spore color can be summarized as cold, dark browns in opposition to warm browns, like the rusty or light brown group described above. The most famous edibles are in the chocolate-brown-sporing genus *Agaricus* and the black-spored Shaggy Mane (*Coprinus comatus*). Also black spored are Mica Cap (*Coprinellus micaceus*) and Alcohol Inkcap (*Coprinopsis atramentaria*), as well as gilled members of the bolete order, like pinespikes (*Chroogomphus*) and slimespikes (*Gomphidius*); also described in this section is the nonconforming Western Gilled Bolete (*Phylloporus arenicola*).

In between on this spectrum, we find dark-brown-spored Spring Fieldcaps (*Agrocybe praecox*), purple-brown to deep-purple-gray-spored Conifer and Sulfur Tufts (*Hypholoma*), dark-purplish-brown-spored brittlestems (*Candolleomyces / Psathyrella*), and the dark-purplish- to nearly-black-spored roundheads (*Stropharia*) and magic mushrooms (*Psilocybe*).

Agaricus Species

Everyone is familiar with *Agaricus* even if they don't realize it. The genus contains America's most widely eaten mushroom, *Agaricus bisporus*. When young, it is the button mushroom, when older and opened up, the portobello, and as a different, darker-capped strain, the crimini. *A. bisporus* is the one mushroom we are all allowed to eat without contemplating the finality of human existence. It somehow outsmarted Anglo-fungophobia. All *Agaricus* are saprobic; they feed on dead biomass. The Pacific Northwest is rich in edible *Agaricus* species, with relatively few (but abundant) mildly toxic ones that will upset a human digestive system—

PROCESSSING AND COOKING AGARICUS

Agaricus go bad quickly when mature, so check them for larvae and use them as soon as possible. Remove the dark gills if you don't like blackening your food. In Tibet, people love their Drukdak Shamo or Ka Sha, as they know *Agaricus*, either the "dragon thunder mushroom" or just "white mushroom." In Tibetan mythology, dragons fighting in the sky cause thunderstorms and result in juicy field mushrooms that fruit at elevations of up to 17,000 feet (5,200 m) high on windswept highlands!

Tibetans love to cook whole caps. They remove the stems, and then wash the outside of the cap but keep the gills dry. The cap is placed on the fire, gill-side up; the extra moisture from washing it protects it from burning. The chopped stem is piled onto the gills with some butter, salt, and a couple spoons of *tsampa* (roasted ground barley). The moisture in the cap boils all the ingredients nicely, making them *shim po du*—tasty! This recipe is great for campfire barbecues. Washing the cap is a helpful hack, and *tsampa* is not needed to enjoy the Dragon Thunder Mushroom.

the "lose your lunch bunch," as David Arora calls them.

Agaricus is easily recognized by a few key features. The gills are free, they are not attached to the stem, and the stem breaks cleanly from the cap. They usually have a prominent ring. *Agaricus* never have a sac or volva on the stem base as Amanitas do. Many *Agaricus* have a robust, often stocky stature. Its caps are never sticky or slimy.

Most telling and unique is their change in gill color; young gills are off-white or pink and mature as the cap opens to dark chocolate-brown (sometimes with hues of purple or blackish-brown), the color of their spores. Observing this color change is a dead giveaway that you do not have a lethal *Amanita* but instead are looking at an *Agaricus*. Many people could have saved themselves a wicked poisoning by giving their falsely identified "*Agaricus*"

the time to mature and brown its gills with spores (east of the Rockies, there are toxic pink-gilled Amanitas). Identification down to the species level often remains challenging in this large and diverse genus of mushrooms, ranging from small to big and with more than fifty species in the Pacific Northwest.

Luckily, a few choice edible species are common and fairly easy to identify, at least to their common names. This genus allows foragers to be daring while staying safe! Still, not every *Agaricus* poisoning is a walk in the park. Most sane people do not crave an experience of intense vomiting and extended dry heaving, even if they know there is no serious long-term health risk.

There are two key criteria to picking edible *Agaricus* if you cannot identify them to species. First, it has a pleasant almond or anise odor, often most pronounced when young or arising from a crushed stem base. Second, the stem base does *not* turn strongly yellow. Many toxic *Agaricus* turn bright yellow in the stem base and/or often have an unpleasant "bad science lab" odor, also referred to as phenolic or like creosote. The odor is often detectable in the stem base or emerges after cooking (but before eating!).

If your *Agaricus* has the pleasant almond odor, it is safe to eat, even if its stem base changes slightly yellow. It gets a bit murkier if it does *not* have the sweet almond odor; then at least the stem base should stain reddish. Your alarm bell should ring if you see intense yellow staining in the stem base or detect an unpleasant chemical smell. Once you are completely sure you have an *Agaricus* and you have detected the crucial characteristics, you can be confident that you have an edible one, without having to identify it to the species level. Be aware that there are plenty of *Agaricus* species that lack clear signs—

you need to know your odors! However, there are a few unlucky people who have an allergic reaction to these otherwise great mushrooms.

To add another twist, many of the scientific names used in the past have been proven wrong based on DNA analysis. Luckily, we can still use their names to group them, and applying European mushroom names to our Pacific Northwest species works well for pot hunters. The exception is the Prince (*Agaricus augustus*), which retains its name to rule around the Northern Hemisphere, an exceptional creature in many ways!

THE PRINCE
Agaricus augustus

This majestic mushroom is loved by humans and loves to grow around human settlements. Its grandness, almond odor, scaly cap and stem, yellow staining of the cap edges, and subtle dull orange staining of the flesh facilitate easy identification. One challenge for beginning mushroom hunters is its fruiting season. The Prince likes it warm and wet. When the fall mushroom season excites most fungivores, the Prince has already retired.

Its almond flavor is a real treat! It can be enjoyed fried, steamed, grilled, and raw in a salad. However, the Prince has a tendency to attract heavy metals— no, not gold and silver, but the more toxic ones like cadmium and mercury. So it's best not to overindulge or pick from suspect sites.

The Prince, Agaricus augustus

A basket of Princes

Be sure not to confuse it with either Buck's Agaricus or the Flat-topped Agaricus, both slightly toxic, grayish-hued mushrooms that love the same habitat but stain yellow or ochre-brown in the stem base and lack the almond odor.

IDENTIFICATION

- Midsize to huge mushroom, mature cap diameter starts at 8 cm but can reach 40 cm! Color is warm light brown with gold overtones, conspicuous brown scales.
- Gills free, changing from pale white via pink to chocolate-brown, the color of its spores.
- Stem (1.5–6 cm wide x 8–35 cm tall) with smooth upper part and shaggy (with white or brown-tipped) scales in its lower part.
- Partial veil dropping into a prominent skirtlike ring after cap opens.
- White flesh changing to yellow in cap, especially at cap edge; strong almond odor.

- Grows singly, in groups, or in clumps, mostly in disturbed, park, or suburban landscapes.
- Fruits in summer and fall, especially before the weather cools; widespread and common.

NAME AND TAXONOMY: *Agaricus* Latin, is derived from the Agaril, Scythian people of Sarmatia who lived northeast of the Black Sea, where the medicinal conk Agarikon (*Laricifomes officinalis*) was sourced for the market in Greece in ancient times. (It is not clear how and why the derivate name of a woody conk was applied to gilled mushrooms.) *Augustus*, also Latin, means "noble" and "majestic." *Agaricus augustus* Fr. 1836. Agaricaceae, Agaricales.

HORSE MUSHROOM
Agaricus "arvensis" group

Traditionally, this is a midsize whitish *Agaricus* bruising warm yellow when rubbed (but not in the stem base!) with a pleasant almond fragrance that grows in grass. However, recent DNA has messed with the idea, and what has been known to date as *A. arvensis* in the Pacific Northwest is now called *A. fissuratus*, named for cracks and fissures in the aging cap surface. What was known as *A. osecanus*, a much bigger grassland mushroom variety, also now belongs in *A. fissuratus*. Also similar is the very enjoyable Amber-staining Agaricus, *A. albolutescens*.

The choice edible Crocodile Mushroom (*A. crocodilinus*) is in many regards similar as well, but it is bigger, with caps reaching up to 35 cm in diameter. While the young cap surface has tiny whitish scales and can appear bald, mature caps can crack into

Horse Mushroom, Agaricus "arvensis" *group*

fruiting late spring to fall; widespread and common.

LOOK-ALIKES: Very similar in coloration and sharing the almond fragrance and choice edibility is the Wood Mushroom, *Agaricus silvicola*. It is a common forest "version" with a whitish cap (Ø 5–12 cm) that bruises yellow, just as the slender stem does, but its stem base does not! DNA analysis has revealed that *A. silvicola* is a species complex with three or four look-alikes in the Pacific Northwest. Remember, these taxonomic developments do not interfere with edibility status.

big scales resembling crocodile skin. It used to be much more common in meadows, but is becoming rarer.

Stay away from all *Agaricus* that smell unpleasant and bruise yellow in their stem base. Beware too of extremely toxic white Amanitas (see Boogeymen of the Fungal Kingdom) that always have white gills and spores and usually have a volva or at least a widened stem base. In the PNW, no known *Amanita* bruises yellow or red and has a fragrance of almond, but elsewhere, some very toxic Amanitas do stain red and can smell of anise, like the Anise and Raspberry Limbed-Lepidella, *Amanita mutabilis*, growing along the American Atlantic coast.

IDENTIFICATION

- Midsize to big mushroom, white to yellowish cap (Ø 10–15 cm, sometimes to 35 cm), first covered with tiny cottony scales, but balding with age and cracking.
- Gills free, changing with spore maturation from white to grayish-pink to dark chocolate-brown.
- Membranous ring with cog-wheel pattern on underside, starting white, then becoming yellow, orange, or buff; small flakes on lower stem.
- White, firm flesh, unchanging in both cap and stem (or only slightly yellowing, but never in stem base).
- Almond or anise fragrance and flavor.
- Grows in groups, fairy rings, and sometimes clusters in grasslands,

NAME AND TAXONOMY: *Fissuratus* "cracked," referring to fissured cap surface, *arvensis* "field," *albo-lutescens* "white-yellowing," and *silvi-cola* "forest inhabitant" (all Latin). *Agaricus fissuratus* F. H. Moeller 1951, *Agaricus arvensis* Schaeff 1778; *Agaricus crocodilinus* Murrill 1912; *Agaricus albolutescens* Zeller 1938; *Agaricus silvicola* (Vittad.) Peck 1872.

MEADOW MUSHROOM
Agaricus aff. *campestris*

The Meadow Mushroom is very similar to the cultivated button mushroom and shares its white color, flavor, and odor. It has long been highly valued and eaten by mushroomers around the world. As a wild food, it is likely more nutritional than its cultivated sister. The transitional intense pink gills, lack of almond odor when young, smaller size, and tapering stem base help to distinguish it.

Since *Agaricus campestris* is a European species and most of the time misapplied in this region, we will soon have new scientific names for our Pacific Northwest Meadow Mushrooms. So far, the most commonly sequenced Meadow Mushroom is *Agaricus porphyrocephalus* var. *pallidus*. However, several species loosely fit the description of the very tasty Meadow Mushroom, which fruits in grasslands whenever precipitation and temperature enable these fungi to send up their reproductive organs. Meadow Mushroom fruitings have seriously declined, most likely due to synthetic fertilizer.

IDENTIFICATION

- Midsize mushroom, white cap (Ø 5–10 cm), inrolled at first, often with partial veil tissue hanging from cap edge.
- Free gills, bright pink when young, turning dark chocolate-brown from spores of same color.
- White stem features pointed or tapering base and thin, not skirtlike (often fleeting) ring.
- White, thick flesh, not bruising but sometimes staining brownish or reddish when old or soggy.
- Flavor and odor pleasant, when aged smelling faintly almondy.
- Fruits in groups or fairy rings in grasslands and pastures and around livestock in spring, summer, and fall; widespread and common.

LOOK-ALIKES: The Pavement Mushroom, *Agaricus bitorquis*, also has a tapering stem and a white cap (Ø 5–20 cm), but often there is dirt on the cap when this very squat mushroom pushes up, not always revealing

Meadow Mushroom, Agaricus aff. campestris

Buck's Agaricus, Agaricus buckmacadooi

itself, along roads and hardpacked, disturbed soils. It fruits from spring to fall and is common and widespread, especially inland, though much rarer along the coast.

The also similar, but briny smelling Salty Mushroom, *A. bernardii* (cap Ø 5–15 cm), claims a similar ecological niche. Both are good, firm edibles and have thick sheathing veils, but while *A. bitorquis* stains only a little red (its cap might yellow when old), *A. bernardii* stains quickly and clearly red.

NAME AND TAXONOMY: *Campestris* Latin "field" and *bi-torquis* Latin "twice" and "necklace." *Agaricus campestris* Fries 1821. *Agaricus bitorquis* (Quel.) Sacc 1887.

ⓧ BUCK'S AGARICUS
Agaricus buckmacadooi

This nonedible, slightly poisonous stately mushroom is one of the most common *Agaricus* you will find in manipulated human landscapes and along roads. It was formerly known as either *Agaricus praeclaresquamosus* in numerous guides or *Agaricus moelleri*, a European species not yet found in the Pacific Northwest. DNA studies reveal that our West Coast species

are actually two and deserve their own names: Gray Agaricus, *A. deardorffensis*, and Buck's Agaricus, *A. buckmacadooi*. Buck's Agaricus is often bigger, has a darker cap (Ø 10–25 cm), and slowly stains yellowish-brown in the stem base, while the Gray Agaricus is often a bit smaller (cap Ø 5–15 cm) and offers quick bright yellow staining in the stem base as a warning. These two mushrooms are often confused with the Prince, but the Prince will charm you with its sweet almond odor, warm brown scaly cap, and scaly stem.

Getting close to the "lose your lunch bunch" will reveal the presence of at least one of these unpleasant characteristics: stinky phenolic odor, cold silver-gray cap, lack of almondy notes, and yellow staining in the stem base. These signs should convince you not to culinarily engage with these albeit attractive imposters. While on your quest for the Prince or other good edible *Agaricus*, you will become familiar with these stately mushrooms and soon be able to tell them apart from a distance.

IDENTIFICATION

- Midsize to large mushroom, with flattened cap (Ø 10–25 cm) even when unopened, often shaped like a marshmallow.

- Whitish cap covered by gray fibrous scales, much darker in center.
- Close gills, free from stem; pale pink when young, maturing to dark brown from spore coloration.
- Finely fibrous, whitish stem; often pinkish above the thick ring and somewhat bulbous.
- Flesh is white and unchanging; only stem base is staining, possibly first showing some slight yellowing, turning to a pale rust-brown with a pinkish tinge.
- Often grows in groups in disturbed areas, at the edge of woods with Douglas-fir and hemlock, and along roads.
- Fruits from wet summer into early winter in the coastal region from southern BC to California.

LOOK-ALIKES: Another common mildly toxic *Agaricus* in the PNW is the white- to grayish- or tan-capped (Ø 5–15 cm) Yellow Stainer, *A. xanthodermus*—patron of all toxic *Agaricus*—offering quick bright yellow staining in the stem base and everywhere else (*xanthos* "yellow" and *derma* "skin," both Greek).

It shares a felty ring with the common, also slightly toxic, Felt-ringed Mushroom (*A. hondensis*, originally described from La Honda, California), which signals its inedibility with an often strong, phenolic odor. Its cap (Ø 6–20 cm) has pale red-brown scales. The smooth stem stains faintly yellow in its base and is ringed by an impressive felty partial veil. It can be found deep in the woods where there is much litter accumulation.

NAME AND TAXONOMY: *Duckmŋçadooi* is named for Northwest mycologist Buck McAdoo, and *deardorffensis* is for Deardorff Road in Calaveras County, California, where it was first described. *Agaricus buckmacadooi* and *Agaricus deardorffensis* Kerrigan 2016; previously falsely subsumed under *Agaricus moelleri* Wasser 1976 and *Agaricus praeclaresquamosus* A. E. Freeman 1979.

Black-Spored Inkcaps and Shaggy Manes

Inkcaps seemed like a happy family until the revolutionary spirit of molecular genetics terminated this illusion. DNA analysis has revealed that these similar-looking, cylindrical to conical capped mushrooms are not all closely related though they share black spores and liquefying caps. The genus *Coprinus* was the first famous "casualty" of the DNA revolution, smashed into three genera in 2001: *Coprinus*, *Coprinellus*, and *Coprinopsis*. At least the mycologists Redhead, Vilgalys, and Moncalvo were easy on the stunned fungal community by adding genus names that can be found together in any guidebook index, though their two families sit on different twigs of the fungal tree of life.

Furthermore, many sources still warn about the risks of consuming Shaggy Manes with alcohol, a plausible concern before the DNA revolution, since Shaggy Mane and its look-alike, Alcohol Inkcap (*Coprinopsis atramentaria*) seemed closely related. However, *Coprinus* species are free of the mycotoxin coprine—*cheers to that!*

Shaggy Manes, Coprinus comatus

SHAGGY MANE
Coprinus comatus

Shaggy Manes are distinctive with their longish, bullet-shaped caps covered in white fluffy scales, which set them up to be known in the UK as Lawyer's Wig! They love growing in often big groups on lawns, in pastures, along roadsides or trails, or near compost heaps, so wherever you live you might get lucky. They are some of the easiest gilled mushrooms to recognize and hence quite safe. If it wasn't for their short shelf life, saprobic Shaggy Manes would be offered in stores, since they can be cultivated.

Tough yet fragile: they offer an interesting combination of juicy consistency and firm texture and a subtle earthy flavor. The late, great mycologist Gary Lincoff recommended steaming young fresh Shaggy Manes and enjoying them like asparagus with butter or cream sauce. I suggest going fancy with sauce bearnaise. Make sure they don't get crushed in transport. In their white stage, they break easily, though they have been observed pushing up through pavement! Drying risks deliquescence, which kicks in in four to six hours. Interestingly Chef Chad Hyatt insists liquefied black spore mass has a more intense fungal aroma than the fresh tissue, and he loves to use it as a tasty pigment. There are several ways to avoid deliquescence. Store caps upside down or underwater, both of which slow down maturation substantially. Or blanch the fresh mushroom, pureeing and freezing the mass in measured amounts to add later to soups or sauces.

When finding Shaggy Manes, you have to evaluate how pristine or polluted the site is. Shaggies are excellent heavy metal accumulators, as are the Prince and puffballs, all members of the Agaricaceae, so you should not eat any you find growing along busy roads or in industrial sites.

Traditionally, the black pigment has been used to make ink. Some artists have created beautiful, usually fungally themed drawings with it. The founding documents of the Puget Sound Mycological Society in 1964 were signed with ink derived from Shaggy Manes!

- Cylindrical to oval white cap (Ø 4–10 cm and 7–12 cm high) maturing to bell shape.
- Light brown top, smooth at first, rips into brownish segments and scales; many more fibrous white scales cover lower parts of the cap.
- Densely crowded, free white gills mature through pinkish-brown phase to soot-black, liquefied spore mass.
- Smooth, slender stem (8–25 cm tall), hollow core containing white strands; stem base widens or is bulbous.
- White membranous partial veil forms a small, movable ring on lower stem that can fall down or off.
- Grows alone, scattered, or in big groups from disturbed ground.
- Fruits in spring and summer, but mostly in fall in the Pacific Northwest; global distribution.

MEDICINAL: Shaggy Manes have been used in Chinese medicine to support digestion and cure hemorrhoids. Many lab studies have shown in vitro inhibition of ovarian, breast, and leukemic cancer cells, and it may help prevent or treat prostate cancer, but clinical studies are lacking. Also, Shaggy Manes may aggravate eczema in those with atopic dermatitis.

LOOK-ALIKES: Most similar to *Coprinus comatus* is the Inkcap, *Coprinopsis atramentaria*, which lacks the shags and shows more leg, since its caps are shorter. People also confuse Shaggy Manes with shaggy parasols (*Chlorophyllum* spp.), as long as they are in their drumstick phase (see Shaggy Parasols, Dapperlings, and Allies). However, parasol caps are roundish, not conical-cylindrical. They are white spored and never autolyze.

NAME AND TAXONOMY: *Coprinus* Greek "growing in dung" and *comatus* from Greek *coma* "hair," a reference to the hairy shags. *Coprinus comatus* (O. F. Muell.: Fr.) Pers. 1797. Agaricaceae, Agaricales.

⚠ ALCOHOL INKCAP
Coprinopsis atramentaria

Appreciated as a good edible in Middle Eastern societies, where traditionally people abstain from alcohol, these gracious mushrooms with the odd habit of liquefying themselves for the next generation do not get to shine their culinary light in our alcohol-soaked society. In Great Britain, this mushroom is also known as Tippler's Bane! It contains the mycotoxin coprine, a disulfiram-like substance that causes antabuse syndrome. Antabuse is used to help alcoholics resist the temptation of drinking by making them feel miserable upon indulging. If a person has recently consumed alcohol, within five to ten minutes Alcohol Inkcap can cause facial flushing, nausea and possibly vomiting, a fast heartbeat and palpitations, severe headache, sweating, anxiety, vertigo, confusion, and (rarely) collapse. However miserable that sounds, no deaths have been reported.

Coprine inhibits acetaldehyde dehydrogenase, an enzyme that breaks down alcohol in our stomach. Coprine is neither neutralized nor weakened by cooking. Symptoms occur only after drinking alcohol, but sensitivity can last for three full days after eating Inkcaps. Apparently, in Italy right after World War II, when wine was rare, some people would eat Inkcaps and drink just a thimble of *vino rosso* so everyone could enjoy the buzz. For teetotalers or people who drink rarely, Alcohol Inkcaps are fine edibles and quite similar to Shaggy Manes.

There are several Inkcaps in the Pacific Northwest. *Coprinopsis atramentaria* var. *atramentaria* (as described below) tends to be the biggest of them, with wider caps, and is more often found growing in grass. Streaked Inkcap, *C. striata*, has a bit smaller but narrower cap and a scaly lower third of its stem, and it grows in forest habitats. Scaly Inkcap, *C. romagnesiana*, has a finely scaly cap and small dark scales on the lower stem.

IDENTIFICATION

- Lead-gray to brownish cap (Ø 2–8 cm), bald (sometimes finely scaly), lined with grooves.
- Oval cap matures to conical to bell-shaped; flattens out (retaining umbo) when liquefying.
- Free, densely crowded, mature to brown, then to an ink-like black spore mass.
- Hollow, slender, smooth white stem—with ring or traces—lower part with small brown scales.
- Fragile soft flesh, mild flavor.
- Grows alone, in groups, or in dense clusters, often in soil but feeding on dead wood below lawns and gardens and along roads.
- Fruits from spring through summer, especially in fall.

TOP: *Scaly Inkcap*, Coprinopsis striata; **BOTTOM:** *Alcohol Inkcap*, Coprinopsis atramentaria

Mica Cap, Coprinellus micaceus; Mica cap specks

LOOK-ALIKES: Edible Shaggy Mane (*Coprinus comatus*) has a more elongated, cylindrical, and very scaly white cap. The edible and smaller Mica Cap (*Coprinellus micaceus*) has a yellow-brown, mica-flecked cap and grows directly on dead wood. And then there are many more, all smaller, Inkcap type mushrooms.

NAME AND TAXONOMY: *Coprin-opsis* Greek "like-*Coprinus*" (see above); *atramentaria* from Latin "ink." *Coprinopsis atramentaria* (Bull.: Fr.) Redhead, Vilgalys, and Moncalvo 2001; previously *Coprinus atramentarius* (Bull.) Fr. 1838. Psathyrellaceae, Agaricales.

MICA CAP
Coprinellus micaceus

Mica Caps are a nice culinary pleasure in spring when there are limited other options for fungal feeding. It often fruits heavily right after a rain. They are called Mica Caps for the fine, sparkly, mineral-like particles spread over the streaked cap when fresh, but those particles can wash off. Harvesting these smallish, fragile mushrooms can be a bit tedious, but often they grow in big congregations on dead wood, enough for a nice meal. Their flavor is richly fungal.

IDENTIFICATION

- Yellow-brown cap (Ø 2–5 cm) first oval-conical, opening to bell shape, and nearly flattening with age; young cap covered in fine, shiny particles.
- Pale gills mature gray or brownish to black, producing liquefied spores.
- Fragile white flesh, with mild fungal flavor and odor.
- Grows in tufts or groups around old hardwood stumps or buried wood.
- Fruits year-round, but mainly in spring and fall; a widespread cosmopolite.

LOOK-ALIKES: The smaller nontoxic Firerug Inkcap, *Coprinellus domesticus*, is similar and can be told apart by a mat of coarse, orange-brown hairs around its stem base. Inkcap (*Coprinopsis atramentaria*) is grayer, bigger, and fleshier with a fatter stem. Pale Brittlestem (*Candolleomyces candolleanus*) is also similar but differs in several key ways (see below).

NAME AND TAXONOMY: *Coprin-ellus* Latin for "little" *Coprinus*. *Coprinellus micaceus* (Bull.: Fries) Vilgalys, Hopple, and Jacq. Johnson 2001; *Coprinus micaceus* (Bull.) Fries 1838. Psathyrellaceae, Agaricales.

Other Dark-spored Gilled Mushrooms

This group includes gilled mushrooms that produce cold dark brown or gray spores, mostly with a purple tinge. Most of them grow on wood, such as woodtufts (*Hypholoma*) and brittlestems (*Candolleomyces* and *Psathyrella*), or on wood chips or dead debris, like roundheads and Wine Cap (*Stropharia*), as well as Spring Fieldcap (*Agrocybe*), its dark brown spores devoid of purple tones. In addition, the wood chip–worshiping magic Wavy Cap and the grass-feeding Liberty Cap (both *Psilocybe*) are presented.

!! PALE BRITTLESTEM
Candolleomyces candolleanus

When young, this small mushroom looks so clean, bright, and happy with its obvious appendiculate cap margin—the little white specks originating from the partial veil that soon hangs from the cap edge. The cap, however, quickly changes color, drying out to pale white. This striking change in tissue is described as hygrophanous, a term capturing the different appearance of a darker, well-hydrated area and an already drier area; the cap center often loses its moisture first.

Most people ignore these fragile little mushrooms, or rather they shy away from them, since Pale Brittlestems take a high degree of mushroom mastery to identify; a life lived without them in your frying pan is not missing much but a tiny, tasty LBM. However, when harvested in quantity at their prime and after you have conclusively identified them, they quickly turn into an enjoyable treat fried in butter or mixed into whatever dish needs a nice fungal touch.

IDENTIFICATION

- Smallish mushroom (Ø 3–8 cm), honey colored, often quickly becoming pale; hygrophanous.
- Cap edge slightly striate, often with white partial veil remnants dangling.
- Crowded whitish, adnate gills that turn violet-tinged dark brown (retaining their white edges) from spores.
- Fragile, hollow, white stem with no ring but possibly veil remnants; stem often 1.5 times as long as cap is wide.
- Very thin, fragile, white flesh; mild fungal flavor and odor.
- Grows in tufts or groups on old hardwood stumps or buried wood in disturbed places.
- Fruits from spring to fall after good rains; widespread cosmopolite.

LOOK-ALIKES: At first glance, there are many LBMs that could be mistaken for Pale Brittlestem, including brethren that

are, or used to be, *Psathyrella*, which are mostly regarded as harmless. Mica Caps (*Coprinopsis micaceus*) have similarly colored caps, but also feature mica grains and are without ring remnants; when mature, their cap margins start to liquefy. The dangerous look-alikes are the very toxic fiberheads (*Inocybe* spp.), Deadly Skullcap (*Galerina marginata*), and Ringed Conecap (*Pholiotina rugosa*): all have much lighter, more rust-brown spores than the nearly black ones of *Candolleomyces candolleanus*.

Pale Brittlestem, Candolleomyces candolleanus

NAME AND TAXONOMY: *Candolleomyces candolleanus* honors the great Swiss botanist Augustin de Candolle (1778–1841). *Candolleomyces candolleanus* (Fr.) D. Wächt. and A. Melzer 2020; formerly *Psathyrella candolleana* (Fries) Maire 1937. Psathyrellaceae, Agaricales.

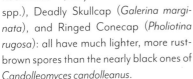

CONIFER TUFT
Hypholoma capnoides

Conifer Tuft, the widely overlooked little brother of the charismatic but poisonous Sulfur Tuft (*Hypholoma fasciculare*; see below), is common and often fruits in abundance, while offering a pleasant flavor and good structure. Since it is rarely collected, it is an interesting edible for the advanced mushroom hunter in otherwise overpicked forests. Conifer Tuft fruits in fall on dead conifers, the same habitat appreciated by

Sulfur Tuft, Deadly Skullcap (*Galerina marginata*), and several mostly bitter, brown-spored scalycaps (*Pholiota*). Use great care—this is not a mushroom for beginners! Its rusted-iron flavor, which disappears when cooked, is very helpful for identifying it. Usually, I enjoy it fried in butter or olive oil—it has a surprisingly firm texture and a good fungal flavor.

IDENTIFICATION

- Yellow-brown to ochre-brown, smooth (but sticky when wet), hygrophanous cap (Ø 2–7 cm) with low umbo and often lighter edge, which often retains thin patches of the partial veil, including a thin dark brown line of caught spores.
- Close, narrow gills, usually adnate but often separating from stem, at first pale gray developing to smoky-purple-brown, reflecting the purple-brown to deep purple-gray spore color.
- Long (4–12 cm), slender stem with obscure ring zone, colored like cap, but darker toward base.

- Ring finely fibrous, often disappearing with age but if present, marked by dark spore deposit.
- Flavor mild, fungal, and slightly metallic.
- Thin flesh, tougher in stems.
- Grows in groups or clusters on dead conifers.
- Fruits in sometimes large clusters from wood, in fall and early winter; widespread and common.

LOOK-ALIKES: Toxic Sulfur Tuft (*Hypholoma fasciculare*) has yellow-greenish gills and tastes intensely bitter. The inedible Snakeskin Brownie (*Hypholoma dispersum*) is much smaller and usually fruits solitarily or in groups on the ground; rarely, it grows clustered on wood. Deadly Skullcap (*Galerina marginata*) is smaller and lacks yellow tones and if clustered, they are usually small clusters. Scalycaps (*Pholiota* spp.) share the same habit, growing clustered on wood, but many have scaly caps, all are brown spored, and even those that are not toxic are usually not eaten.

NAME AND TAXONOMY: *Hypho-loma* Greek means "thread-fringe," a reference to the threadlike veil; *capnoides* Greek "smoky," as in the gill color. *Hypholoma capnoides* (Fr.) P. Kumm. 1871; synonymous with *Naematoloma capnoides* (Fr.: Fr.) P. Karst. Hymenogastraceae, Agaricales.

SULFUR TUFT
Hypholoma fasciculare

The very bitter Sulfur Tuft (*Hypholoma fasciculare*, also called Clustered Woodlover) is similar to the Conifer Tuft (*Hypholoma capnoides*) but easy to tell apart by flavor and gill colors. When young, the gills are bright light yellow to electric green, then they mature to dark purple-brown, often with a greenish hue.

Conifer Tuft, Hypholoma capnoides

Sulfur Tufts, Hypholoma fasciculare

This wood-decaying, eye-catching mushroom, which often produces impressive fruitings, can seemingly fruit in soil, but if you investigate, you will always find a dead root or wood underground that it is growing on.

It is difficult to imagine how anyone would eat a mushroom with such an awful flavor. Poisoning is delayed, often kicking in five to ten hours after ingestion, causing nausea and a messed-up digestive system, blurred vision, and even paralysis.[85] Fresh and dried Sulfur Tufts reflect a spooky green light under UV light. This feature is displayed at the Puget Sound Mycological Society's annual mushroom show in a dark-viewing tent revived by Derek Hevel, author of the nifty 2019 PSMS cookbook.

WINE CAP
Stropharia rugosoannulata

The Bordeaux-colored Wine Cap can grow to be impressively large, sometimes reaching one foot across. You would not want to gulp down a cup of wine the size this mushroom can reach. Also known as the King Stropharia, it has expanded its domain to the Pacific Northwest in recent decades, supported by loyal subjects who worked its commercially available mycelium into a 4-inch layer of wood chips and straw spread between beds, usable as a garden path. Here this saprobic mushroom reliably fruits by feeding on all kinds of organic debris, improving the soil and sending its spores off to conquer the hinterland. When young, it sports a wine-red cap, solid stem with a cog-like ring and mycelial strands (rhizomorphs) on the wide base, and grayish-purple gills.

With a few white tufts, it resembles a comic book version of the more slender, white-gilled, and white-spored Fly Agaric. Cultivators praise the "Garden Giant" as a very enjoyable edible, but others with less commercial interest are not so impressed. Many people compare them to button mushrooms for their culinary qualities and suggest using *Strophoria rugosoannulata* in the same fashion. Mushroom Mountain's

Tradd Cotter and Olga Katic describe it appreciatively: "It tastes like potato cooked in a light wine marinade, and the stringy stem is reminiscent of steamed asparagus."

IDENTIFICATION

- Wine-red to reddish-brown, smooth cap (Ø 5–20 cm) fades in age to pale yellow-brown or gray.
- Cap can be sticky when wet, and when young features irregularly whitish tufts of veil remnants.
- Adnate or notched (with age sometimes free), crowded gills starting whitish and maturing via gray to purple-black with sterile gill edges remaining whitish; produces purple-black spores.
- Hefty whitish stem (8–25 cm tall), wider at base, darkening yellowish or browning with age; base often with white mycelial threads.
- Thick, firm, unchanging white flesh
- Ring grooved with lines on both sides, cog-like, often segmented.

- Grows scattered and in groups on wood chips and other mulch in cultivated areas.
- Fruits mainly in spring and fall, but also during wet summers.

LOOK-ALIKES: Very similar is the probably toxic but smaller Conifer Roundhead (*Stropharia hornemannii*). It has a sticky cap when wet, and its ring lacks the cog pattern on the underside and is less likely to break into segments. Also similar is the rare, probably harmless, and often bitter Scalloped Scalecap, *Hemistropharia (Pholiota) albocrenulata*, which fruits on hardwoods and has whitish scales on cap and stem. Young Wine Caps can resemble several *Agaricus* species, but *Agaricus* has chocolate-brown spores, while *Stropharia* has purple-black ones.

NAME AND TAXONOMY: *Stropharia* from Greek *strophos* "band" or "rope," a reference to the partial veil, and *rugoso-annulata* Latin for "wrinkled ring," describing the stem annulus. *Stropharia rugosoannulata*, Farlow ex Murrill (1922). Strophariaceae, Agaricales.

Wine Cap, Stropharia rugosoannulata

AMBIGUOUS ROUNDHEAD
Stropharia ambigua

There is not much ambiguity and nothing questionable about this shining, elegant roundhead, which is also called the Questionable Stropharia, besides its appreciation as an edible. It is maligned for tasting like rotten leaves (one of its favorite foods) by some people, while others appreciate its earthy flavor. Most people regard them as marginal, but so are toaster pastries, which are surely worse for your health.

That fluffy, white margin around a sticky, color-saturated yellow cap makes *Stropharia ambigua* easy to recognize. Add the fluffy stem and the purplish, nearly black spores (sometimes caught on the fluffy ring tissue), and you can be confident about its ID. It should not surprise foragers that the Ambiguous Roundhead tastes best in midwinter—lean times for mushroomers—sautéed in a generous dollop of organic butter. Fair warning: omitting the butter might render this *Stropharia* inedible.

Ambiguous Roundhead, Stropharia ambigua

- Grows solitarily, in small groups, or in big troops, in rich soil or wood chips, and in mixed and conifer forests.
- Common all over the Pacific Northwest, but especially in coastal areas; fruits mostly in fall but also in winter and spring.

LOOK-ALIKES: The Mulch Maid (*Leratiomyces percevalii*, a.k.a. *Stropharia percevalii*)—edibility harmless—has been taking over wood chips by storm. David Arora, comparing *L. percevalii* (then known as *L. riparius*) to *Stropharia ambigua*, notes that it has a (1) slightly smaller, slimmer stature (stem less than 1 cm thick); (2) thinner veil (like facial tissue) that often leaves cap remnants near the margin rather than dangling from the margin itself; (3) stem that is not nearly as shaggy; and (4) cap more likely to be slightly umbonate.

The black spores of *Stropharia* make a clear break from all the toxic, white-spored Amanitas. The stockier Conifer Roundhead (*Stropharia hornemannii*) has a chestnut-brown cap (soon paling) and felty

IDENTIFICATION

- Yellow to yellowish-brown cap (Ø 4–15 cm), sticky or slimy when moist, paling with age.
- Cap margin hung with cottony, fluffy white veil remnants.
- Adnate crowded grills, at first white, turning quickly gray to purplish-black; dark purple-brown to black spores.
- Off-white stem (6–18 cm tall x 0.5–2 cm wide) features a fleeting ring; lower stem has tufted white scales and often orangish-brown stains; bulbous base has white rhizomorphs.
- Thick, soft to fairly firm, white flesh with earthy fungal flavor and odor.

Spring Fieldcap, Agrocybe praecox

Spring Fieldcaps have a smooth, light brown cap, brown gills, and a thin ring. They have good structure and a unique odor and flavor, very farinaceous or cucumber-like but with their own bent. Some people enjoy the flavor, and others will be upset that they wasted butter on them. Most will not give them a second chance, though their abundance may tempt them to try again!

skirtlike ring, and it often grows from rotten wood. Spring Fieldcap (*Agrocybe praecox*) has a paler cap that cracks easily in irregular patterns and is brown spored.

NAME AND TAXONOMY: *Stropharia ambigua* (Peck) Zeller 1914. Strophariaceae, Agaricales.

SPRING FIELDCAP
Agrocybe praecox

When spring is peaking, the Spring Fieldcap is all over: alone, in small groups, or in the hundreds covering wood chip fields. Dry and sunny spring weather makes the cap crack into irregular polygons or blocks, a phenomenon known as areolate, helping cinch their identification. Before the cap cracks, many people have a hard time figuring out what mushrooms they are looking at, since it can grow in so many shapes, from slender to stocky, and there are plenty of brown-spored mushrooms that sprout from the ground.

IDENTIFICATION

- Smooth and dry, cream to tan cap (Ø 2–8 cm), pale yellow-brown toward center, cracking irregularly as it matures.
- Adnexed or adnate, crowded, broad gills are at first pale to light brown, darkening to a dark rich brown, the color of the spores.
- Whitish or cap-colored stuffed stem, mealy and lined at top; thin mycelial cords on a sometimes wider stem base.
- Persistent and drooping felty white ring relatively high up on stem.
- Unique, farinaceous odor and flavor, described by some as mild and pleasant; flavor sometimes bitter, which will not cook away.
- Grows in groups or troops and small tufts in barren soil, grass, fields, on wood and forest edges, and especially in wood chips.
- Fruits mostly in spring, but a few pop up in summer or fall; widespread.

MAGIC MUSHROOMS

The magic of the picture-book LBMs (little brown mushrooms) has been gaining more and more momentum! There is a clear trajectory from the obscure, sacred substance of Indigenous cultures to a fascination in fringe subcultures to now knocking on the doors of perception of the medical mainstream, well chronicled in Michael Pollan's bestseller *How to Change Your Mind*. Research spearheaded by Johns Hopkins University and Imperial College London in the past couple decades has documented incredible potential for using psilocybin to treat clinical depression, anxiety disorders, nicotine and alcohol addiction, and cluster headaches and migraines, as well as devastating PTSD (post-traumatic stress disorder).

Magic mushrooms are not food, but they are proving to be medicinal. In 2020, Oregon's voters approved Measure 109, allowing regulated use of psychedelic mushrooms in a therapeutic setting. Warning: The high induced by magic mushrooms can be very disorienting, and in rare cases it has been known to trigger psychosis, especially in individuals where there is a family history of serious mental disease. Also, there have been cases of serious, even lethal, poisoning of young children. However, according to a growing body of literature and study, in the right setting, the psilocybin-induced state of mind can lead to profound feelings of unity with nature and the universe and produce a more open, welcoming attitude.

LOOK-ALIKES: Almost indistinguishable is the harmless Bearded Fieldcap, *Agrocybe dura* (syn. *A. molesta*), which grows in grass, tends to be more robust, and has an ephemeral ring zone and partial veil tissue hanging from the cap margin, hence the name Bearded Fieldcap. The often bitter, probably harmless Mulch Fieldcap (*A. putaminum*) has a club-shaped, usually darker stem and no visible veil.

Several roundheads can be similar, like the inedible Garland Roundhead, *Stropharia coronilla*, which is smaller and has the Stropharian dark purple-brown spores. Also quite similar is the Ringed Webcap (*Cortinarius caperatus*), but it grows in fall in conifer forests and has brighter, rusty-brown spores and a wrinkled cap surface.

NAME AND TAXONOMY: *Agro-cybe* Greek "fieldhead," *prae-cox* Latin "early ripening." *Agrocybe praecox* (Fr.) Fayod 1889. Strophariaceae, Agaricales.

WAVY CAP
Psilocybe cyanescens

Distinguishing Psilocybes from other little brown mushrooms would be a total challenge were it not for the revealing blue staining of most psychoactive fungi. (While psilocybin is assumed to be the active ingredient, it is rapidly metabolized into psilocin, the main active substance. It is psilocin that enzymes oxidize to a blue color. Indeed, if a magic mushroom is very blue, much of the psychoactivity is "rusted"—oxidized—

Wavy Cap, Psilocybe cyanescens; *note the bluing on the stems and cap edges.*

away.) Because there are quite similar deadly and other toxic LBMs, a key line to remember is "no bluing, no magic."

In addition to the bluing, all Psilocybes produce a dark purple-brown spore print and have a semigelatinous separable pellicle, a sticky, cohesive cap skin that can be peeled off. *Psilocybe cyanescens*, the Wavy Cap, has caps that become wavy and fleeting, fibrous stem veils; it grows on dead wood, especially wood chips, the new fungal fast food.

Very similar to Wavy Caps are Flying Saucers (or Azzies, *P. azurescens*), which tend to be a bit bigger (cap Ø 3–10 cm, stem 9–20 cm tall), with flat (not wavy) caps when opened that have obvious, often pointed umbos. Azzies fruit in dune grass along our Pacific coast from late September through April. Two smaller *Psilocybe* species, growing on wood chips and somewhat common, are Blue-ringers (*P. stuntzii*), which have a felty membranous ring and fruit in fall and early winter, and Ovoids (*P. ovoideocystidiata*), which fruit in spring. A fall-fruiting forest dweller, especially in years after logging and in wood debris fields, is the Conifer Psilocybe, *P. pelliculosa*.

IDENTIFICATION

- Caramel-brown cap (Ø 1–6 cm), sticky when moist, peelable; matures from rounded bell to broadly convex to flat with wavy margins.
- Hygrophanous cap color varies by water content, fading from full orangish/reddish-brown tones to pale yellowish-straw. Often blue or blue-green stains develop with age, especially near margin.
- Adnate to somewhat decurrent broad gills at first mottled cinnamon-brown, becoming deep smoky-brown with maturity (keeping paler gill edges); spores dark purple-brown.
- Stiff, bright white stem (2–8 cm long x 0.2–0.5 cm wide), silky to finely hairy, slightly browning with age and bluing from touch or other impacts.
- Partial veil made of white spider-web-like fibers that catch the dark spores and soon disappear.

- Thin flesh, colored like cap (whiter on stem); quickly bruises bluish.
- Mild, mealy, or earthy odor and similar flavor, sometimes bitter.
- Grows in troops on hardwood chips, sawdust, or soil with woody debris in yards, parks, and landscaped areas.
- Widespread and common especially around human settlements from Southwest British Columbia to California west of the Cascades; fruits mostly in late October and November with colder weather.

LOOK-ALIKES: The Deadly Skullcap (*Galerina marginata*) shares stature and habitat with *Psilocybe cyanescens*; they are known to grow side by side. Both mushrooms have hygrophanous caps. But skullcaps have rusty-brown (not dark purple-brown) spores and never stain blue. Skullcaps have a darker stem and a more pronounced white ring zone.

The similar and ubiquitous but inedible Scurfy Twiglet (*Tubaria furfuracea*) gets plucked again and again by those seeking magic, but no matter how much you squeeze it, you will not provoke bluing! The benign Mulch Maid (*Leratiomyces percevalii*) also grows on wood chips; it is darker spored and bigger and lacks bluing, as does the tiny orange-brown and deadly Ringed Conecap (*Pholiotina rugosa*). Then there are dozens of poisonous fiberheads (*Inocybe* spp.) that do not stain blue, have brown spores, and often feature a fibrous cap surface that cannot be peeled. And the list of LBMs goes on.

NAME AND TAXONOMY: *Psilo-cybe* from Greek *psilos* "naked, smooth" and *kybe* "head"; *cyan-escens* Latin "becoming blue." *Psilocybe cyanescens* Wakef, 1946. Hymenogastraceae, Agaricales.

LIBERTY CAP
Psilocybe semilanceata

As I crossed the US-Canada border at the Peace Arch, a customs officer asked about my visit. I honestly replied, "I am coming for the mushrooms." The officer responded, "Magic mushrooms!" Startled, I had to think for a second. "Although I think all mushrooms are magic," I said, "no, ma'am, I am not visiting for magic mushrooms, which make up less than 0.01% of the known mushrooms, but rather for chanterelles and lobsters." Content with my answer, she welcomed me to British Columbia. To my surprise, two days later,

Liberty Cap, Psilocybe semilanceata

I found myself on my knees with a dozen people, most of them of retirement age, all attendees of the awesome Sicamous Fungi Festival. We were combing through the tallish grass of a wet pasture to find a few tiny Liberty Caps and admire these magic mushrooms in their natural environment—without breaking the Crown's law.

The bell-shaped to conical caps of Liberty Caps do not flatten, and typically they have a little extended nipple in the center, hence the name, a reference to the headgear, also called the *bonnet rouge*, popular during the French Revolution. Their whitish stems are very tall (3–10 cm) in relation to cap diameter (0.5–2.5 cm). The cap is smooth and slightly sticky; when fresh, the cap skin or pellicle can be peeled off. Being hygrophanous, the chestnut-brown or olive-brown caps fade to olive-buff or straw-yellowish. Close and narrow adnate to adnexed gills are pale whitish to gray. Maturing spores turn the gills dark purple-brown; the edges stay whitish. Cap, stem, and flesh bruise

a much less pronounced blue than many other magic mushrooms.

NAME AND TAXONOMY: *Semi-lanceata* Latin "half spear-shaped," referring to the shape of the overall mushroom. *Psilocybe semilanceata* (Fr.) P. Kumm. 1871.

Gilled Bolete Relatives

Several genera of gilled mushrooms are part of the Boletales and not Agaricales. A few are edible but not impressive, such as slimespikes (*Gomphidius*) and pinespikes (*Chroogomphus*). Some are of highly questionable edibility like False Chanterelle (*Hygrophoropsis aurantiaca*), which is presented as a chanterelle look-alike in a sidebar earlier (p. 104). None reach the culinary heights of their pored cousins.

Another interesting gilled bolete is the Rollrim (*Paxillus involutus* group)—its gill layer can be peeled off like the pores of a bolete! *Paxillus* has been cooked and eaten for centuries without any perceived incidents, but it can turn deadly—an antigen can trigger the immune system to attack red blood cells, causing hemolysis and kidney failure. This birch (and aspen) associate is infamous for being the only mushroom known to have fatally poisoned a professional mycologist, Julius Schäffer. The Rollrim was not known as toxic before this tragedy. However, Schäffer's death in October

Rollrim, Paxillus involutus *group*

1944 in Bavaria must also be attributed to the complete breakdown of medical support during the final months of World War II.

BLACKENING SLIMESPIKE
Gomphidius oregonensis

Blackening Slimespikes are cool, especially when you find one with a fully stretched and translucent (like a clear umbrella) slimy partial veil! And all that slime, which is actually a surprisingly cohesive protective layer, is easily peeled off. When collecting them, keep the sticky slimespikes separated from other mushrooms. The sticky slime and loud yellow of the lower stem make them easy and safe to identify.

There are several edible species in the Pacific Northwest. The easiest to ID is the Rosy Slimespike, *Gomphidius subroseus*, which sports cute, slimy, pinkish to red caps. Slimespikes are members of the bolete order and are closely related to pinespikes (*Chroogomphus*) and jacks (*Suillus*)! They are not simply related to jacks, but appear to depend on them; slimespikes seem parasitic on the mycelia of *Suillus*. For too many years, slimespikes got away with that behavior while being mistaken as mycorrhizal with the very trees *Suillus* (not *Gomphidius*) are trading nutrients and water for sugars!

To enjoy slimespikes, first peel away the sticky layer. When peeled, they are all right edibles—or is it edible, alright? Young specimens have a better texture; older slimespikes soften, and their developed black spores darken the final culinary product. They are best processed like *Suillus* species: dried to firm up and concentrate flavor (rehydrate them later or grind them for spice powder). When cooking them fresh, make sure to dry sauté the peeled mushrooms, firming up the flesh and bringing out the earthy fungal flavor with a hint of bolete. Avoid harvesting slimespikes in areas treated with processed sludge from water treatment facilities ("dilution is the solution"). Slimespikes can be loaded with lead and arsenic,[86] so it is better to shun such areas, which are usually posted with warning signs.

IDENTIFICATION

- Very slimy, smooth cap (Ø 3–15 cm) ranging widely in color from whitish to dull pinkish or salmon when young, turning purplish to reddish-brown with age.
- Overall shape first peg-like with rolled-in margin, and then expanding to nearly flat, often with depressed center.

Blackening Slimespike, Gomphidius oregonensis

- Deeply decurrent, thick, distant, waxy gills produce brownish-black spores.
- Straight to curved, firm, solid stem, moist and whitish in upper portion, lower part staining blackish with age, always yellow at the base.
- Ring made of whitish layer of slime with fibrous layer beneath; forms a slightly hairy slime tissue blackened by caught spores.
- Softish, white to grayish flesh, with pink tinge under cap surface; stem base has yellow flesh.
- Grows alone or often in small clusters on *Suillus* species in pine, Douglas-fir, or other conifer forests.
- Common from Alaska Panhandle to California and in the Rockies; fruits in fall.

LOOK-ALIKES: Another slimespike, *Gomphidius glutinosus*, grows less clumped and usually has a darker or more purple cap when young, but other colors are similar, and it can taste sour. The larger spores of *G. glutinosus* are the only definitive characteristic to tell it apart from Blackening Slimespike. Rosy Spike (*G. subroseus*) can be similar, but its cap is pink or rosy-red to red, its thicker stem is yellower, and it occurs occasionally in clumps.

NAME AND TAXONOMY: *Gomph-idius* from Greek *gomphos* "peg" or "wedge" and *-idios* "resembling," named for its shape, especially of young fruiting bodies. *Gomphidius oregonensis* Peck 1898. Gomphidiaceae, suborder Suillineae, order Boletales.

WINE PINESPIKE
Chroogomphus vinicolor

Where there are pines, there are pinespikes! But without jacks (*Suillus* spp.) or false truf-

Wine Pinespike, Chroogomphus vinicolor

fles (*Rhizopogon* spp.) in the forest, there are hardly any, since these mushrooms "spike" into the mycelium of their cousins and feed off them. It seems they do have direct ectomycorrhizal relations as well,[87] at least the European species. Since several *Suillus* are mycorrhizal with Douglas-fir and larch, pinespikes can also be found in their habitat. The Pacific Northwest has several species of pinespikes, all of which are edible.

When young, *Chroogomphus vinicolor* is easily recognized by its sticky dark brown cap, widely spaced and deeply decurrent gills, pale orange flesh, and pine habitat. Similar in stature, but orange and never reddish, is the Woolly Pinespike (see below). Two darker species are harder to differentiate. The Giant Pinespike (*C. pseudovinicolor*) has a reddish-orange to burgundy-colored cap (Ø 5–15 cm) that lacks an umbo and an orangish-tan, stubby stem covered with reddish chevrons. The more slender Olive Pinespike (*C. ochraceus*), besides having a purplish-brown/grayish-brown to olive-tan cap (Ø 3–7 cm), often sports a lovely pointed light brownish umbo.

Pinespikes offer bland fungal flavor with a firm enough texture when they are young. Overall, they are rather mediocre. As with many of their relatives, such as the other members of the gilled bolete order or many of the jacks, they benefit greatly from dry sautéing to firm up their tissue. It's probably better to dry them to use as whole dry pieces or as mushroom powder in soups and sauces.

IDENTIFICATION

- Dark red-brown cap (Ø 2–10 cm), sticky when wet and shiny or silky when dry, conic with a slight umbo, becoming convex in age.
- Distant, deeply decurrent, broad gills; dull orange turning olive-gray to smoky-black from spores.
- Pale orange, finely fibrous (quickly disappearing) veil leaves a slight hairy ring.
- Firm, solid stem (3–15 cm tall x 0.5–2 cm), pale orange-tan to yellowish-brown, taking on red tones in age; narrowing downward.
- Thick, fibrous, pale orange flesh.
- Fruits alone or in groups (with *Suillus* spp.) from soil under pines and other conifers.
- Widespread and fairly common, fruiting mostly in fall.

LOOK-ALIKES: Besides slimespikes (*Gomphidius* spp.), there are no close look-alikes. Pinespikes are unique with their well-spaced, broad, decurrent gills and dark spores.

NAME AND TAXONOMY: *Chroo-gomphus* from Greek *chros* "skin" or "skin-color" and *gomphus* "peg" or "wedge"; *vini-color* Latin "vine-colored." *Chroogomphus vinicolor* (Peck) O. K. Mill. 1964. Gomphidiaceae, Boletales.

WOOLLY PINESPIKE
Chroogomphus tomentosus

A very common Golden Chanterelle look-alike, the Woolly Pinespike receives most of its attention when mistaken as a chanterelle. It is quite easy to do, when mesmerized in the hunt picking chantie after chantie, until you feel the woolly cap and see the true gills below, which breaks the spell. Woolly Pinespikes are also paler and more orangish-brown than chanterelles. They

Woolly Pinespike, Chroogomphus tomentosus

turn wine-red when cooked, which makes them easy to tell apart from chanterelles at this stage, should you miss all other signs while collecting and cleaning.

Woolly Pinespikes often grow in the same habitat as chanterelles. From the coast of Northern California, Christian Schwarz and Noah Siegel report that *C. tomentosus* appears to associate with the Admirable Bolete (*Aureoboletus mirabilis*) in western hemlock habitat.[88] Buck McAdoo likes to dry his Woolly Pinespikes and then reconstitute them later for stuffing because, as he says, "the magenta color really brightens up the bird."

NAME AND TAXONOMY: *Tomentosus* Latin "woolly." *Chroogomphus tomentosus* (Murrill) O. K. Mill. 1964.

WESTERN GILLED BOLETE
Phylloporus arenicola

The gilled bolete was always a thumb in the eye of small-minded taxonomists and a refreshing reminder that classifying nature is a human obsession and not something Mother Nature is invested in. From the cap surface, stipe, flesh, bruising, and spores, it is evidently a bolete, and very close to the suede bolete, *Xerocomus*. Look below, though, and the attractive yellow gills make it seemingly impossible to identify. Looking even closer reveals intervening between the gills, even rarely showing as pores, clearly demonstrating that pores are in its genes! DNA and microscopy indicate that in the case of gills versus pores, what can seem a world of difference is actually just a small variation of the evolutionary mushroom theme striving for maximum spore-producing surface area under a protective cap.

The gilled bolete is easily recognized by its bright golden gills that slowly bruise blue; a velvety, greenish-olive-brown cap; and yellow mycelium at the stem base. Flavor and texture are comparable to *Xerocomellus* boletes, offering a nice fungal aroma and good structure, especially when young. The main challenge when feeding

Western Gilled Bolete, Phylloporus arenicola

on gilled boletes is that it is hard to find this gorgeous and striking mushroom in sufficient numbers to serve as a decent meal.

IDENTIFICATION

- Velvety, dry cap (Ø 2–12 cm) ranges from grayish-olive-green to tan and reddish-brown with age.
- Golden-yellow gills are thick, wide, well-spaced, sometimes forking and have some cross-veins; decurrent, notched, or even nearly free.
- Gills are sometimes bluing, and produce dull orange-brown spores.
- Yellowish stem with brown streaks and brown dots at top; tapering stem base with yellow rhizomorphs.
- Firm, yellowish flesh, rarely staining; flavor mild.
- Grows solitarily or in small groups, often in sandy soil with conifers, especially pines and Douglas-fir.

- Fruits in spring, summer, and fall from southern British Columbia to California.

LOOK-ALIKES: Very similar and also edible is the Golden Gilled Bolete (*Phylloporus rhodoxanthus*), which has a dull reddish to brown cap and gills that are more conspicuously decurrent. However, this eastern North American species might not grow out West. Molecular studies will tell. For other yellow-gilled mushrooms, see Man on Horseback and its look-alikes (in Tricholoma: Knights or Trichs earlier).

NAME AND TAXONOMY: *Phylloporus* from Greek *phyllon* "leaf," as in gill, and *poroi* "pores"; *areni-cola* Latin (as in arena) "growing on sand." *Phylloporus arenicola* A. H. Sm. and Trappe 1972. Boletaceae, Boletales.

PART THREE
THE RECIPES

While this handbook is predominately a field guide to finding and identifying mushrooms, I offer inspiration in this section for how to cook and preserve mushrooms, including detailed recipes to enhance your culinary use and enjoyment of wild mushrooms. To avoid sending you into the woods with a cookbook in your pack, many of these suggestions are merely jumping-off points; use your creativity to make them your own. For general notes on processing, cooking, and eating mushroom species, see Cooking with Mushrooms in Part One: The Hunt.

OPPOSITE: *Candy Cap Butter Cookies*

THE RECIPES

APPETIZERS, SNACKS, AND SIDE DISHES

SOUPS AND SAUCES

HEARTY MEALS

Golden Chanterelles, Winter Chanterelles, and Hedgehogs cleaned and ready to be cooked

DRINKS AND DESSERTS

Bolete powder to make Bolete Butter (p. 333) or add to other dishes as you wish

Cooking with mushrooms is not a precise art like baking, where, when you deviate from the recipe, the project easily fails, like a cake not rising. Rather, a mushroom forager has to be much more observant and tailor their approach to the mushrooms' condition. Though button mushrooms from the supermarket have a reliable consistency, even the same species of foraged wild mushrooms can vary widely in structure and/or taste depending on their condition when found. Not all wild-collected mushrooms arrive in the kitchen in their perfect state: young and firm. We don't have to shun more mature specimens or seemingly waterlogged chanterelles collected after five days of rain. We do,

however, have to adjust our processing and cooking. If the texture is on the softer side, be it from age or weathering or intrinsic to the species, dry sautéing your mushrooms first is always a great choice, because while a softer structure is no problem in sauces or soups, it might disappoint in a stir-fry.

Also, don't forget that cooking or drying are important preparational steps in rendering many species edible. For most mushrooms, frying or boiling them five to seven minutes will do the detox; some others like scaberstalks (see Boletes in Part Two: The Mushrooms) might need more like twenty minutes. When preparing mushrooms in ways that do not involve heat, pay attention to the mushroom species you want to use. Are they edible raw, or do they need detoxification? Is drying at 130°F (55°C) sufficient for detox? Some volatile toxins may mostly dissipate; others absolutely do not. If you intend to eat a mushroom without cooking, please double-check your species in this guide—or, if necessary, elsewhere—before taking on the extra risk. And while we all wish we had a precise answer for each specific toxin, so much is still unknown. I have not powdered dried fresh shaggy parasols for spicing popcorn because they may cause digestive upset; they generally need to be cooked (not just dried) to be rendered edible. It's best to stick to species that are known to be harmless raw if you plan to use their powder in preparations you aren't going to cook first.

REGARDING FOOD CHOICES AND LIMITATIONS

As some readers may quickly realize, none of these recipes include meat. Not that mushrooms won't enhance most meat dishes—they surely do! However, I am of the opinion that when you are preparing dishes with delicious mushrooms, there is no real need for extra meat. Mushrooms alone contribute so much flavor, texture, and healthy nutrients (proteins, fibers, minerals, and vitamins) while having a tiny carbon footprint that they are perfect without meat. Still, I am a conscious omnivore, and have made suggestions in some recipes about how to integrate some meat as well. And yes, any mushroom duxelles, cream sauce, or gravy will enhance a festive roast or piece of meat.

Also, many recipes have been prepared gluten-free, since my wife's digestive system, like that of many other people, cannot break down gluten. Luckily, at this point there are great alternative ingredients that make substitution easy; you do not need to sacrifice taste when cooking with mushrooms. Furthermore, several recipes use cream, milk, and cheese, but you can easily substitute cashew cream, soy milk, or your preferred alternatives. If you adhere to a specific diet, I trust you know how to find your way in making the most of your cooking.

Finally, never eat wild mushrooms without a confident identification! *When in doubt, throw it out.* If you think you have eaten a poisonous mushroom, call Poison Control: 1-800-222-1222.

Bolete Butter

Pickled Chanterelle Buttons, p. 334

APPETIZERS, SNACKS, AND SIDE DISHES

BOLETE BUTTER

One way to make the most of your life is to have bolete butter ready to enjoy 24/7 year-round! Bolete butter is very easy to make, holds up well refrigerated and frozen, and is extremely versatile. It is delicious on a slice of fresh sourdough bread or a hot piece of toast, either by itself or with some *Penicillium camemberti* in the form of a ripe camembert cheese or any charcuterie. Bolete butter melts to spread its yumminess onto steamed veggies and adds savory flavor to steak as well. The twin cooks Jamie and Dennis Nothman, members of Seattle's Puget Sound Mycological Society, turned me on to the idea of this scrumptious butter, which, following my mother's herb butter recipe, I've fine-tuned a bit.

Makes several months' worth

1 pound (4 sticks) butter
1 ounce powdered porcini
1/2 cup finely chopped parsley (about 1/2 bunch)
2 teaspoons Dijon mustard
1 teaspoon lemon juice (or more, to taste)
Salt, to taste, when using unsalted butter

In a small mixing bowl, set the butter out to soften. Mix in the powdered porcini (preferably *Boletus* species), combining well. Stir in the finely chopped parsley and Dijon mustard and add the lemon juice and salt to taste.

Let it sit for a while—the butter becomes better-infused with flavor at room temperature. Store in a closed container in the refrigerator, or label it with the date and freeze it.

Note: Before adding the powder, you can mix it with bit of water to make a paste, but normally there is enough water in commercial butter to rehydrate it.

Bolete butter is not limited to king boletes. It is a great way to make good use of your dried jacks (*Suillus*)—many of them offer a great bolete taste, best brought to shine after drying and powdering! I enjoy adding a good dose of pressed fresh garlic, minced finely, to *Suillus* butter without worrying about overpowering the precious and delicate porcini aroma of king boletes.

PICKLED CHANTERELLE BUTTONS

Jumping through the hoops to pickle your mushrooms is so worth it! What a great way to preserve them in their glory and enjoy them throughout the year. Mostly I pickle mushrooms to share them! They make great presents and are so versatile. You can cook with them or use them as garnish in salads, but I love to eat them to mark special moments, like major holidays or a forest picnic while mushrooming.

Makes 8 half-pint jars

2 pounds chanterelle
   buttons
1 tablespoon olive oil,
   divided

PACKING
8 (8-ounce) glass canning
   jars with sealing lids
8 cloves garlic
8 sprigs rosemary
8 sprigs thyme
8 leaves sage
8 leaves bay laurel

MARINADE
1 1/2 cups extra-virgin
   olive oil
1 1/2 cups apple cider
   vinegar
1/4 cup lemon juice
2 teaspoons minced garlic

Clean mushrooms if necessary and slice into bite-size pieces.

PACK THE JARS

Sterilize clean jars and lids for 10 minutes in boiling water. Remove and let drain, jar mouth down, on a clean towel.

Prepare ingredients for each jar. Cut the garlic cloves into quarters, and reserve four quarters for each jar. Prepare the herbs for each jar: a sprig of rosemary, a sprig of thyme, and a leaf each of sage and bay laurel.

PREPARE THE MARINADE

In a medium-sized saucepan, mix the pickling marinade of olive oil, apple cider vinegar, lemon juice, minced garlic, salt and pepper, and the finely chopped herbs and bring to a boil. Let simmer for 10 minutes to infuse with herbs.

In a large sauté pan on high heat, add 1/2 tablespoon of the olive oil and half the mushrooms. Sauté until they begin to release their water. Remove the cooked mushrooms from the pan and set aside. Repeat the same sautéing process for the other half of the mushrooms.

2 teaspoons sea or kosher salt

1 teaspoon crushed black pepper

2 tablespoons finely chopped herbs, like sage, rosemary, thyme, and/or bay laurel

Pack mushrooms into prepared jars, but leave about a half inch of head space. Pour enough hot marinade into jars to cover packed mushrooms, and then seal each jar with a sterilized lid.

If you want to keep your pickled mushrooms for many months without refrigeration, boil the tightly closed hot jars in a water bath for 15 minutes. Make sure you use a rack in the pot so the jars don't touch the bottom, which could cause them to crack. For the best flavor, wait for at least several weeks after pickling before eating the mushrooms. They should keep fine for up to a year.

Variations

The species best suited for pickling are firm, dense mushrooms like young chanterelles, boletes (not just kings but others, including young jacks), matsutake, brittlegills, and milkcaps. Since the way something looks affects how it tastes, using small, complete specimens or slicing them in an attractive way makes for beautiful pickled mushrooms. This is not the time to use larvae-infested mushrooms!

Instead of frying your mushrooms, you could parboil them for 5 to 10 minutes in the pickling liquid. Then separate the mushrooms from the liquid before packing the jars and topping them off with hot marinade.

You can add onion slices or whole pearls and colorful slices of pimento peppers for color. People with a sweet tooth may like to add up to a half cup of sugar to the pickling marinade. Others like it spicy and add chiles or, more commonly, a few peppercorns and several mustard seeds per jar. If you use kosher salt in the recipe, make sure to avoid anticaking agents.

A note on sterilization: You can also place your filled jars into a preheated oven (350°F) for 20 minutes or so until they bubble. For this method, do not close your jars too tightly, so that the bubbles can rise up. Remove jars from the oven and tighten the lids—this time tight, but be careful because the jars will be boiling hot! Lids should seal in the cooling process as the hot pickled contents cool and contract. Keep jars that don't seal properly in the refrigerator and eat them first. Boiling will make them stable for weeks.

Some people just pickle mushrooms in their favorite vinegar-based salad dressing for a couple of minutes. They can then be quick-fried for immediate enjoyment or stored in the refrigerator for a week or two of fridge shelf life.

CHANTERELLE APRICOT CHUTNEY

The fruitiness of the chanterelle is not just one of its important identification traits—and the root of its Japanese name, anzutake, "apricot mushroom"—but is also instrumental in its versatile culinary uses. In *The Mushroom Hunter's Kitchen*, Chad Hyatt shares great recipes for Chanterelle Lemon Marmalade and for a mildly spicy, sweet-and-sour Chanterelle Apricot Chutney. Chanterelles' fruitiness is nicely enhanced with apricots. This chutney recipe can be used in many ways, but it shines best with crackers, cheese, and/or charcuterie.

*Makes 4 to 5
(8-ounce) jars*

1 1/2 pounds chanterelles, cleaned and cut into 1/8-inch dice

1/2 pound dried apricots, cut into 1/8-inch dice

1 jalapeño or serrano pepper, chopped into very small pieces

1/2 cup sugar

1/2 cup apple cider vinegar

1 1/2 teaspoons salt

Zest of 1 lemon

Combine all ingredients and 1 cup of water in a heavy-bottomed pot and mix well. Over medium heat, bring the mixture to a simmer, stirring occasionally. Lower the heat to maintain a slow simmer, stirring frequently, until the liquid has become syrupy.

Let it cool down to room temperature before pouring it into jars. This chutney will keep for several weeks in the refrigerator.

Variations

If you want to store it for longer periods, use sterilized jars, fill them with the hot product, and immerse them in a hot water bath for 15 minutes. This way they can keep for many months! I enjoy adding a cup of finely chopped onions. You can find all kinds of inspiration from other chutney recipes too.

RUSSULA HUMMUS

Short-stemmed Brittlegills are ignored by most mushroom hunters—because they have never tried Russula Hummus! Chef Chad Hyatt first came up with the idea of using these common and often very large mushrooms as a chickpea substitute. Chef

Hyatt soaks the *Russula* overnight in very salty water, followed by several changes of fresh water to leach out salt. My method is simpler and quicker: parboiling or frying fresh slices. Both these approaches make great hummus.

Serves 2

1 cup sliced cooked *Russula brevipes*

1/4 cup extra-virgin olive oil (high-quality oil is key)

2 large cloves garlic, peeled and crushed

1 1/2 tablespoons tahini

Juice of 1/2 lemon

Salt to taste

Fresh cilantro leaves, for garnish

Clean your mushrooms (often the cap needs washing) and cut them into ¼-inch-thick slices. Parboil in lightly salted water for 10 minutes. Drain mushrooms.

Using a food processor if you have one, blend together the mushrooms, olive oil, garlic, tahini, and lemon, adding salt to taste. Top with cilantro leaves and drizzle everything with more olive oil. It is great with chips and crackers and as veggie dip or a sandwich spread!

Variations

You can also fry your sliced mushrooms in olive oil before blending them. Cooking the mushrooms improves digestibility. We still don't have a lot of information on whether eating Russulas raw is safe.

WITCHES' BUTTER FRUIT LEATHER

More than just a weird forager snack in the dark winter woods, witches' butters provide a great base for a very enjoyable fruit leather. I came across this recipe posted by Serena Juarez on Facebook's Pacific Northwest Mushroom Social Club group, tried it, and loved it. Turning a jelly fungus into an enjoyable snack, which when refrigerated stays edible for months, gives a whole new twist to the magic of witches' butter!

Makes 3 big leathers

1 1/2 cups fresh witches' butters

2 cups orange juice

1/2 cup lemon juice

2 pears, peeled and cubed

Clean, wash, and drain the witches' butters. Mix the orange and lemon juices in a small bowl. Add the witches' butters, letting them marinate for 8 to 24 hours. After marination, drain the fruiting bodies (reserving the juice) and set aside.

In a small saucepan, combine the reserved juice, pears, sugar, and cinnamon. Bring to a simmer, stewing on low

Five varieties of mushroom jerky

Witches' Butter Fruit Leather

1/2 to 1 cup sugar (to taste)

1 to 2 teaspoons ground cinnamon

1 tablespoon finely sliced dried ginger or grated fresh ginger

heat for 20 minutes until thickened. Combine the fruit stew and marinated mushrooms in a blender and blend well.

Spread the blended mushroom and fruit mix thinly on parchment paper, and sprinkle the top with dried ginger. Dry it in a dehydrator (or bake in an oven) at 150°F for 6 to 8 hours until a leathery consistency is achieved. It's best to check frequently for doneness after the first 5 hours. While the leather is warm, cut it into slices or roll it up first before slicing.

Variations

Here I have matched the orange color of the slippery fungus to the fruit of the same color, but you can experiment with other types of fruit pulp or juice to take the witches' butter in many new directions.

WOOD EAR MUSHROOM JERKY

Turning your mushrooms into tasty, zesty, and long-lasting jerky is a great way to enjoy them year-round. Jerked wild mushrooms can feed you while you are hiking, traveling, or snacking on the sofa. Starting with a recipe I found on the awesome site the3foragers.blogspot.com, I developed several variations depending on the type of mushroom one is jerking around.

Makes about 1 cup of marinade; good for 3 ounces (about 4 cups) dried mushrooms

1/3 cup low-salt soy sauce or gluten-free tamari

2 1/2 tablespoons maple syrup (or 2 tablespoons brown sugar)

2 cloves garlic, pressed

Salt, to taste

1/4 teaspoon ground pepper

1 to 2 teaspoons chili-garlic sauce (optional)

For the marinade, combine the soy sauce, maple syrup, garlic, salt, and pepper in a blender. Add the chili-garlic sauce if you like it spicy. Puree well. Add the juice or cider and blend a bit more.

Place the dried wood ears into a 1 quart or larger lidded jar or container and cover them with the marinade. A large French coffee press dedicated to mushrooms is perfect for holding down a maximum amount of mushrooms in the marination sauce. Let mushrooms steep 4 to 6 hours in the refrigerator.

Drain your marinated mushrooms, arrange on a drying rack, and dry for 2 to 4 hours at 140°F in a food dehydrator or in the oven with the fan going. Check frequently for

1/2 cup apple juice or sweet apple cider

4 cups (3 ounces) dried wood ear mushrooms

the degree of dryness you like: some people love crunchy marinated mushrooms, others prefer chewy. Stored in the freezer, airtight and out of the light, your mushroom jerky should keep for at least a year.

Note: If you have collected the wood ears yourself, you do not need to parboil them before use. However, with store-bought mushrooms, parboiling is important to kill possible bacterial contamination: boiling them for 1 to 2 minutes should work.

Variations

There are endless possibilities for marinating and drying mushrooms. The character of the mushroom species has a huge impact on the final result. Wood ears shine thanks to their thin yet rigid structure and benefit mightily from a flavorful marinade. Dried till crispy, the fibrous Chicken of the Woods makes a surprisingly great jerky. Oysters, edible corals, button mushrooms, and shiitake are all good candidates. The natural apricot taste of chanterelles fades from their dried taste profile, but their rubbery texture works well: chanterelle jerky is best enjoyed when still a bit moist and flexible. Similarly, King Boletes marinate and dry well, but their innate deliciousness gets overpowered by strong marinades, and their culinary qualities might be used in better ways.

Important: Most mushrooms should not be eaten uncooked (with a few exceptions: porcini, matsutake, button mushrooms, and several jelly fungi; see Advice for Staying Safe When Eating Mushrooms in Part One). Parboil your mushrooms before marinating and drying them. Cover ⅛-inch-thick slices or small whole caps in water and boil for 10 minutes. Drain the mushrooms, let cool, and squeeze out any extra moisture (this increases marinade absorption and reduces drying time). Save the mushroom water as broth for other cooking projects!

TERIYAKI ROASTED CAULIFLOWER MUSHROOM

Cauliflower Mushrooms are big and so firmly textured that while they might not be famous for any intrinsic taste beyond a lovely fungal one, they are a great base for many types of dishes. In this preparation, shared with me by Andrew MacMillan of the Kitsap Peninsula Mycological Society, their mushroom goodness is paired with a sweet and spicy, nutty-flavored umami sauce, making for a memorable snack or appetizer. I just wish there were more Cauliflower Mushrooms out there!

Preheat oven to 350°F. Whisk soy sauce, honey, ginger, and oils together in a large mixing bowl.

8 ounces (2 pints)
Cauliflower Mushroom
1 tablespoon soy sauce
1 tablespoon honey
1 teaspoon grated fresh
ginger
1 tablespoon vegetable oil
A few drops toasted sesame
oil

Tear the Cauliflower Mushroom into 2-to-3-inch pieces, discarding any of the thicker base blades. Thoroughly coat each mushroom piece in the sauce, and spread the mushroom pieces in a single layer onto a rimmed baking sheet.

Roast in the middle of the oven for 30 to 45 minutes until the frilly edges of the mushrooms are crisp but not burned and the centers are still chewy. Let cool a bit. Enjoy while they are still hot or once they have cooled off.

Variations

Many other firm mushrooms, like oysters or corals, can be prepared the same way, to everyone's gustatory delight!

SHAGGY PARASOL EGGPLANT BALLS

Tired of frying your mushrooms every single time? Try these baked mushroom eggplant balls! These awesome and versatile mushroom balls need no meat. The mushrooms take care of our protein and savory needs. These umami balls can be used as the main course in a tomato sauce or as appetizers to be dunked into a delicious dip. Adjust ball size to your intended use.

Serves 4

1 medium-sized eggplant
2 tablespoons olive oil,
divided
1 teaspoon kosher salt
1/4 teaspoon black pepper,
divided
3 shaggy parasol mush-
rooms
3/4 cup crushed cashews

Preheat oven to 425°F. Peel and dice the eggplant. In a medium-sized mixing bowl, drizzle the diced eggplant with 1 tablespoon of the oil, add the salt and half of the pepper, and mix well. Spread in a single layer on one half of a rimmed baking sheet.

Slice the mushrooms. Add them to another medium-sized bowl to drizzle them with the remaining oil and sprinkle on the remaining pepper. Spread them evenly on the other half of the baking sheet. Bake for about 15 minutes in the preheated oven. Once the mushrooms and eggplant are

Teriyaki Roasted Cauliflower Mushroom, p. 340

Shaggy Parasol Eggplant Balls

1 cup bread crumbs
1 large egg, lightly whisked
2 cloves of garlic, minced
2 tablespoons minced fresh
cilantro
1 tablespoon dried sweet
basil
1 tablespoon dried oregano

finished cooking, pull them out and let them cool a bit. Reduce the oven heat to 400°F.

Pulse the cashews separately in a food processor or crush them. Combine the baked mushrooms and eggplant in a food processor and pulse until they are nicely minced, but not a paste. If you do not have a food processor, mash the mixture. Transfer the mash into a large bowl and add the rest of the ingredients: cashews, bread crumbs, whisked egg, garlic, and spices. As needed, substitute the fresh cilantro with dried cilantro or cilantro paste, cutting the amount to 1 tablespoon. Using a large spoon, mix well. Cover and chill in the refrigerator for 15 minutes.

Hand-roll 1 to 3 tablespoons of mushroom mixture (measured using a tablespoon or small ice-cream scoop) into evenly sized balls, and place them onto a baking sheet lined with parchment paper. Oven bake for 20 to 25 minutes until the balls are golden-brown.

Variations

You can substitute shaggy parasols with all kinds of mushrooms; however, this recipe works best for other firm-textured species like Pig's Ears, Lobsters, chanterelles, boletes, or Horse Mushrooms. Selecting your spices, you can either go more Italian (substituting thyme for the cilantro and optionally adding about ½ cup ground Parmesan) or more Asian (swapping the oregano and sweet basil for mint and holy basil).

Cheese Fondue with Fried Chanterelle; Matsutake and Goat Cheese Salad

FURTHER INSPIRATION

These bite-size recipes are intended to inspire you to get creative with the mushrooms you find while foraging—the possibilities are endless!

SAVORY MUSHROOM LEATHERS

Producing your own mushroom leathers is surprisingly simple and provides you enduring and delicious savory snacks. Prepare your mushrooms as you like them, be it as a sauce or fried alone and well spiced. Cook off most of the moisture and pour everything into a blender, blending until you have a smooth paste.

Spread the paste evenly and thinly on parchment paper and dry in a dehydrator or an oven at 130°F. Drying time varies greatly, but the end result should be pliable and come off in one piece when lifted from the parchment paper.

Adding dried huckleberries or cranberries adds a nice fruity twist—mushrooms, pemmican-style!

CHEESE FONDUE WITH FRIED CHANTERELLE

A favorite fall special of mine, warm and hearty, is adding fried chanterelles or other pre-cooked mushrooms to cheese fondue. The fried mushrooms lovingly enhance the fondue and add the fun of having to fish them out by spearing them or trying to catch them up onto a chunk of bread, or a piece of potato or other vegetable, on your fondue fork. Just to clarify, mushrooms are added for enjoyment, not necessarily to up the protein intake, which is already secured by the cheese without the fungal boost.

CROSTINI WITH GRISETTE AND ARUGULA

The beauty of serving a sautéed mushroom on crostini is that the mushroom's unique flavor and texture can shine unmasked. It is a great way to introduce people to mushrooms they

Honey Mushroom Pizzette; Matsutake Deglazed in Sake and Tamari

might not have tasted before. Here, the delicate taste of a grisette (as with all mushrooms, but especially Amanitas, be 100% sure of your identification!) mates well with the piquancy of fresh arugula.

MATSUTAKE AND GOAT CHEESE SALAD
Matsutake belong to the small and elite club of edible choice mushrooms than can be enjoyed raw. They are perfect when thinly sliced as an accent on a tomato, onion, basil, and goat cheese salad—or for that matter on many other salads.

HONEY MUSHROOM PIZZETTE
Honey mushrooms add great pizzazz to pizzette. Because you are substituting whole mushroom caps for the dough, first fry the caps in a little olive oil until slightly browned. Transfer cooked caps to a baking sheet and spread each cap with sun-dried tomatoes and thinly sliced garlic. Dust the pizzette with oregano, thyme, salt, and pepper, and sprinkle on a grated cheese of your choice. For 8 midsize honey mushrooms, I use ½ to 1 cup of cheese. In a toaster oven or baking oven, cook till the cheese is melted, and serve hot.

You can use all kinds of mushrooms as your "pizza" base. While smaller caps are more attractive for appetizers, bigger caps like shaggy parasols and Ringed Webcaps make delicious main dishes. And regarding toppings: stay classic, add bacon chunks, or just run wild!

Be aware that there are people whose digestive systems will greatly appreciate a detox of the honey mushrooms by a 10-minute parboil followed by dry frying. You can also freeze mushroom caps after a quick frying, or oven-sweat them to use later as a pizzette base, long past peak mushroom season.

MATSUTAKE DEGLAZED IN SAKE AND TAMARI
The distinctive aroma of matsutake seems at first to integrate poorly into the flavors and traditions of most European cooking. However, a simple and tasty approach that highlights

Suillus Quesadillus; Fire Coral with Blue Potatoes

the pine mushroom's special aroma as well as its amazingly firm texture is slicing and frying them with a bit of oil first. Once the slices are nicely caramelized, deglaze the slices (that is, add moisture to capture the cooked bits) with a 3:1 mix of sake (or white wine) and soy sauce (or tamari). Don't overdo the soy sauce: it will make the mushrooms too salty and overpower the matsutake taste and aroma.

Also, instead of frying your sliced matsutake, you can steam them on top of the meal in a closed pot. Read more about this process in the matsutake section of Part Two: The Mushrooms.

FIRE CORAL WITH BLUE POTATOES
Bright red coral mushrooms add beautiful color to your cooking. However, adding them to a dish can be a bit hit-and-miss. Sometimes these corals are too bitter for my taste, but adding caramelized sweet onions can add a nice twist.

SUILLUS QUESADILLUS
A mixture of Purple-veiled Slippery Jacks and Matte Jacks (*S. luteus* and *S. lakei*) truly jazz up a lunchtime quesadilla. I regard any mushroom as an enjoyable edible if it adds flavor to a quesadilla. You'll want to first free the Slippery Jacks from their stickiness by peeling away the slippery cap layer (the cap cuticle), which comes off in one easy go. Next, slice the jacks, dry sauté them, and finish with olive oil and a bit of salt and pepper. Some days I fancy up the mushrooms by deglazing them in the pan with sherry or a port wine.

GREEN BEANS WITH SHRIMP *RUSSULA*
The firm structure of Shrimp Brittlegills and their subtle, fishy taste (a taste missing in some otherwise texturally well-suited Russulas) make it a nice addition to many stir-fried veggies, as here, with a simple string bean dish. Adding a black bean sauce brings the green beans back to earth with an added umami taste.

LOBSTER CHOWDER

What better use of our Pacific Northwest Lobster Mushroom than in a chowder? The Lobster's slightly fishy aroma paired with firm, bright orange flesh makes a fragrant and colorful core ingredient to this super-tasty soup. Mushroom Mountain man Tradd Cotter insisted I try this recipe. He first got it from Raeleen Wilson and Chris Tullar of the Asheville Mushroom Club as a blewit-focused dish, which they called "clam-less chowder." But I am more of a "cup half full" person and refer to it as Lobster Chowder.

Makes 4 servings

1 cup chopped onion
Butter or oil for sautéing
3/4 pound Lobster Mushrooms, cleaned and sliced
2 cloves garlic, minced
1 1/2 cups peeled and cubed russet potatoes (2 medium-sized potatoes)
1 cube chicken or vegetable bouillon
1 teaspoon Worcestershire sauce
1/2 teaspoon fresh chopped thyme
1/2 teaspoon pepper
2 cups milk (or oat milk)
1 cup light cream

Sauté the onions in oil or butter in a large frying pan. Remove them and set aside. To the same pan, add the sliced mushrooms and minced garlic and sauté them together in a bit of butter or olive oil.

In a large saucepan, boil the potatoes in 1½ cups of water, adding the cooked onion, bouillon cube, Worcestershire sauce, thyme, and pepper. Boil for about 10 minutes until soft.

In a medium-sized mixing bowl, stir together milk, cream, flour, and salt. Add this mixture, plus the sautéed mushrooms and kelp powder, to the potatoes and cook for 10 to 15 minutes until thick and bubbly. Serve with oyster crackers as a garnish.

2 tablespoons flour
1 teaspoon salt
2 tablespoons kelp powder
Oyster crackers, for garnish

Note: If you don't have a bouillon cube, substitute 1 cup of real broth and reduce water for boiling the potatoes accordingly.

Variations

You can use all kinds of mushrooms: Shrimp Brittlegills keep the seafood theme nicely afloat, as do oyster mushrooms in a more subtle way. Whichever mushroom you use, sautéing them first adds greatly to the taste and texture of the chowder. Substitute nondairy cream, such as coconut, as you prefer.

Lobster Chowder

MUSHROOM DUXELLES

Foraging writer extraordinaire Langdon Cook introduced me to duxelles when we cooked chanterelles for a Seattle public radio show. It turns out I had made this sauce often, but was simply unaware of the fancy French name, probably like many other mushroom lovers out there who use a version of this recipe, which quickly and easily turns mushrooms into a delicious, versatile sauce. Use duxelles as a topping for bruschetta, as a stuffing for meat or pasta like tortellini, or as a sauce for veggie dishes. It is also excellent tossed with pasta, polenta, rice, or dumplings. Duxelles will keep fine for a week or longer in the refrigerator. And if you store it in your freezer, you can use it in many, versatile dishes throughout the year.

Serves 2 to 3

2 pounds mushrooms
1/2 tablespoon oil or butter
2 shallots (or 1 small onion),
    finely diced
3 cloves garlic, minced
1/2 cup white wine, or a
    fortified wine like sherry
1 tablespoon chopped fresh
    herbs, like parsley and/
    or thyme
Salt and pepper, to taste
Parsley, chopped (enough
    for garnish)

Clean your mushrooms well and finely chop them. Dry sauté the mushrooms in a large skillet for 5 to 10 minutes until they have released their water.

Making space in the center of the pan, add the oil or butter and sizzle the shallot and garlic for 10 seconds (taking care not to burn them). Stir them into the broader pan with the cooked mushrooms.

Deglaze the skillet with the white wine. Sherry, cognac, marsala, or port wines are also marvelous for duxelles. Scrape all pieces of goodness from the frying pan and continue cooking until moisture is almost fully reduced. Add fresh herbs (including 3 to 5 sprigs of thyme) and season to taste with salt and pepper. When serving, garnish with fresh chopped parsley.

Variations

Vary the consistency of the mushrooms in your duxelles, chopping by hand or using a food processor. The alcohol you choose for deglazing will strongly flavor your sauce. Marsala is sweeter than dry white wine. Red wine works fine too but will darken your sauce. Turn the duxelles into a cream sauce by adding ½ cup or more of heavy cream after deglazing.

Matsutake Miso Soup; Nettle Soup with Oyster Mushrooms

FURTHER INSPIRATION

These bite-size recipes are intended to inspire you to get creative with the mushrooms you find while foraging—the possibilities are endless!

NETTLE SOUP WITH OYSTER MUSHROOMS

While picking spring nettles for this soup, keep an eye on dead alders or cottonwoods for a fresh crop of oyster mushrooms! Outside nettle season, oysters add great flavor to make soups yummy.

MATSUTAKE MISO SOUP

A hot miso soup on a cold fall or winter day can't get much better when it includes a few slices of matsutake. It is up to you how long you want to "steam" your matsutake, since they can also be enjoyed raw. However, adding the pine mushroom to the soup early softens the pieces and infuses the miso base with the rich aroma of the matsutake.

MUSHROOM HOT POT

A collectively fun way to enjoy a diversity of wild-harvested or store-bought mushrooms is to prepare them in a hot pot (a.k.a. Chinese fondue), which always results in a memorable dinner. Originally, we enjoyed hot pots as special family meals around the holidays. During Mushroaming tour adventures in Tibet, however, I started bringing freshly collected wild mushrooms to Chinese restaurants. Contributing them to the hot pots, I realized, was a fantastic way to taste and appreciate each mushroom species individually!

Heat a large pot of flavorful broth to just below a simmer. It's nice to have mushroom broth on hand to add to this. Spice the broth with garlic, cilantro, a small amount of sugar (if you like) or extra salt, and all kinds of other ingredients. It is called "hot pot" since traditionally the broth is very spicy. So spice up your pot to your heart's desire or provide chili sauce on the side.

Have ready a wide range of bite-size food items including mushrooms, veggies, meat, or tofu to immerse in the broth. A range of different dipping sauces, including crushed fresh garlic in oil, spicy prepared garlic sauce, crushed peanut–based dip, or hoisin sauce, are usually prepared for the cooked ingredients as well.

The pot is placed in the center of the table, ideally with a heat source under it; otherwise it needs to be reheated several times during the meal. For group enjoyment, add the pieces in stages to the hot pot, fishing them out with a slotted spoon or chopsticks, or dip them in one at a time using individual fondue forks. The leftover broth makes for some of the most delicious soup you have ever had.

Mushroom Hot Pot

Chanterelle and Chèvre Galette

Bold Bolete Quiche, p. 356

HEARTY MEALS

CHANTERELLE AND CHÈVRE GALETTE

My daughter Sophia introduced us to this tasty galette, and my wife, Heidi, and I make it frequently. Before Heidi had to eat gluten-free, I would buy puff pastry and fancy up my mushrooms in cream sauce dishes by serving them in a crust. For years, not having a great gluten-free alternative, we were cut off—not anymore!

Serves 4

CRUST
1 1/4 cups regular flour
    (or Bob's Red Mill 1-to-1
    gluten-free flour)
1/4 teaspoon salt
1/2 cup cold butter (plus 1
    teaspoon for rubbing on
    crust)
1/4 cup Greek yogurt or sour
    cream
1 to 4 tablespoons cold
    water
1 egg (for spreading on
    crust)

FILLING
4 cups fresh chanterelles
1 tablespoon olive oil
1 yellow onion

MAKE CRUST
Combine the flour and salt. Cut ½ cup cold butter into small cubes, combine with the flour, and add in the yogurt, mixing it in well, using your hands. Once the butter and yogurt are incorporated, add small amounts of cold water (up to 4 tablespoons) until you are able to form the mixture into a nonsticky ball of dough. Do not overwork the dough; this should take only a few minutes. Let it rest in a bowl (covered with a clean dish towel) for 1 hour in the refrigerator, or 20 minutes in the freezer.

Note: It is important that the dough be cold! This ensures it doesn't get soggy in the cooking process.

While you wait for the dough to chill, clean and cut or rip your chanterelles into strips; some people prefer ripping to cutting for texture and looks. Julienne the onion (cut it into long, thin slices). Chop the garlic. Core and dice the apple, and remove leaves from the thyme sprigs.

1 clove garlic

1/3 cup sherry wine (or white wine)

5 1/2 ounces chèvre cheese, divided

1/2 large apple (or pear)

5 sprigs fresh thyme (or 1 tablespoon dried thyme)

2 teaspoons lemon juice

1 teaspoon maple syrup

1/4 teaspoon Dijon mustard

Salt and pepper, to taste

Heat a large saucepan on high heat, add the chanterelles, and "sweat" them for 3 to 5 minutes, adding the olive oil as they start to stick. Turn the heat down to medium, add the onions, and sauté together for another 5 minutes. Add chopped garlic and continue the sauté until onions are soft. Return the pan to high to deglaze it with sherry or wine, stirring to free anything sticking to the pan.

Turn off the heat and mix in 5 ounces of chèvre and the apple or pear, thyme, lemon juice, maple syrup, and Dijon mustard. Add salt and pepper to taste. Let filling cool to room temperature.

Remove chilled dough from fridge, sprinkle flour onto a clean surface, and roll the dough into a 12-by-14-inch or so rectangle. Lift and place the crust onto a baking sheet lined with parchment paper or a silicon mat. Rub a thin layer of soft butter on the crust where the filling will be placed—this helps prevent the crust from getting soggy—then spread the filling evenly over the crust, leaving clear a 2-inch perimeter for folding over. Cutting 1-inch pleats into the crust edge at 3-inch intervals makes the folding process easier. Fold over the five to seven pleats, leaving the center of the galette exposed. Sprinkle remaining chèvre on top. In a small bowl, beat the egg with a fork and use a brush to wash the egg onto the galette crust.

Preheat oven to 400°F. While the oven heats, place finished galette in the freezer or fridge for 10 to 15 minutes. The colder the galette is before cooking, the better!

The galette will take about 30 minutes to bake. If you have a convection oven, check it around 15 minutes and add time as necessary. When the crust is golden-brown, remove the galette from the oven and let it sit for 5 minutes before serving. Believe me, the insides will still be hot!

Variations

The galette dough can be used for all kinds of fungal (and non-fungal) dishes demanding an attractive and delicious crust. Mushrooms in cream sauce, or any other way you like to prepare your mushrooms, can be enclosed—just make sure the filling is not too moist or the bottom dough will be soggy and not bake well. The fruit and cheese, as well as the thyme, mustard, and lemon, can be omitted or exchanged for other ingredients.

KING BOLETE POTATO GRATIN

My mother used to make potato gratins, but I do not recall which mushrooms she used. When my sister praised her own porcini gratin, I had to make it, and I loved it. Gratins are outstanding, hearty dishes for a fall or winter evening. Dried mushrooms come back alive nicely, and king boletes add their lovely, special nutty sweet aroma to potatoes and cream. The main recipe is the fastest, simplest approach. Leaving off the cheese and using nondairy substitutes for the milk and cream makes it easily preparable for vegans.

Serves 4 to 6

1/2 ounce dried porcini (5 ounces fresh)
2 pounds (Yukon gold) potatoes
1 to 2 cloves garlic
1 cup milk or nondairy milk
1 cup cream (or nondairy substitute)
1/2 teaspoon salt
Ground pepper, to taste
Ground nutmeg, to taste
1 teaspoon extra-virgin olive oil or butter
1/2 cup soft cheese or cashew cheese (optional)
1/4 cup hard cheese (optional)

Preheat oven to 400°F. Cut or break the dried porcini into small pieces, or slice them if fresh, roughly ⅛ inch thick. Peel the potatoes and, using a mandoline or sharp knife, cut them into slices about ⅛ inch thick. Peel and press the garlic.

In a medium saucepan, combine the milk, cream, and garlic, adding pinches of salt, pepper, and nutmeg. Bring to a low boil over medium heat, then remove from the heat and set aside.

Grease an 8-inch square baking dish with olive oil or butter. On the bottom of the baking dish, spread a layer of potato slices, slightly overlapping one row with the next. Sprinkle a third of the mushrooms (dried or fresh) on top. Add another two layers of sliced potatoes and mushrooms, covering the topmost mushroom layer with the nicest-looking remaining potato slices. Pour the cream and spice mixture evenly over the potatoes and mushrooms, and let it soak in. (To create a nondairy cream, combine ¾ cup nondairy milk and ¼ cup olive oil.)

Bake the gratin for 45 to 60 minutes, positioning the pan in the lower half of the oven (otherwise, protect the top from burning with aluminum foil). You know the gratin is done when the potatoes are soft when poked with a knife.

Variations

I often use dried porcini, which rehydrate deliciously during baking. There's no need to go to the trouble of precooking them. If you use other, less special mushrooms, like oysters or Meadow Mushrooms, sautéing them first pumps up their umami quotient.

Including dairy-based cheese or cashew cheese (¼ cup of finely grated aged cheese or ½ cup of soft cheese) in the layering and on top of the gratin adds a lot of savory deliciousness! If using a soft cheese, reduce the amount of liquid appropriately. Parsley and thyme add nice notes to the gratin.

BOLD BOLETE QUICHE

My daughter Sophia loves baking quiche, and we love her quiches, especially when they are enhanced by wild mushrooms. Here the royal aroma and flavor of king boletes, embedded in the goodness of eggs, cheese, onions, and kale—and accented by spices—is packaged in a sumptuous crust.

Makes 1 quiche

CRUST

1 1/2 cups whole wheat
   flour or gluten-free flour
1/2 teaspoon salt
1/2 cup cold butter

FILLING

1 1/2 cups dry boletes (1.5
   ounce powdered) or 8
   ounces fresh boletes
1 medium onion, julienned
4 cloves garlic, minced
2 tablespoons butter or
   olive oil
2 cups chopped kale
1 cup grated cheese
   (Parmesan and cheddar)
2 teaspoons thyme (fresh
   or dry)
1 teaspoon oregano (fresh
   or dry)
Salt and pepper, to taste
Dash of nutmeg
5 eggs
1 cup milk (or milk substitute)

MAKE CRUST

Preheat the oven to 375°F. Add flour and salt to a medium mixing bowl. Slice cold butter and cut it into the flour until the dough is crumbly. Slowly add cold water (4 to 6 tablespoons) until dough forms a ball but isn't too sticky. The goal is a flaky pie crust.

Roll out the crust with a rolling pin and transfer it to a 9-inch pie pan.

MAKE QUICHE

If using fresh boletes, slice them. If using dried boletes, rehydrate them in ½ cup hot water; mix well to saturate all mushroom pieces. Once rehydrated, drain and squeeze the boletes, saving the mushroom water to deglaze the fried ingredients below.

Pan fry the onion, garlic, and boletes in the butter or olive oil. Transfer pan-fried ingredients into a large mixing bowl, adding the kale, cheese, thyme, oregano, salt, pepper, and nutmeg. Stir to mix evenly. Crack eggs into the bowl and beat together well. Add the milk and stir to blend.

Pour the final mixture into the pie crust. Bake for about 45 to 50 minutes. Starting at 30 minutes, check to see if the crust is browning. If so, cover the quiche with aluminum foil for the remaining cook time.

It does not need to be porcini to make a great quiche. Other boletes, jacks, or whatever mushroom you have at hand will work well in quiche, though they might require some special attention while pan frying. (For example they might benefit from dry sautéing to firm them up.)

Feel free to use your favorite cheese or combination of cheeses (or substitute cashew cheese). The kale can be replaced with spinach or other greens.

ESPRESSO PORCINI RISOTTO

Porcini risotto is a favorite mainstay of many mushroom lovers. The rich and creamy umami rice extravaganza is hard to resist. Yet traditional risotto takes a lot of time to prepare, dissuading many of us. My sister Eva, who lives in Switzerland and collects wild mushrooms as a side hustle, introduced me to using a pressure cooker for risotto cooking—a total game changer! Now you, too, can churn out a complex culinary delight in fast-food time without losing any deliciousness. Espresso risotto!

Serves 4

1 cup (1 ounce) dried porcini
  mushrooms
1 onion, chopped
1 clove garlic, chopped
1 tablespoon olive oil
1 1/2 cups arborio risotto rice
1/2 cup white wine
2 cups mushroom, veggie, or
  chicken broth (including
  mushroom-soaking water)
1 1/2 cups heavy cream
  (or your preferred liquid
  substitute)
Salt and pepper, to taste
1 teaspoon parsley, finely
  chopped
1 teaspoon fresh thyme,
  finely chopped (optional)
1/2 cup grated cheese (one
  that melts well)

Soak porcini mushrooms (possibly after crumbling them a bit) in lukewarm water for 5 to 10 minutes, while chopping the onions and garlic.

In an instapot, fry onions and garlic in oil until glassy (1 to 2 minutes) on the sauté setting. Add the risotto rice, stirring continuously to coat the grains with oil. Sauté for 30 seconds, stirring constantly. Deglaze the pan, adding the white wine. Stir in the rehydrated mushrooms, broth, and cream. Add salt and pepper to taste. Mix well.

Close the instapot and cook on high pressure for 5 minutes (instapot might take another 5 minutes until fully pressurized). When the timer goes off, use the quick pressure release, remove the lid, and stir. If the risotto looks overly wet or runny, press "sauté" and stir until it loses enough moisture to have the desired consistency. (It is better to undercook the rice and add a bit of sautéing than to end up with mushy rice.) Add the mixed herbs and grated cheese. *Finito!*

Chanterelles in Cream Sauce

Espresso Porcini Risotto

One option is to forgo or reduce the amount of cream but balance the loss of liquid with broth. There is also the slippery slope of upping cream and cheese content or adding in butter when mixing in the cheese. Be warned, risotto can get decadently rich! Some people like to add vegetables like frozen peas, which can be mixed in after pressure-cooking.

When porcini are in short supply, you can use other mushrooms like dried jacks or shaggy parasols. When made with other, less strongly flavored mushrooms like oysters or *Agaricus*, the risotto might benefit from a couple tablespoons of porcini or jack powder "magic" (see Powdering Mushrooms in Part One).

Note: Using a traditional pressure cooker is similar; however, you might have to reduce the heat as soon as high pressure is reached, and you will need to remove the pot from the heat at the end of the 5 minutes of cooking time. Cool the pot down before opening, possibly by running cold water on the side of the pot. Never try to open a still-pressurized pot!

CHANTERELLES IN CREAM SAUCE

My favorite way to prepare mushrooms is to serve them in a cream sauce, the way my mother taught me when I was a child. Sweet onions add flavor and body, but you can skip them if you prefer. Heavy cream lends delicious flavor and subtle sweetness, while the broth and wine help lighten the sauce so that you don't feel as if you need to sleep it off! This recipe works great with most "regular" mushrooms other than chanterelles, including all kinds of boletes, hedgehogs, Angel Wings, oysters, *Agaricus*, and many others.

Serves 3 to 4

4 cups sliced chanterelles
1 cup sliced sweet onions
3 to 4 tablespoons olive oil
   or butter, divided
1 1/2 cups veggie, mush-
   room, or chicken broth
1/2 cup dry, fruity white
   wine
1/2 cup heavy cream

Cut the chanterelles into pieces, 1 to 2 inches long by ½ inch wide; you should end up with about 4 cups. Slice the onions into ½-inch-thick slices to get about 1 cup. Keep the two ingredients separate.

Divide the olive oil or butter into two different medium to large saucepans, one for the mushrooms and one for the onions, since their cooking times vary. Olive oil allows you to cook the mushrooms at a higher temperature; butter adds a richer flavor.

Salt and black pepper, to taste
1 teaspoon sweet mustard (optional)
Finely chopped parsley, for garnish

Sauté the sliced mushrooms to caramelize them, being careful not to overload the frying pan. If your mushrooms are freshly washed, softish, or at all soggy, dry fry (dry sauté) them first, by starting the mushrooms without cooking fat and adding the oil or butter just as the mushrooms start sticking to the pan. If the mushrooms sweat a lot, pour that liquid into a bowl to use later together with the broth when making the sauce.

While the mushrooms are cooking, sauté the onions in the second pan until they look glassy.

When the mushrooms are done, deglaze them with white wine or broth to capture all of the delicious cooked bits and juices. In the large saucepan, combine the onions, the rest of the white wine and broth, and the heavy cream. Add salt and black pepper to taste and stir well. Sweet mustard adds a welcome dose of acidity, but avoid adding too much to keep the mustard flavor subtle, balancing the sweetness of the cream.

Let the sauce simmer for at least 10 minutes: a few minutes longer allows the sauce to thicken nicely. If too much liquid boils away, add more broth, wine, or mushroom water. To add color and a nicely subtle herbal note, remove from heat and stir the parsley into the finished sauce.

Serve Chanterelles in Cream Sauce on pasta, rice, or any other grain you like. My favorite base carbs to use besides fettuccini are *Semmelknödel*, airy and tender Bavarian bread dumplings made from bread, milk, onion, eggs, and herbs. This cream sauce is also perfect as a gravy for meat dishes.

Note: Dry frying (dry sautéing with little to no cooking fat) is a common treatment for many mushrooms, allowing them to release much of their moisture first. See Cleaning and Preparing Mushrooms in Part One.

Variations

Dried, sweetened cranberries add a fruity twist to this sauce, working especially well when used as a filling for puff pastries or galettes. Finely chopped fruits, fresh or dried, like apple, pear, persimmon, or apricot, add a rich, fruity aroma. Many people love to add a sprinkle of cardamom as well. For a dairy-free version, substitute cashew cream for the heavy cream. For animal protein, fold in sliced and fried strips of chicken, pork, or veal.

LEMONGRASS CHANTERELLE SKEWERS

I kept seeing such beautiful mushroom cooking from Davy Kim online and knew I wanted to include one of her tasty Cambodian dishes in the recipes. Here she contributes a versatile preparation of a lovely, zesty "BBQ" sauce (*kroeung* paste) that is awesome on chanterelles and other mushrooms firm enough to endure being lined up on a skewer. *Kroeung* can also be used with all kinds of fresh and cooked vegetables as well as fried meats.

Serves 4

KROEUNG SAUCE
3 to 6 stalks lemongrass
1 ounce galangal (or 1/2 ounce ginger)
4 makrut lime leaves
5 to 8 cloves garlic
2 shallots or 1 small onion
1 ounce fresh turmeric (or 1 teaspoon powdered)
4 tablespoons oyster sauce
3 tablespoons sweet red chili sauce
1 tablespoon fish sauce
1/2 cup vegetable oil
1 to 2 tablespoons chile flakes (optional)

CHANTERELLE SKEWERS
Bamboo skewers
1 tablespoon vegetable oil
1 pound fresh chanterelle mushrooms, cut into bite-size sections
1 tablespoon fish sauce
1 teaspoon sugar
1/2 cup *kroeung* basting sauce

PREPARE THE SAUCE

Using the bottom 3 to 5 inches of each stalk only, thinly slice the lemongrass (the upper parts are too fibrous but can be used for tea or soups). Peel and mince the galangal. Thinly slice, then mince, the lime leaves, first removing their hard center leaf ribs.

Finely chop the garlic, shallots or onion, and turmeric. Combine the lemongrass, galangal, lime leaves, garlic, shallots, and turmeric in a big bowl, adding in the oyster, chili, and fish sauces—and chile flakes if you want more heat. Mix thoroughly, then add enough oil to coat the mix with at least a ½ inch above.

A food processor works well for making the paste. Include a teaspoon or two of water to get the mixture to blend well. Some very small chunks of lemongrass or galangal are okay, but if the paste is still too coarse, finish up using a mortar and pestle, grinding in a circular motion until it is well blended.

In a pan, using medium heat, add the *kroeung* paste and enough vegetable oil to again cover the ingredients. Let it simmer lightly, stir well, and taste. Adjust to your preferred saltiness and sweetness by adding more oyster sauce, fish sauce, or sweet chili sauce. Add another tablespoon or two of chile flakes if you want an even bigger kick. Simmer for 5 minutes, stirring as needed. Then the sauce is ready to use.

This basting sauce can be frozen for months or kept in the fridge for several weeks. Transfer it to a jar with a screw lid. Add more oil to the top of the paste to keep it from drying out.

Note: If fish sauce is scarce in your cupboard, you can replace it with salt to taste. One tablespoon of fish sauce contains about ½ to 1 teaspoon of salt.

PREPARE THE SKEWERS
Soak the bamboo skewer sticks in water for 30 minutes to prevent burning when using a barbecue grill. Add 1 tablespoon oil to a large frying pan, and turn the heat to medium-high. Once the pan is hot, sauté the sliced chanterelles for a couple of minutes. Add the fish sauce and sugar, mixing well but gently so the pieces stay whole.

When the chanterelle mushrooms have just softened, immediately remove them from the pan. Once they are cool to the touch, skewer the pieces centrally from bottom to top. Make sure to leave 2 inches at the bottom of each skewer stick empty for a place to hold the skewers.

To cook the chanterelle skewers, you may grill them on a barbecue grill over medium-high heat, cooking for about 2 to 4 minutes on each side and brushing liberally with *kroeung* sauce. When the chanterelles are a nice caramel color with a bit of char, they are ready to serve. You may add more *kroeung* paste right before serving.

If you don't have access to a grill, these chanterelle skewers may be cooked in a large cast-iron pan or on a stovetop griddle over medium-high heat.

Lemongrass Chanterelle Skewers

LEBANESE PUFFBALL STEAKS

Giant Puffball was the one wild mushroom my wife Heidi's Midwestern, outdoors-loving father foraged for. Twenty-five years into our relationship, we were finally in the right place at the right time with Giant Puffballs dotting a meadow near Telluride, Colorado, where Heidi used to live. We found ourselves participating in the marvelous Telluride Mushroom Festival, where everyone insisted that marination was the key to a culinary puffball blast. Their mild fungal taste is enjoyable, and luckily, we had plenty of raw material to experiment with! Our hallelujah moment came when we tried a garlic-lemon marinade with turmeric and cumin—a recipe we developed through trial and error in our effort to copy the secret recipe of a local Lebanese restaurant.

Serves 4 to 6

PUFFBALL STEAKS
1 pound fresh Giant Puffball
2 lemons, juiced
1/4 cup olive oil
4 cloves garlic, pressed
1 tablespoon turmeric
   powder
2 teaspoons cumin

RICE
2 cups water
1 cup uncooked white rice
Leftover marinade

TAHINI SAUCE
1/4 cup warm water
3 tablespoons tahini paste
1 fresh lemon, juiced
3 cloves garlic, minced
1/2 teaspoon cumin
Salt, to taste

Cut the Giant Puffball into ¾-inch-thick slabs, and set aside. Mix the lemon juice, olive oil, garlic, and spices to make the marinade. Immerse the puffball slices in the marinade for at least an hour.

After an hour, remove the slabs and set them aside. Retain the excess marinade and add it to a 2-quart saucepan. Add in 2 cups of water and the rice, bringing everything to a simmer. Cook the rice, covered, for 15 to 25 minutes or until done.

Cook the marinated puffball slices by grilling them on a barbecue grill, baking them for 20 minutes at 350°F, or pan-frying them, both sides, until nicely browned.

Mix ingredients for the tahini sauce in a jar with a tight-fitting lid, and shake it vigorously until homogenous. Adjust consistency and taste by adding more water or upping the lemon.

Serve Lebanese Puffball Steaks with tahini sauce over rice.

Lebanese Puffball Steaks, p. 363

Coral Cakes

CORAL CAKES

Coral cakes (and *Hericium* cakes) are tasty and not as well-known as they deserve to be. We talk about "substituting" crab meat with mushrooms, as if mushrooms are lesser. But there is no sacrifice or downgrading when using White Coral Fungus (*Clavulina coralloides*), Ash Coral (*C. cinerea*), Wrinkled Coral (*C. rugosa*), or *Hericium* species like Lion's Mane or Bear's Head—just the opposite, in fact! These mushrooms have a firm texture that can be sliced or ripped into longish, stringy pieces, with a bit of fishy aroma as an added bonus.

They create patties very similar to crab cakes. Alas, in the Pacific Northwest, we find Bear's Head and Lion's Mane too rarely. Our forests, however, overflow with petite *Clavulina* species—foolishly ignored by most people, though they are closely related to chanterelles and lend themselves marvelously to these fried mushroom cakes.

Makes two 4-inch cakes

3 cups coral fungi
2 tablespoons olive oil, divided
1/3 cup minced onion
1/4 cup dried bread crumbs
2 tablespoons regular (high fat) mayonnaise
1 egg
2 tablespoons chopped fresh herbs, like chives, cilantro, and/or Italian parsley (more for garnish)
1/2 tablespoon Worcestershire sauce
Salt and pepper, to taste
Dash each of paprika and cayenne pepper
All-purpose flour for dredging
1/4 cup mild oil for cooking the cakes
Fresh salad greens or spinach leaves

Pull your corals apart into pieces of three to five strands each, removing all dirty ends, debris, and purple-gray areas (infected by a parasitic fungus). Preheat a medium-sized frying pan over medium-high heat with a tablespoon of olive oil, then toss in the corals (if they are wet, dry fry them first) and cook them till their water is released and evaporated. Now add the minced onion and another tablespoon of olive oil, frying the mixture until the corals are slightly browned. Remove from heat and transfer the cooked mushrooms into a bowl to cool. Meanwhile, preheat the oven to 350°F.

In a medium bowl, combine and mix together the bread crumbs, mayonnaise, egg, chopped herbs, Worcestershire sauce, salt and pepper, and spices. Let sit for 15 minutes to allow the bread crumbs to absorb the moisture. Add the now-cooled mushrooms and onions. Mix well.

Take half of the mixture and form your first cake. Use a 4-inch ring form to shape the cakes. (Cutting the bottom out of a round plastic food container makes a decent alternative.) Shape the remainder into a second cake. With both cakes formed, gently dredge each in the flour, coating both sides.

Reheat your frying pan, adding your mild-flavored, high-temperature oil of choice. Place your coral cakes into

Tartar sauce, remoulade, or aioli (optional)

1/2 lemon, cut into wedges

the hot oil (trying not to break the cakes and being careful not to burn yourself from splashing oil), and fry each side until golden-brown.

Move the cooked cakes first to absorptive paper towels, then to a baking sheet in the preheated oven. Bake for 10 minutes to finish cooking the centers of the cakes.

Serve the mushroom cakes on salad greens or spinach leaves, garnishing them with fresh chives, cilantro, and parsley. Serve with tartar sauce, remoulade, or aioli, with lemon wedges to squeeze on top.

Variations

You can make these cakes with other corals (*Ramaria* spp.), Lobster Mushrooms (*Hypomyces lactifluorum*), or many other mushrooms. Don't limit your creativity! Whichever mushroom you choose, to create firm cakes, slice the mushrooms first into lengthy strips and precook them to reduce the water. You can also add small amounts of chopped, cooked veggies to the recipe. As long as you have the right mix of mushrooms, bread crumbs, and egg, the possibilities are endless.

Coral cakes can also be served on a bun, burger-style.

BREADED SAFFRON MILKCAPS

Breading (or battering) your mushrooms is a favorite of many American wild mushroom hunters. A Midwest standard is breading Giant Puffballs. It seems to be especially popular with people who don't eat many mushrooms, and the new air fryers have excited many mycophiles because the mushrooms come out less greasy. Don't expect a mild mushroom to shine its culinary light once breaded; there might be a reason food is called "battered" after deep frying. However, if you love breaded food, and, for example, you have plenty of porcini (or, wisely, and miserly, you want to use this technique for less extravagant mushrooms), you will enjoy the outcome. I love to bread milkcaps for their crunchy structure and shaggy parasols for their rich, savory flavor, retained even when battered!

Serves 4

12 Saffron Milkcaps or 4 midsize shaggy parasols

1 cup all-purpose flour

Clean mushrooms, if necessary. Slice mushroom stems into slices between ½ and ¾ inch thick. Keep caps whole, unless their cap flesh is thicker than ¾ inch.

Spread some of the flour in a flat dish. In a small bowl, whisk

2 eggs

1 tablespoon milk or water

1 1/2 cups fine dried bread
  crumbs

2 tablespoons finely
  chopped Italian parsley
  (or 1/2 teaspoon dried)

1/2 teaspoon garlic powder
  (or 2 cloves garlic,
  pressed and minced)

1/2 teaspoon fine salt
  (and more for serving if
  desired)

1/4 teaspoon black pepper

Flavor-free oil, like grapeseed, for frying (enough to
  cover your pan base well)

About 1/2 cup mayonnaise,
  tartar sauce, or aioli (for
  dipping)

Several lemon wedges

together the eggs with the milk or water and cover another plate with some of the beaten egg. In a medium-sized bowl, combine the bread crumbs, parsley, garlic, fine salt, and black pepper and mix well. Spread the bread crumb mixture onto a third flat dish or plate.

Start by dipping a mushroom slice in flour, coating both sides, then dip it in the whisked egg. Finally, dip the mushroom in the bread crumb mix. Make sure each mushroom is completely covered in bread crumbs. Arrange the breaded mushrooms in a single layer on a big plate.

Setting the heat to high, cover the bottom of a frying pan (I use cast-iron pans; nonstick coatings will need less oil) generously with oil. I recommend mild, flavor-free cooking oils like grapeseed, peanut, avocado, or sunflower. When the oil is hot, place the mushrooms in the pan and lower heat to medium. Cook each side for 3 to 5 minutes or until the breading is golden-brown. Once a mushroom is cooked, carefully transfer it to a large plate lined with paper towels to absorb excess oil. Repeat until you have cooked all the mushrooms.

Note: If you work with thick slices or big caps, put fried mushrooms into a preheated oven (350°F) for another 10 minutes to fully cook the mushroom centers.

Variations

Breading suits many mushrooms, from boring button mushrooms to plentiful oysters and (deep) Fried Chicken Mushrooms. Some mushrooms, however, like the King Bolete, should be sautéed first to help bring out its rich, subtle aroma through the breading.

STUFFED MORELS

A dish of stuffed morels is one of our family favorites! And it is so easy. All you really need are morels big enough to be stuffed, plus several smaller morels for the filling, chopped and cooked, to mix with diverse ingredients, including grains like rice or polenta. Mixing in cheeses that you fancy (or nondairy cheese substitutes) adds flavor and a bit of cohesiveness.

Serves 4

8 big morels (4-inch-tall caps or larger)
Several small morels
1/3 cup port (sherry or white wine)
2 teaspoons olive oil or butter, divided
1 small onion, chopped

Cut off the hollow stems of the big fresh morels, setting aside their caps. (If using dried morels, rehydrate them first.) Separate out the cutoff stems and a few smaller morels and chop them well. In a large pan, fry them in 1 teaspoon of oil or butter, uncovered, until all the moisture has cooked off. Deglaze with port.

Remove cooked morels from pan and set aside. Add another teaspoon of oil or butter, and cook the onions and garlic till fragrant.

1 clove garlic, pressed (optional)

1 cup precooked polenta, rice, or other grain

1 tablespoon chopped fresh herbs

1/4 cup reconstituted dried tomatoes (optional)

Salt and pepper

Remove from heat and stir in your chosen precooked grain base of polenta, rice, quinoa, or another grain. Add some nicely chopped herbs like parsley, thyme, or oregano, possibly some dried tomatoes, and salt and pepper to taste, and you have yourself a perfect stuffing.

Using a small spoon, carefully fill your morel hulls, and then in a large cast-iron pan, fry them in olive oil or butter on all sides until nicely browned. Since the stuffing ingredients, including the chopped morels and the grains, are already cooked, the frying is needed just for cooking the morel caps and to heat up the filling.

Note: Stuffed morels can also be baked in the oven. You can enjoy your beautiful umami-loaded morels as an intriguing appetizer or as the main dish. Stuffed morels freeze well and when reheated make a memorable instant feast!

Variations

Do you need some extra protein? Crack an egg into the filling, or jazz it up with your favorite cheese blend, such as a soft mild cheese like ricotta mixed with a grated, well-aged cheese like Parmesan or sharp cheddar. Nondairy cheeses work well too. For meat lovers, add cubes of ham or crisp crumbled bacon. Only one stuffing recipe we tried did not excite us. It suggested we use store-bought sausages as the sole stuffing, which sounds easy enough. It made good sausages, all right, but the poor morel was completely overpowered and drowned in grease.

Chanterelle Phyllo with Persimmon; Giant Puffball Pizza

FURTHER INSPIRATION

These bite-size recipes are intended to inspire you to get creative with the mushrooms you find while foraging—the possibilities are endless!

CHANTERELLE PHYLLO WITH PERSIMMON

Prepare your mushrooms in your favorite way—be it in a cream sauce or with white wine, lemon, and thyme—and fill up baked puff pastry or roll them up in puff pastry! Chanterelles with persimmon are just one of many options.

GIANT PUFFBALL PIZZA

Giant Puffballs make a sturdy and delicious pizza base. To optimize taste and consistency, bake your puffball slab in the oven or sauté it first in a roomy frying pan. Don't overbrown the lower surface, as it will continue to cook in the oven and you don't want it to burn; the upper side is well protected by the choice of your toppings. The base is absolutely gluten-free, and if no cheese is used, as in pizza marinara, it is also dairy-free.

BBQ BOLETES

Grilling mushrooms might be as ancient as humans' use of fire! Although the fall Northwest mushroom season harvest often starts after the traditional summertime barbecue season is shut down by rain and cold, thankfully there are Spring Kings and Princes around in early summer. Barbecuing is simple and straightforward: the main thing is not burning your mushrooms, and deciding whether you prefer to start them with a bit of an oil coating or add the fat later in the game. Sprinkle on a little salt and pepper, and enjoy a very pleasant primordial fungal feast!

Pasta with Pig's Ears; BBQ Boletes

SAFFRON MILKCAP POLENTA CAKE

The crunchy texture of Saffron Milkcap and other milkcaps blends very well with a polenta cake. Slices of fried onions and some fresh small sage leaves add more deliciousness.

PASTA WITH PIG'S EARS *SUGO DI TOMATE*

Pig's Ears are long-lasting, firm mushrooms that survive well in the woods for weeks on end. More mature Pig's Ears often develop a subtle, astringent note that is easily obscured by a strong sauce. They are perfect in a *sugo di tomate*, Italian for tomato sauce, substituting for minced meat. Fry your finely chopped mushrooms separately in olive oil until they are nicely browned to give them more body in the sauce. The pasta sauce also benefits from herbs like thyme and oregano. Fancy up your Pig's Ears pasta with pine nuts.

Candy Cap Butter Cookies

Chanterelle Vodka, p. 374

DRINKS AND DESSERTS

CANDY CAP BUTTER COOKIES

As kids, we loved to be enlisted to help make *Butterplätzchen*, the simplest Christmas cookies my mother used to bake. Armed with cookie cutters, we had great fun producing stars, trees, moons, and mushrooms. And then we were nibbling away on the raw cutoff dough!

The cookies' crispy texture and rich, buttery taste made them very popular. However, no one back then had an idea that the mushroom shape wasn't just a good luck symbol but represented an awesome fungal ingredient. Here on the West Coast, we can use Candy Cap mushrooms and their delicious maple syrup flavor for butter cookies.

Makes one cookie tray or 50 smallish cookies

1/4 to 1/2 cup (0.2-0.4 ounce) dried Candy Cap mushrooms)
1/2 cup butter (4 ounces)
1/2 cup sugar
1 egg
1 1/2 cups all-purpose flour
1/4 teaspoon salt

Preheat oven to 375° F. Powder the dried Candy Caps in a spice grinder; you want about 2 tablespoons of powder. In powder form, they infuse their aroma into the cookie dough much better.

Cream together butter and sugar. Beat in the egg. Slowly add flour and Candy Cap powder while stirring. Fold the dough a few times by hand to make sure it is well combined.

Sprinkle a clean work space with flour, and roll the dough out to about ¼ inch thick. A mushroom-shaped cookie cutter adds a nice twist, but any cookie cutter shape will do—and if none are in the house, just cut squares or rhombi!

Using a spatula, lift each cookie carefully from your working surface onto an insulated baking tray. (The high butter content in this dough does not require buttering the tray.) If you don't have an insulated tray, adding a second baking sheet below it will prevent cookies from burning on the undersides.

Bake cookies 10 to 15 minutes until they are a rich golden-brown. Underbaked cookies crumble easily and lack taste; caramelization of butter and flour must happen. Let cookies cool before removing them from the tray.

Variations

The dough can be easily amended with ½ teaspoon vanilla or almond extract and enriched with pecans, but the Candy Caps shine brightly without them.

CHANTERELLE VODKA

A nice chanterelle drink that gets you through the doldrums of winter, this infused vodka shared by Yellow Knife fungophile Velma Sterenberg can be enjoyed when chanties are out of season! Otherwise, it could be consumed as an antiparasitic medicine, as traditionally done in the Baltic states, or to prevent an intestinal worm infection. Prevention is key—cheers!

Makes 32 shots

2 to 4 cups chanterelles
1 quart vodka

Clean and slice your mushrooms. Put them into a tightly sealable quart jar and fill with vodka, making sure to fully cover the chanterelles. The amount you use will directly affect the strength of the infused aroma. Seal the jar and let it stand in a dark, cool place for 3 to 5 days.

Remove the chanterelles, squeezing out the vodka—some people like to filter it with a fine mesh sieve or cheesecloth—and bottle your Chanterelle Vodka.

Variations

You can infuse vodka or other alcoholic spirits with other mushrooms as well. I have enjoyed using matsutake as well as Aniseed Funnelcaps. However, keep tasting your infusion and don't let it go too long. My Aniseed Funnelcap Vodka was great after three days, but after a week, it exhibited a whole range of weird chemical tastes that ruined the infusion. And then there are endless possibilities to use your mushroom-infused vodka as a base for a liqueur by adding sugar and fruit extracts.

ACKNOWLEDGMENTS

How can I begin to express my gratitude for being able to write this book? Perhaps I should first thank the kingdom of fungi and all its mushrooms. Throughout my life, I have had the pleasure of getting to know so many of its fascinating members, photographing them and enjoying their delicious flavors. I hope they will forgive my narrow perspective on edibiles and toxic look-alikes! Also, the vicious virus helped a great deal by keeping me home where I could focus on writing.

So many human beings inspired and helped me. First, I thank my parents for instilling my siblings and me early on with a love of nature, an excitement for foraging, and the joy of cooking mushrooms. My wife, Heidi Schor, was an invaluable asset for helping the project come to fruition. I am indebted to her for encouraging me when the challenges seemed insurmountable and I could not see the forest for its mushrooms. Furthermore, she patiently endured critical taste-tests of yet another "edible" mushroom, accommodated me when I proclaimed, "Sorry, I need to take pictures of our meal before we eat it," and let me ramble on about mushrooms far beyond her level of fascination with fungal facts. Our daughters, Sophia and Lilith, were always happy to join in on the fun of the hunt and the joy of cooking the mushrooms we found. Sophia also assisted a great deal with writing and testing the recipes.

My transformation from "mushroom hunter" to "expert" was possible only through the support and inspiration of many people. The mutual support and camaraderie in our mushroom community is absolutely remarkable and very touching. When we moved from Munich, Germany, to Seattle in the mid-1990s, I joined the Puget Sound Mycological Society, a crucial stepping stone in my education about Pacific Northwest mushrooms. My participation in PSMS helped me become familiar with a new set of somewhat familiar edible mushrooms, but more importantly, it exposed me to the science of mycology.

While working as a consultant on projects for community reforestation and rural income generation, I was inspired to research economically important wild edible and medicinal mushrooms in Eastern Tibet. Studying edible species in the Americas and Eurasia impressed upon me how various cultures appreciate a wide variety of mushrooms, as well as how their taste around what is edible diverge—a mushroom considered a delicacy in one community may be frowned upon or even shunned by another culture. Understanding these divergent perspectives made me realize how common it is to narrowly focus on a handful of "choice" mushrooms, a practice that does not do justice to the culinary diversity and quality Pacific Northwest fungi offer curious mushroom hunters.

There is not enough space to thank all the ambassadors of the kingdom of fungi who tirelessly spread the fungal gospel. They all sparked my curiosity and helped me broaden my knowledge. I am grateful for the contributions of all the members of the Pacific Northwest Key Council in helping understand, identify, and study our funga. David Arora's contributions to our understanding of western edible mushrooms are invaluable.

For manuscript revisions, as well as clarifications around taxonomy, toxicologic, medicinal, and culinary specifics, I am deeply grateful to Alan Rockefeller, Amie Broadsword, Andy MacKinnon, Buck McAdoo, Chad Hyatt, Chris Hobbs, Christian Schwarz, Colin Meyer, Danny Miller, Davi Kim, Denis Benjamin, Drew Parker, Elinoar Shavit, Else Vellinga, Erica Cline, Eva Flückiger, Gary Gilbert, Ian Gibson, James "Animal" Nowak, Langdon Cook, Jonathan Frank, Karin Montag, Larry Evans, Leah Bentlin, Michael Beug, Noah Siegel, Paul Kroeger, Rick Kerrigan, Robert Rogers, Shannon Adams, Sophia Winkler-Schor, Steve Trudell and Tyson Ehlers. You all contributed to make this book a reality!

I am grateful for the generous contribution of several mushroom photos I did not have in my collection, including several images by Alan Rockefeller, Bruce Newhouse, Jonathan Frank, and Rich Tehan, as well as images by Daniel Kamberelis, Heidi Schor, James "Animal" Nowack, Joanne Schwartz, Libby Wu, Mike Potts, Thea Chesney, Tim Sage, Warren Cardimona, Wolfgang Bachmeier, and Yi-Min Wang. Thank you as well to Roo Vandegrift for artwork.

Special thanks as well to Alana McGee, Art Goodtimes, Brian Luther, Britt Bunyard, Chris Melotti, David Campbell, Denis Oliver, Dorjee Tsewang, Eugenia Bone, Helen Lau, Jakob Winkler, Jeremy Collison, Jim Erckmann, Jim Gouin, Jinpa Gyatso, Joe Ammirati, John Sirjesse, Kate Mohatt, Kim Brown, Larry Evans, Leslie Scott, Luise Asif, Marianne Maxwell, Milton Tam, Molly Widmer, Olga Katic, Paul Hill, Paul Stamets, Tatiana Sanjuan, Tony McMigas, Tradd Cotter, as well as the late Gary Lincoff, and Patrice Benson. Thanks to all the generous PSMS volunteers, as well as those for all the other mycological societes who have supported me along the way. Such organizations fulfill a vital mission in sharing the fascination of fungi far and wide.

Last but not least special thanks to the marvelous people at Skipstone and Mountaineers Books. We had no idea how complex this project would be. Emily White initiated the project, and Laura Shauger and Erin Moore cheered me on and worked tirelessly to edit my writing and make it shine in this book, while Jen Grable crafted the beautiful design. Special thanks also to Langdon Cook who sent Mountaineers Books my way! I alone am responsible for any mistakes or shortcomings in this field guide.

GLOSSARY

adnate Attached broadly, more or less in a straight line, to stem; used for gill attachment in gilled fungi. See Figure 4 for gill attachments.

adnexed Attached narrowly to the stem; used for gill attachments.

aff. Short for Latin *affinis,* "having an affinity." Inserted between genus and species names when a species is still undescribed but very similar to the referenced species.

agaric A mushroom with gills.

amatoxins Compounds (cyclid octopeptides), several deadly to humans, commonly found in *Amanita phalloides* and other mushrooms.

amorphous Without a clearly defined shape or form.

amyloid Condition of cells staining blue to gray to black when exposed to solutions of iodine (for example Melzer's or Lugol's reagents).

annulus Ring around a mushroom's stem left from the partial veil.

Ascomycetes Fungi belonging to the phylum Ascomycota, a higher taxon that includes the largest group of fungi: those that produce their spores in sacs called asci. Morels are Ascomycetes.

attached Gills that connect to the stem, as opposed to free gills, which do not connect.

Basidiomycetes Fungi belonging to the phylum Basidiomycota, a higher taxon that includes most gilled mushrooms as well as jelly fungi, chanterelles, tooth fungi, boletes, polypores, puffballs, and stinkhorns.

beta-glucans β-D-glucose polysaccharides, soluble fibers with many health benefits found in mushroom cell walls.

buff A common mushroom color: pale tones of yellow with tinges of brown or gray.

bulbous Used to describe a stem with an enlarged, bulging base similar to the shape of a bulb.

centrally depressed Used to describe a cap, when the cap's center is below the cap surface; concave.

chitin A tough and structural polysaccharide containing nitrogen; a primary component of cell walls in fungi. Insect exoskeletons are also composed of chitin.

conical Cone-shaped; used for the whole mushroom or the cap. Also conic.

convex Curved outward like the surface of a ball.

convoluted Intricately folded or twisted.

cortina Weblike or silky veil connecting cap margin to stem and found in some mushrooms; especially evident when the mushroom is young, and often soon disappearing or leaving remnants on stem or cap margin.

decurrent Running down the stem; gills grow down the stem and arch downward; used to describe gill attachments.

eccentric Off-center. Used to describe attachment of stem to cap.

ectomycorrhizal Type of symbiosis between certain fungi and plants, especially trees in the pine, birch, and oak families. The fungal mycelium (*myco*) connects to individual plant outside (*ecto*) of their rootlet (*rhizo*) cells, exchanging nutrients (like phosphorus and potassium) and water with sugary compounds that the plant produces by photosynthesis.

farinaceous Odor (or flavor) of fresh ground meal; also similar to the smell of cucumber or watermelon rind.

fibril A small fiber, thin and threadlike.

fibrillose Composed of long, delicate fibers, evenly arranged on the surface.

fibrous Composed of tough, stringlike tissue.

free gills Gills attached only to the cap, not growing onto the stem. The stem sometimes detaches from the cap in "ball and socket" fashion.

fruiting body A spore-producing macrostructure or organ of a fungi. Also "fruit body."

funga All fungi in a certain area, a.k.a. mycota or mycoflora; funga completes the trinity of flora, fauna, and funga.

furrowed Marked with lines, ridges, or grooves.

gill Bladelike structure that produces spores on basidia on its surface; often aligned spoke-like under the cap.

glandular dots Resinous spots that occur on the stems of some *Suillus* species.

gregarious Growing in close groups but not fused or clustered.

guttation Tissue formation of small drops of liquid; seen on outside of mushroom.

hygrophanous Used to describe tissue, especially the cap, that changes color as it dries. From Greek *hygro* "moisture" and *phainein* "to show."

hymenium Fertile, spore-producing tissue of a mushroom, like gills, pores, tubes, or spines, or a relatively smooth surface cell layer as on the cup fungi and morels.

hypha (pl. hyphae) Threadlike, tube-shaped fungal cells that make up the body (mycelium) of most fungi.

macrofungi Fungi visible to the naked eye.

marbled Marked with streaks or veins of different colors, like marble.

membranous Having skinlike tissue, somewhere between tissue paper and thin felt.

morphology Form and structure of organisms without consideration of function.

mottled Marked with patches or smears of different shades or colors.

muscarine Common toxin found especially in *Inocybe*, *Clitocybe*, *Omphalotus*, and *Entoloma* species. Quick onset, within half hour after ingestion, causing sweating, salivation, slow heart rate, low blood pressure, blurry vision, irregular pulse, and difficulty breathing. People who recover usually do so within twenty-four hours, but severe cases may result in death from respiratory failure. Atropine is a specific antidote.

mushroom Visible fleshy fruiting bodies of fungi, often with caps and stems.

mushrumps Humps produced in the duff layer from a mushroom pushing up without having broken through the needles, leaves, or twigs.

mycelium Network of complex branched hyphae, making up the body of many fungi.

partial veil Membranous, fibrous, or weblike tissue protecting the immature gills or pores and connecting from cap edge to stem. As the cap matures, the partial veil breaks open, leaving remnants on stem, and sometimes tissue at cap edge, i.e., a ring.

perithecia Small, flask-shaped fruiting bodies in certain ascomycetous fungi that contain the ascospores.

phylogeny The evolutionary history of a species or other taxon, showing its line of descent and thus relationships among organisms.

polysaccharide A complex sugar. Chitin, cellulose, and starches are all polysaccharides.

pore Tiny hole on the underside of a cap. Pores are at the mouths of spore-producing tubes.

pot hunter A mushroom forager primarily motivated by mushroom edibility.

primordium An organ, structure, or tissue in the earliest stage of development; a "mushroom fetus."

radial Arranged like rays or lines outward from a common center.

reticulated Having a netlike pattern, characteristic of the stems of some boletes.

rhizomorph Strands of hyphae forming rootlike strings and present around the base of stem and nearby substrate. Honey mushrooms are characterized by rhizomorphs.

ring Often fibrous or membranous tissue circling the stem; see also *partial veil*.

saprobic Feeding on dead and decaying organic matter. Also saprotrophic.

scaber Short projecting scale or tuft.

scale Erect, flattened, or recurved piece of tissue on a surface that projects outward or is torn and uplifted from matrix below.

seceding gills Formerly attached gills that separate from the stem (as fruiting body ages), leaving vertical lines on the stem where the gills once connected.

sequence A patterning of the four "base" amino acids making up the genetic code, identified through the molecular study of DNA.

serrated Sawtoothed to almost ragged, often referring to gill edges.

sheath A close-fitting covering, often used in the context of a ring, clasping or clinging to the stem and opening upward.

skirtlike Of a ring (annulus), hanging down like a skirt.

sporulate To produce and release spores.

stem Stalk; see *stipe*.

stipe Stalk of the mushroom, the structure that bears the spore-producing cap and raises it from the substrate; see also *stem*.

striated Marked with delicate lines, ridges, or grooves, which may be parallel or radiating.

striations Fine parallel or radiating lines or grooves.

subgill Short gill that does not reach all the way from the cap margin to the stem.

taxon (pl. taxa) Taxonomic group, such as species, genus, family, order.

teeth Spinelike structures of some fertile surfaces. Tooth fungi produce their spores on the surface of small, toothlike structures that point straight down; the teeth are tapered, allowing spores to fall straight down.

tomentose Woolly, densely covered with matted soft hairs.

tubes Cylindrical structures of the hymenium of polypores and boletes, lined with spore-producing basidia. En masse in the boletes, referred to as a "sponge."

tuft Short cluster of elongated strands.

umbo Raised knob or mound at the center of the cap. *Umbonate* means to have such a knob.

universal veil Enveloping tissue that covers the whole young mushroom from bottom of stem base to top of cap. The tissue rips as the mushroom expands, often leaving fragments (specks, warts, or patches) on the cap or stem, and a volva (cup) at the stem base.

viscid Sticky but not slimy or lubricous.

volva Saclike membranous tissue or cup at base of stipe, remnant of the universal veil.

RESOURCES FOR MUSHROOM HUNTING

Mushroom identification is too often a tricky business, especially if you are new to foraging. Although we have dozens of choice edibles that can be safely identified, how do you get to the point where you know enough to trust your knowledge? Get good field guides for the Pacific Northwest, join your local mushroom club or mycological society, find support online, and spend time in the field. And always remember when identifying any mushroom for the cookstove: **When in doubt, throw it out!**

PACIFIC NORTHWEST FIELD GUIDES

These days, there are many good mushroom field guides focused on our region. Look for them in the References, where they are highlighted in **bold**.

DANIEL WINKLER'S MUSHROAMING

The Mushroaming website includes listings of upcoming presentations, online classes, and presentations to local mycological societies. Here online you will find Daniel's descriptions, illustrated with a lot of photos, of some of the best and most common edibles: visit mushroaming.com/pacific _northwest). Also offered are high-quality laminated foldouts of field guides to edible mushrooms and medicinal fungi, as well as up-to-date western bolete flash cards, mushroom knives, and emergency whistles for forays. To join in a private mushroom hunt, please send an email to me@danielwinkler.com.

PACIFIC NORTHWEST MUSHROOM CLUBS

Local mycological societies are the best place to find expert support in person! Membership fees are modest, and usually included are club forays with expert identifiers, mushroom presentations, and other events. Many clubs have their own email lists just for members. It's a great way to meet fellow mushroom lovers.

Membership dues are worth every penny. No club where you live? Start your own informal club, maybe first online, to connect to people! It's so much easier to learn from people in the field than from books and the internet alone! Below are a few organizations in our region. Or find a local club by visiting: mushroaming.com/fungal_links.

British Columbia

South Vancouver Island Mycological Society, svims.ca, Victoria, British Columbia

Vancouver Mycological Socety, vanmyco.org, Vancouver, British Columbia

Western US

Cascade Mycological Society, cascademyco.org, Eugene, Oregon

Colorado Mycological Society, cmsweb.org, Denver, Colorado

Humboldt Bay Mycological Society, hbmycologicalsociety.org, Arcata, California

Kitsap Peninsula Mycological Society, kitsapmushrooms.org, Bremerton, Washington

North American Mycological Association, namyco.org

North Idaho Mushroom Club, facebook.com/IdahoWild Mushrooms, Coeur d'Alene, Idaho

North American Truffling Society, natruffling.org, Corvallis, Oregon

Northwest Mushroomers Association, northwestmushroomers.org, Bellingham, Washington

Olympic Peninsula Mycological Society, mmuller@olympus.net, Chimacum, Washington

Oregon Mycological Society, wildmushrooms.org, Portland, Oregon

Puget Sound Mycological Society, psms.org, Seattle, Washington

Snohomish County Mycological Society, scmsfungi.org, Everett, Washington

South Sound Mushroom Club, southsoundmushroomclub.com, Olympia, Washington

Southern Idaho Mycological Association, idahomushroomclub .org, Boise, Idaho

Southwest Washington Mycological Society, swmushrooms.org, Chehalis and Centralia, Washington

Western Montana Mycological Association, facebook.com /montanamushrooms, Missoula, Montana

Willamette Valley Mushroom Society, wvmssalem.org, Salem, Oregon

Yakima Valley Mushroom Society, yvms.org, Yakima, Washington

FREE ONLINE TOOLS

MycoMatch—mycomatch.com—is an awesome mushroom ID application covering thousands of Pacific Northwest species and images compiled by Ian Gibson; it is now available for all devices. Through it you can search using one (or more) features: for example, all mushrooms that smell of garlic or any white-spored fungus that grows in clusters.

The very helpful PNW Mushroom Pictorial Key by Danny Miller combines images with brief descriptions. Taxonomically, it is always up to date. alpental.com/psms/PNWMushrooms /PictorialKey

iNaturalist and Mushroom Observer are the two leading online citizen science communities for fungi. Very helpful! You can upload photos of your mushrooms with the location, the date, and notes, and other people might help you with your mushroom IDs. You also can search for certain mushrooms and see when they fruit and where else they are found.

FACEBOOK AND MEETUP

Mushroom groups on Facebook and other social media sites can be very helpful for seeing what is fruiting right now and for getting ID help; but be careful, a lot of people will volunteer mushroom names and IDs, and some are totally wrong. Be aware and never go with one identification only. A virtual misidentification can cause a real poisoning!

- Pacific Northwest Mushroom Identification Forum
- Pacific Northwest Mushroom Social Club
- Vancouver Island Mushroom Identification and Info Group
- Pacific Northwest Mushroom Foraging and Identification
- Meetup site for Jeremy Collison of Salish Sea Mushrooms, www.meetup.com/SalishMushrooms

SOURCE NOTES

INTRODUCTION

1. Egli et al., 2006.
2. Hobbie and Hobbie, 2006; Sheldrake, 2020.
3. Borchers et al., 1999.
4. Friedman, 2016.
5. Friedman, 2016.
6. Hobbs, 2021.
7. Sari et al., 2017.
8. Feng et al., 2019.
9. Zhang et al., 2009.

PART ONE

10. Katalyse Institute, 2021.
11. Pajak et al., 2020.
12. Meisch et al., 1977.
13. Trappe et al., 2014.
14. Galgowska and Pietrzak-Fiećko, 2017.
15. Heshmati et al., 2019.
16. Cline, 2013.
17. Shavit and Shavit, 2010.
18. Schellmann et al., 1980.
19. Diehl and Schlemmer, 1980.
20. Norvell and Roger, 2016.
21. Paoletti et al., 2007.
22. Hu et al., 2020.

PART TWO

23. Hobbie et al., 2016.
24. Saviuc et al., 2010.
25. Piqueras, 2021.
26. Loizides, 2017.
27. Beug, *Mushrooms of Cascadia*, 2021.
28. Davoli and Sitta, 2015.
29. Dirks, 2023.
30. Lagrange et al., 2021.
31. Niskanen et al., 2018.
32. Nelson, 2010.
33. Sari et al., 2017.
34. Lemieszek et al., 2017.
35. Falandysz, 2008.
36. Frank et al., 2020.
37. den Bakker et al., 2005.
38. Siegel and Schwarz, 2016.
39. Beug, *Mushrooms of Cascadia*, 2021.
40. Hobbs, 2021.
41. Colak et al., 2009.
42. Gilmore, 1919.
43. Stepney and Goa, 1990.
44. Burk, 1983.
45. Exeter et al., 2006.
46. Jayasuriya, 1998.
47. Nuytinck and Ammirati, 2014.
48. Wood et al., 2012.
49. Rubel and Arora, 2008.
50. Zechmann, 2011.
51. Berch, Kroeger, and Finston, 2017.
52. Bunyard, 2015.
53. Hong et al., 2018.
54. McAdoo, 2020.
55. McAdoo, 2020.
56. Laubner and Mikulevičeinė, 2016.
57. Nieminen et al., 2005.

58. Rzymski et al., 2018.
59. Turner et al., 1987.
60. Jones, 1983.
61. Blanchette et al., 1992.
62. Kuhnlein and Turner, 1991; Turner et al., 1987.
63. Turner et al., 1990; Turner and Cuerrier, 2022.
64. Turner et al., 1990.
65. Anderson and Lake, 2013.
66. Arora, 1990.
67. Falandysz and Rizal, 2016.
68. Falandysz and Rizal, 2016.
69. Teaf et al., 2010.
70. Li et al., 2021.
71. Oswald et al., 2021.
72. Caspar and Spiteller, 2015.
73. Barron, 1977.
74. Tomita et al., 2004.
75. Rogers, 2020.
76. Rogers, 2020.
77. Halford, 2011.
78. Schmitt and Tatum, 2008.
79. Patton/Oregon Public Broadcasting, 2015.
80. Kawahara et al., 2016.
81. Rogers, 2020.
82. Arora, 1986.
83. Yan et al., 2015.
84. Winkler, 2009.
85. Benjamin, 1995.
86. Cline, 2013.
87. Agerer, 1990.
88. Siegel and Schwarz, 2016.

REFERENCES

The best mushroom field guides for the Pacific Northwest are highlighted in **bold**.

Agerer, Reinhard. "Ectomycorrhizae of *Chroogomphus helveticus* and *C. rutilus* (Gomphidiaceae, Basidiomycetes) and Their Relationship to Those of *Suillus* and *Rhizopogon*." *Nova Hedwigia* 50 (1990): 1–63.

Anderson, M. Kat, and Frank K. Lake. "California Indian Ethnomycology and Associated Forest Management." *Journal of Ethnobiology* 33, no. 1 (2013): 33–85.

Arora, David. *All That the Rain Promises and More: A Hip Pocket Guide to Western Mushrooms*. Berkeley: Ten Speed Press, 1990.

——. *Mushrooms Demystified: A Comprehensive Guide to the Fleshy Fungi*. 2nd edition. Berkeley: Ten Speed Press, 1986.

Bachmeier, Wolfgang. Pilzsuche 2022. www.123pilze.de. Accessed March 13, 2022. In German.

Barron, G. L. *The Nematode-Destroying Fungi*. Guelph, Canada: Canadian Biological Publications, 1977.

Benjamin, Denis. *Mushrooms: Poisons and Panaceas: A Handbook for Naturalists, Mycologists, and Physicians*. New York: W. H. Freeman, 1995.

——. "Neurological Effects of *Morchella* sp." *Fungi Magazine* 8, no. 3 (2015): 24–25.

Berch, Shannon M., Paul Kroeger, and Terrie Finston. "The Death Cap Mushroom (*Amanita phalloides*) Moves to a Native Tree in Victoria, British Columbia." *Botany* 95, no. 4 (2017): 435–440.

Beug, Michael. *Mushrooms of Cascadia: An Illustrated Key*. Batavia, IL: The Fungi Press, 2021.

——. "North American Mushroom Poisonings and Adverse Reactions to Mushrooms 2018–2020." *Fungi Magazine* 14, no. 2 (2021): 17–22.

Blanchette, Robert A., et al. "Nineteenth Century Shaman Grave Guardians Are Carved *Fomitopsis officinalis* Sporophores." *Mycologia* 84, no. 1 (1992): 119–124.

Borchers, A. T., et al. "Mushrooms, Tumors, and Immunity." *Proceedings of the Society for Experimental Biology and Medicine* 221 (1999): 281–293.

Bunyard, B. "The Real Story Behind the Increasing Number of Amanita Poisonings in North America." *Fungi* 8, no. 3 (2015): 5–9.

Burk, William R. "Puffball Usages among North American Indians." *Journal of Ethnobiology* 3 (1983): 55–62.

Caspar, Jan, and Peter Spiteller. "A Free Cyanohydrin as Arms and Armour of *Marasmius oreades*." *ChemBioChem* 16, no. 4 (2015): 570–573.

Cline, Erica. "Toxic Metals Uptake by Edible Wild Mushrooms: Why It Happens and Which Species to Avoid." University of Puget Sound, 2013. www.pugetsound.edu /news-and-events/arts-at-puget-sound /thsms/erica-cline. Accessed April 27, 2019.

Colak, A., Ö. Faiz, and E. Sesli. "Nutritional Composition of Some Wild Edible Mushrooms." *Turkish Journal of Biochemistry* 34, no. 1 (2009): 25–31.

Davoli, Paolo, and Nicola Sitta. "Early Morels and Little Friars, or a Short Essay on the Edibility of *Verpa bohemica*." *Fungi* 8.1 (2015): 4–9.

den Bakker, H. C., and M. E. Noordeloos. "A Revision of European Species of *Leccinum* Gray and Notes on Extralimital Species." *Persoonia—Molecular Phylogeny and Evolution of Fungi* 18, no. 4 (2005): 511–574.

Diehl, J. F., and U. Schlemmer. "Identification of Bioavailability of Fungal Cadmium Based on Feed Trial with Rats: Relevance for Humans." *Z. für Ernährungswiss* 23 (1984): 126–135. In German with English summary.

Dirks, Alden C., et al. "Not All Bad: Gyromitrin Has a Limited Distribution in the False Morels as Determined by a New Ultra High-Performance Liquid Chromatography Method," *Mycologia* 115(1) (2023):1-15.

Egli, S., M. Peter, C. Buser, W. Stahel, and F. Ayer. "Mushroom Picking Does Not Impair Future Harvests—Results of a Long-Term Study in Switzerland." *Biological Conservation* 129 (2006): 271–276.

"Empfehlungen zur Verzehrseinschränkung von Speisepilzen. Mitteilungen aus dem Bundesgesundheitsamt." ("Recommendations to Limit Ingestion of Edible Mushrooms.") *Bundesgesundheitsblatt* (*Federal Health Administration*) 21 (1978): 204. In German.

Exeter, Ronald L., Lorelei Norvell, and Efrén Cázares. Ramaria *of the Pacific Northwest*. USDA, BLM, Salem District. 2006.

Falandysz, Jerzy. "Selenium in Edible Mushrooms." *Journal of Environmental Science and Health*, Part C, 26, no. 3 (2008): 256–299.

Falandysz, Jerzy, and Leela M. Rizal. "Arsenic and Its Compounds in Mushrooms: A Review." *Journal of Environmental Science and Health*, Part C, 34, no. 4 (2016): 217–232.

Feng, Lei, et al. "The Association Between Mushroom Consumption and Mild Cognitive Impairment: A Community-Based Cross-Sectional Study in Singapore." *Journal of Alzheimer's Disease* 68, no. 1 (2019): 197–203.

Frank J. L., et al. "*Xerocomellus* (Boletaceae) in Western North America." *Fungal Systematics and Evolution* 6 (2020): 265–288.

Friedman, Mendel. "Mushroom Polysaccharides: Chemistry and Antiobesity, Antidiabetes, Anticancer, and Antibiotic Properties in Cells, Rodents, and Humans." *Foods* 5, no. 4 (2016): 80.

Gałgowska, Michalina, and Renata Pietrzak-Fiećko. "Pesticide Contaminants in Selected Species of Edible Wild Mushrooms from the North-Eastern Part of Poland." *Journal of Environmental Science and Health*, Part B, 52, no. 3 (2017), 214–217.

Genaust, Helmut. *Etymological Dictionary of Botanical Plant Names*. Hamburg: Birkhäuser, 1996. In German, available as pdf.

Gilbert, Gary, and Daniel Winkler. Boletes *of Western North America*. MycoCards, 2021.

Gilmore, Melvin Randolph. "Uses of Plants by the Indians of the Missouri River Region." In *Thirty-Third Annual Report of the Bureau of American Ethnology to the Secretary of the Smithsonian Institution*, 1911–1912 Washington, DC: Government Printing Office, 1919. 43–154.

Gry, Jørn, Christer Andersson, et al. *Mushrooms Traded as Food*. Vol. 2, Sect. 1. TemaNord series. Copenhagen: Nordic Council of Ministers, 2012.

Halford, Bethany. "The Angel's Wing Mystery: An Unstable Amino Acid May Have Something to Do with Deadly Mushroom Poisoning." *Chemical and Engineering News* 89, no. 13 (2011).

Hallock, Robert M. *A Mushroom Word Guide: Etymology, Pronunciation, and Meanings of Over 1,500 Words*. Self-published, 2019.

Heshmati Ali, Mina Hamidi, and Amir Nili-Ahmadabadi. "Effect of Storage, Washing, and Cooking on the Stability of Five Pesticides in Edible Fungi of *Agaricus bisporus*: A Degradation Kinetic Study." *Food Science & Nutrition* 7, no. 12 (2019): 3993–4000.

Hobbie, E. A., et al. "Isotopic Evidence Indicates Saprotrophy in Post-Fire *Morchella* in Oregon and Alaska." *Mycologia* 108 (2016): 638–645.

Hobbie, J. E., and E. A. Hobbie. "N-15 in Symbiotic Fungi and Plants Estimates Nitrogen and Carbon Flux Rates in Arctic Tundra." *Ecology* 87 (2006): 816–822.

Hobbs, Christopher. *Medicinal Mushrooms: The Essential Guide.* **North Adams, MA: Storey Publishing, 2021.**

Hong, Luo, et al. "The MSDIN Family in Amanitin-Producing Mushrooms and Evolution of the Prolyl Oligopeptidase Genes." *IMA Fungus* 9 (2018): 224–225.

Hu, Daihua, et al. "Ultraviolet Irradiation Increased the Concentration of Vitamin D_2 and Decreased the Concentration of Ergosterol in Shiitake Mushroom (*Lentinus edodes*) and Oyster Mushroom (*Pleurotus ostreatus*) Powder in Ethanol Suspension." *ACS Omega* 5 (2020): 7361–7368.

Jayasuriya, Hiranthi, et al. "Note Clavaric Acid: A Triterpenoid Inhibitor of Farnesyl-Protein Transferase from *Clavariadelphus truncates*." *Journal of Natural Products* 61, no. 12 (1998): 1568–1570.

Jones, Anore. *Nauriat Nigiñaqtuat: Plants That We Eat.* Indian Health Service, 1983.

Karlsen, Pål, and Tommy Østhagen. *SpiSopp: 200 Sopper du må Smake før du Dør.* (*Edible Mushrooms: 200 Mushrooms You Have to Taste Before You Die*). In Norwegian. Oslo: Kolofon Forlag, 2020.

Katalyse Institute. Schwermetalle und mehr in Wildpilzen, Berlin, 2021. http://chemie-in-lebensmitteln.katalyse.de/schwermetalle-und-mehr-in-wildpilzen.

Kawahara, Hidehisa, et al. "Antifreeze Activity of Xylomannan from the Mycelium and Fruit Body of *Flammulina velutipes*." *Biocontrol Science* 21, no. 3 (2016), 153–159.

Kuhnlein, Harriet V., and Nancy J. Turner. *Traditional Plant Foods of Canadian Indigenous Peoples: Nutrition, Botany, and Use.* London: Gordon and Breach Publishers, 1991.

Kuusi, T., et al. "Lead, Cadmium, and Mercury Contents of Fungi in the Helsinki Area and in Unpolluted Control Areas." *Z. Lebensm Unters Forsch* 173 (1981): 261–267.

Lagrange, E., et al. "An Amyotrophic Lateral Sclerosis Hot Spot in the French Alps Associated with Genotoxic Fungi." *Journal of the Neurological Sciences* 427 (2021): 117558.

Laubner, Gabija, and Gabija Mikulevičeinė. "A Series of Cases of Rhabdomyolysis after Ingestion of *Tricholoma equestre*." *Acta Medica Lituanica* 23, no. 3 (2016): 193–197.

Lemieszek, Marta, et al. "A King Bolete, *Boletus edulis* (Agaricomycetes), RNA Fraction Stimulates Proliferation and Cytotoxicity of Natural Killer Cells Against Myelogenous Leukemia Cells." *International Journal of Medicinal Mushrooms* 19, no. 4 (2017): 347–353.

Li, H., et al. "Reviewing the World's Edible Mushroom Species: A New Evidence-Based Classification System." *Comprehensive Reviews in Food Science and Food Safety* 20, no. 2 (2021): 1982–2014.

Loizides, Michael. "Morels: The Story So Far." *Field Mycology* 18, no. 2 (2017): 42–53.

MacKinnon, Andy, and Kem Luther. *Mushrooms of British Columbia.* **Victoria: Royal BC Museum, 2021.**

McAdoo, Buck. *Profiles of Northwest Fungi.* **Boulder: GL Design, 2020.**

Meisch, Hans-Ulrich, Johannes A. Schmitt, and Wolfgang Reinle. ("Heavy Metals in Higher Fungi—Cadmium, Zinc, and Copper.") *Z. für Naturforschung* 32e (1977): 172–183. In German, with English abstract.

Nelson, Stephen F. "Bluing Components and Other Pigments of Boletes." *Fungi* 3, no. 4 (2010): 11–14.

Nieminen, P., A. M. Mustonen, and M. Kirsi. "Increased Plasma Creatine Kinase Activities Triggered by Edible Wild Mushrooms." *Food and Chemical Toxicology* 43 (2005): 133–138.

Niskanen, Tuula, et al. "Identifying and Naming the Currently Known Diversity of the Genus *Hydnum*, with an Emphasis on European and North American Taxa." *Mycologia* 110, no. 5 (2018): 890–918.

Norvell, Lorelei, and Judy Roger. "The Oregon *Cantharellus* Study Project: Pacific Golden Chanterelle—Preliminary Observations and Productivity Data (1986–1997)." Conference paper, Mycological Society of America, 2016.

Nuytinck, Jorinde, and Joseph F. Ammirati. "A New Species of *Lactarius* Sect. *Deliciosi* (Russulales, Basidiomycota) from Western North America." *Botany* 92 (2014): 767–774.

Oswald, Iain W. H., et al. "Identification of a New Family of Prenylated Volatile Sulfur Compounds in Cannabis Revealed by Comprehensive Two-Dimensional Gas Chromatography." *ACS Omega* 6, no. 47 (2021): 31667–31676.

Pajak, M., et al. "2020 Risk Assessment of Potential Food Chain Threats from Edible Wild Mushrooms Collected in Forest Ecosystems with Heavy Metal Pollution in Upper Silesia, Poland." *Forests* 11 (2020): 1240.

Paoletti, M. G., et al. "Human Gastric Juice Contains Chitinase That Can Degrade Chitin." *Annals of Nutrition and Metabolism* 51, no. 3 (2007): 244–251.

Parker, Drew, and Teresa Marrone. *Mushrooms of the Northwest: A Simple Guide to Common Mushrooms.* Cambridge, MN: Adventure Publications, 2019.

Patton, Vince. "Oregon Humongous Fungus Sets Record as Largest Single Living Organism on Earth." Oregon Public Broadcasting, February 12, 2015, Video, 7:26.

Peintner, U., et al. "Mycophilic or Mycophobic? Legislation and Guidelines on Wild Mushroom Commerce Reveal Different Consumption Behaviour in European Countries." *PLoS ONE* 8, no. 5 (2013).

Piqueras, Josep. "Morel Mushroom Toxicity: An Update." *Fungi Magazine* 14, no. 2 (2021): 42–52.

Pollan, Michael. *How to Change Your Mind: What the New Science of Psychedelics Teaches Us About Consciousness, Dying, Addiction, Depression, and Transcendence.* New York: Penguin, 2018.

Qizheng, Liu, et al. "Artificial Cultivation of True Morels: Current State, Issues, and Perspectives." *Critical Reviews in Biotechnology* 38, no. 2 (2017): 1–13.

Rogers, Robert. *The Fungal Pharmacy: The Complete Guide to Medicinal Mushrooms and Lichens of North America.* Berkeley: North Atlantic Books, 2011.

——. *Medicinal Mushrooms: The Human Clinical Trials.* Edmonton, Canada: Prairie Deva Press, 2020.

Rubel, William D., and David Arora. "A Study of Cultural Bias in Field Guide Determinations of Mushroom Edibility Using the Iconic Mushroom, *Amanita muscaria*, as an Example." *Economic Botany* 62, no. 3 (2008): 223–243.

Rzymski, Piotr, and Piotr Klimaszyk. "Is the Yellow Knight Mushroom Edible or Not? A Systematic Review and Critical Viewpoints on the Toxicity of *Tricholoma equestre*." *Comprehensive Reviews in Food Science and Food Safety* 17 (2018): 1309–1324.

Sarawi, Sepas, et al. "Occurrence and Chemotaxonomical Analysis of Amatoxins in *Lepiota* spp. (Agaricales)." *Phytochemistry* 195 (2022): 113069.

Sari, Miriam, et al. "Screening of Beta-Glucan Contents in Commercially Cultivated and Wild Growing Mushrooms." *Food Chemistry* 216 (2017): 45–51.

Saviuc, Philippe, et al. "Can Morels (*Morchella* sp.) Induce a Toxic Neurological Syndrome?" *Clinical Toxicology* 48, no. 4 (2010): 365–372.

Schellmann, B., M. J. Hilz, and O. Opitz. "Fecal Excretion of Cadmium and Copper after Mushroom (Agaricus) Diet." *Zeitschrift f. Lebensmittel-Untersuchung und -Forschung* 171, no. 3 (1980): 189–192. In German, with English summary.

Schmitt, C. L., and M. L. Tatum. *The Malheur National Forest: Location of the World's Largest Living Organism (The Humongous Fungus).* Forest Service, US Department of Agriculture, 2008.

Shavit, Elinoar, and Efrat Shavit. "Lead and Arsenic in *Morchella esculenta* Fruit Bodies Collected in Lead Arsenate Contaminated Apple Orchards in the Northeastern United States." *Fungi* 3, no. 2 (2010): 11–18.

Sheldrake, Merlin. *Entangled Life: How Fungi Make Our Worlds, Change Our Minds, and Shape Our Futures.* New York: Random House, 2020.

Siegel, Noah, and Christian Schwarz. *Mushrooms of the Redwood Coast.* Berkeley: Ten Speed Press, 2016.

Siegel, Noah, et al. *A Field Guide to the Rare Fungi of California's National Forests.* Minneapolis: Bookmobile, 2019.

Sitta, Nicola, Paolo Davoli, Marco Floriani, and Edoardo Suriano. *A Reasoned Guide to the Edibility of Mushrooms. (Guida Ragionata alla Commestibilità dei Funghi.)* Italy: Regione Piemonte, 2021. In Italian.

Stepney, Philip H. R., and David J. Goa. *The Scriver Blackfoot Collection: Repatriation of Canada's Heritage*. Edmonton: Provincial Museum of Alberta, 1990.

Teaf, Christopher, et al. "Arsenic Cleanup Criteria for Soils in the US and Abroad: Comparing Guidelines and Understanding Inconsistencies." *Proceedings of the Annual International Conference on Soils, Sediments, Water and Energy*, vol. 15, no. 10, 2010.

Tomita T., et al. "Pleurotolysin, a Novel Sphingomyelin-Specific Two-Component Cytolysin from the Edible Mushroom *Pleurotus ostreatus*." *Journal of Biological Chemistry* 279, no. 26 (2004): 26975–26982.

Trappe, M. J., et al. "Cesium Radioisotope Content of Wild Edible Fungi, Mineral Soil, and Surface Litter in Western North America after the Fukushima Nuclear Accident." *Canadian Journal of Forest Research* 44, no. 11 (2014): 1441–1452.

Trudell, Steve, and Joe Ammirati. ***Mushrooms of the Pacific Northwest.*** **Portland, OR: Timber Press, 2009.**

Trudell, Steve A., P. Brandon Matheny, Andrew D. Parker, Matthew Gordon, Dianatha B. Dougil, and Erica T. Cline. *Pacific Northwest Tricholomas: Are We Using the Right Names?* Published by the authors, February 2022.

Turner, Nancy J., and Alain Cuerrier. "'Frog's Umbrella' and 'Ghost's Face Powder': The Cultural Roles of Mushrooms and Other Fungi for Canadian Indigenous Peoples." *Botany* 100, no. 2 (2022): 183–205.

Turner, Nancy J., Harriet V. Kuhnlein, and Keith N. Egger. "The Cottonwood Mushroom (*Tricholoma populinum*): A Food Resource of the Interior Salish Indian Peoples of British Columbia." *Canadian Journal of Botany* 65, no. 5 (1987): 921–927.

Turner, Nancy J., et al. *Thompson Ethnobotany: Knowledge and Usage of Plants by the Thompson Indians of British Columbia*. Memoir #3. Victoria: Royal British Columbia Museum, 1990.

Vilgalys, R., A. Smith, B. L. Sun, and O. K. Miller Jr. "Intersterility Groups in the *Pleurotus ostreatus* Complex from the Continental United States and Adjacent Canada." *Canadian Journal of Botany*, 71, no. 1 (1993): 113–128.

Watling, Roy, and M. R. D. Seaward. "Some Observations on Puff-balls from British Archaeological Sites." *Journal of Archaeological Science* 3, no. 2 (1976): 165–172.

Winkler, Daniel. "Tales of the Himalayan Gypsy: Wandering Between *Rozites* and *Cortinarius emodensis*." *Mushroom—The Journal of Wild Mushrooming* 102, vol. 27. 1 (2009): 33–40.

Winkler, Daniel, and Robert Rogers. ***Field Guide to Medicinal Mushrooms of North America.*** **Kirkland, WA: Mushroaming Publishing, 2018.**

Wood, William F., et al. "The Maple Syrup Odour of the 'Candy Cap' Mushroom, *Lactarius fragilis* var. *rubidus*." *Biochemical Systematics and Ecology* 43 (2012): 51–53.

Yan, Naihong, et al. "Antiviral Activity of a Cloned Peptide RC28 Isolated from the Higher Basidiomycetes Mushroom *Rozites caperata* in a Mouse Model of HSV-1 Keratitis." *International Journal of Medicinal Mushrooms* 17, no. 9 (2015): 819–828.

Zechmann, Alois. "Auf Pilzpirsch im Bayerischen Wald." ("Hunting Mushrooms in the Bavarian Forest.") *Mycologia Bavarica* 12 (2011): 1–9. In German.

Zhang, Min, J. Huang, X. Xie, and C. D. Holman. "Dietary Intakes of Mushrooms and Green Tea Combine to Reduce the Risk of Breast Cancer in Chinese Women." *International Journal of Cancer* 124 (2009): 1404–1408.

INDEX

ABOUT THE AUTHOR

Photo by Heidi Schor

A mushroom educator, ethnomycologist, ecologist, and award-winning photographer, **Daniel Winkler** grew up collecting mushrooms in the Alps. Winkler has been foraging for the past several decades in the Pacific Northwest, as well as in South America and High Asia, where he is involved in ethnomycological fieldwork. A world-renowned *Cordyceps* expert, he has served as vice president of the Puget Sound Mycological Society, who recognized his lifetime of service with their Golden Mushroom Award.

Through his travel agency, Mushroaming, Winkler runs mushroom-focused ecotours to Tibet, Bhutan, the Amazon, Colombia, and the Austrian Alps and throughout the Northwest. He lives in Kirkland, Washington. Find out what he's up to at mushroaming.com.

ABOUT SKIPSTONE

Skipstone guides explore healthy lifestyles, backyard activism, and how an outdoor life relates to the well-being of our planet. Sustainable foods and gardens; healthful living; realistic and doable conservation at home; modern aspirations for community—Skipstone tries to address such topics in ways that emphasize active living, local and grassroots practices, and a small footprint. Our hope is that Skipstone books will inspire you to celebrate the freedom and generosity of a life outdoors.

All of our publications, as part of our 501(c)(3) nonprofit program, are made possible through the generosity of donors and through sales of 700 titles on outdoor recreation, sustainable lifestyle, and conservation. To donate, purchase books, or learn more, visit us online:

www.skipstonebooks.org

www.mountaineersbooks.org

SKIPSTONE

LIVE LIFE

MAKE RIPPLES

YOU MAY ALSO LIKE:

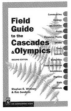 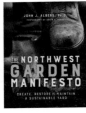   

NONGILLED MUSHROOMS

**MORELS &
FALSE MORELS**
(pp. 60–77)

**LOBSTER
MUSHROOM**
(pp. 88–90)

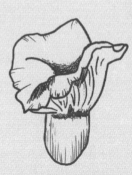

TRUFFLES
(pp. 83–88)

**CORALS, CLUBS
& OTHERS**
(pp. 182–189)

CUP FUNGI
(pp. 78–82)

**BEAR'S HEADS,
LION'S MANES &
BEARDED TEETH**
(pp. 122–126)

STINKHORN
(pp. 180–181)

PUFFBALLS
(pp.174–180)

LION'S MANE